Fractional Integral Transforms

Fractional Integral Transforms: Theory and Applications presents over 25 integral transforms, many of which have never before been collected in one single volume. Some transforms are classic, such as the Laplace, Fourier, etc, and some are more recent, such as the Fractional Fourier, Gyrator, Linear Canonical, Special Affine Fourier Transforms, as well as, continuous Wavelet, Ridgelet, and Shearlet transforms.

The book provides an overview of the theory of fractional integral transforms with examples of such transforms, before delving deeper into the study of important fractional transforms, including the fractional Fourier transform. Applications of fractional integral transforms in signal processing and optics are highlighted. The book's format has been designed to make it easy for the readers to extract the essential information they need to learn about the fundamental properties of each transform. Supporting proof and explanations are given throughout.

Features

- Brings together integral transforms never before collected into a single volume
- A useful resource on fractional integral transforms for researchers and graduate students in mathematical analysis, applied mathematics, physics and engineering
- Written in an accessible style with detailed proofs and emphasis on providing the reader with an easy access to the essential properties of important fractional integral transforms

Ahmed I. Zayed is a Professor of Mathematics at the Department of Mathematical Sciences, DePaul University, Chicago, and was the Chair of the department for 20 years, from 2001 until 2021. His research interests varied over the years starting with generalized functions and distributions to sampling theory, applied harmonic analysis, special functions and integral transforms. He has published two books and edited seven research monographs. He has written 22 book chapters, published 118 research articles, and reviewed 173 publications for the Mathematical Review and 81 for the Zentralblatt für Mathematik (zbMath). He has served on the Editorial Boards of 22 scientific research journals and has refereed over 200 research papers submitted to prestigious journals, among them are IEEE, SIAM, Amer. Math. Soc., Math Physics, and Optical Soc. Journals.

Fractional Integral Transforms

Theory and Applications

Ahmed I. Zayed

DePaul University, USA

CRC Press
Taylor & Francis Group
Boca Raton London New York

CRC Press is an imprint of the
Taylor & Francis Group, an **informa** business
A CHAPMAN & HALL BOOK

Designed cover image: Ahmed I. Zayed

First edition published 2024
by CRC Press
6000 Broken Sound Parkway NW, Suite 300, Boca Raton, FL 33487-2742

and by CRC Press
4 Park Square, Milton Park, Abingdon, Oxon, OX14 4RN

CRC Press is an imprint of Taylor & Francis Group, LLC

Library of Congress Cataloging-in-Publication Data

Names: Zayed, Ahmed I., author.
Title: Fractional integral transforms : theory and applications / Ahmed I. Zayed, DePaul University, USA.
Description: First edition. | Boca Raton, FL : C&H/CRC Press, 2024. | Includes bibliographical references and index.
Identifiers: LCCN 2023046154 (print) | LCCN 2023046155 (ebook) | ISBN 9780367543877 (hardback) | ISBN 9781003089353 (paperback) | ISBN 9780367544485 (ebook)
Subjects: LCSH: Integral transforms. | Transformations (Mathematics)
Classification: LCC QA432 .Z39 2024 (print) | LCC QA432 (ebook) | DDC 519.2/3--dc23/eng/20231031
LC record available at https://lccn.loc.gov/2023046154
LC ebook record available at https://lccn.loc.gov/2023046155

ISBN: 978-0-367-54387-7 (hbk)
ISBN: 978-0-367-54448-5 (pbk)
ISBN: 978-1-003-08935-3 (ebk)

DOI: 10.1201/9781003089353

Typeset in TeXGyreTermes-Regular font
by KnowledgeWorks Global Ltd.

Publisher's note: This book has been prepared from camera-ready copy provided by the authors

Dedication

To

My wife Elena and my daughter Nora for their love and support

Contents

Preface

Integral transforms have existed for two centuries and their historical roots go back to the work of the French mathematicians Pierre Simon Laplace and Joseph Fourier whose work led to the formation of the two seminal integral transforms that were later named after them: the Laplace and Fourier transforms. Integral transforms have proved to be a useful tool in solving many problems in mathematics, applied mathematics, physics and engineering. Hundreds, if not thousands, of articles and books have been written about them.

Although the subject of integral transforms is considered classic with a long history, it has not ceased to grow. Many novel transforms with a plethora of practical applications have been introduced in the last three decades, such as the linear canonical transforms, wavelets and shearlets transforms.

In 1980 an off-shoot of the subject of integral transforms started to emerge in the work of Victor Namias who introduced a new integral transform which he called fractional order Fourier transform and which he used to solve problems in quantum mechanics. Namias's idea of a fractional Fourier transform had appeared implicitly in earlier work by J. Wiener in 1927 and E. U. Condon in 1937, but it was Namias who developed it explicitly and was the first to use the phrase *fractional transform*.

At the beginning, Namias's work received little attention and the one that was received was mainly from mathematicians, such as A. C. McBride, F. H. Kerr and D. Mustard. Almost a decade later, at the beginning of the 1990s, the subject started to take shape and receive more attention when engineers and physicists found numerous practical applications of Namias's transform in signal processing and optics. Some of the pioneers in this field were L. B. Almeida, M. Kutay, A. Lohmann, D. Mendlovic, H. M. Ozaktas and Z. Zalevsky, just to mention a few.

After the publication of his paper on the fractional Fourier transform, Namias extended his work to another integral transform, namely the Hankel transform, and he called the new transform fractional Hankel transform. Namias's work opened a window on new possibilities of extending the fractional transform idea to other integral transforms. Currently there is a slew of fractional integral transforms, such as fractional Hilbert, fractional Stockwell, fractional wavelets, fractional Radon transforms, etc. It is no exaggeration to say that nowadays, practically every integral transform has a fractional version. However, so far most of them have not shown to render any useful practical applications to compete with the fractional Fourier transform.

The notion of fractions existed in mathematics for thousands of years as part of the counting systems used by the Ancient Egyptians, Babylonians and Indian Mayans. The term *fractional* appeared in classical mathematical analysis in the work of Riemann on fractional derivatives and fractional integrals. There is a connection between fractional derivatives and fractional integrals on the one hand and fractional integral transforms on the other hand, which will be discussed in Chapter 3 of this volume.

The fractional Fourier transform has distinguished itself among other fractional integral transforms of being a special case of a more general class of integral transforms that arose in quantum mechanics, time-frequency representations and abstract harmonic analysis. The fractional Fourier transform is a special case of a class of integral transforms known as the *Linear Canonical Transforms* which may be viewed as a group of unitary transformations acting on the Hilbert space of all square-integrable functions on the real line $L^2(\mathbb{R})$. Their action on that Hilbert space is represented by the Metaplectic group.

The idea of this book came to me when some of my colleagues and students, who were interested in learning the rudiments of the subject, asked me for references. I then realized that although there are many books in the literature on integral transforms, there is a very limited number on fractional integral transforms. The first and most noticeable one is the book by Ozaktas, Zalevsky and Kutay, *The fractional Fourier transform with Applications in Optics and Signal Processing,* (2001) which is more than twenty years old. Since the publication of that book, new results and advances in the field have taken place which deserved to be compiled in a publication. Another point of departure from the former book is that the former focused mainly on the fractional Fourier transform and its applications, while this one deals with the general topic of fractional integral transforms and then delves into the study of important fractional transforms, including the fractional Fourier transform.

The book is designed to give the reader an overview of the subject from its infancy until the present state of affairs. It is written at a level that a graduate student in mathematics will find accessible. The book is not meant to be a textbook with exercises and class activities, nevertheless, I provided detailed proofs of results that a student can easily follow. I envisioned this book to be a supplementary reference for a course on integral transforms. The first five chapters should also be accessible to scientists with some basic knowledge of mathematical analysis and could be used as an introductory course on fractional integral transforms, in general, and on the fractional Fourier transform, in particular. I hope this book will be useful to scientists in different fields. In fact, more than fifty percent of the references cited in the bibliography were written by physicists and engineers and are published in engineering and optics research journals.

The book is organized as follows: To make the book self-sufficient, I included in Chapter 1 the preliminary material and basic results on special functions, functional analysis and harmonic analysis that will be used later. Some special topics, like sampling theorems, Poisson summation formula, generalized functions and distributions and uncertainty principle, which are discussed in later chapters, were also included in Chapter 1 but without proofs. However, references where the interested reader can find the proofs are given. Chapter 2 is an introduction to the general topic of integral transforms and its history. It includes a list of more than 25 integral transforms, some are more than hundred years old, like the Laplace transform, and some are recent, like the Gyrator transform.

The general concept of fractional integral transforms is presented in Chapter 3 with a number of examples of such transforms, such as fractional Fourier, fractional Hankel and fractional Jacobi transforms. The connection between fractional derivatives and fractional integrals on the one hand and fractional integral transforms on the other hand is discussed in this chapter.

Chapter 4 focuses on the fractional Fourier transform and its basic properties. It may be viewed as a tour of the fractional Fourier transform's journey; it begins by a historical overview of the development of the transform, then proceeds to unwrap some of its unique properties and finally ends with one of its basic applications in optics. This is the longest chapter in the book because in addition to the discussion of the basic properties of the transform, it explores how the fractional Fourier, fractional Hilbert, fractional Wigner and fractional ambiguity transforms intertwine.

Chapter 5 discusses shift-invariant and sampling spaces in the setting of the fractional Fourier transform. This discussion leads to the introduction of a discrete fractional Fourier transform, fractional convolution structure and fractional Zak transform. Chapter 6 introduces the coupled fractional Fourier transform which is a novel extension of the fractional Fourier transform to two dimensions. The chapter discusses the unique properties of this transform, in particular, its sampling theorem and the associated sampling points configurations. The chapter is concluded by a discussion of a relatively new integral transform that is intimately related to the two-dimensional fractional Fourier transform, and which is called the Gyrator transform. The Gyrator transform, which was also discovered in optics in the year 2000, is obtained from the two-dimensional fractional Fourier by coordinate rotations.

The relationship between the two-dimensional coupled fractional Fourier transform and the four-dimensional Wigner distribution is presented in Chapter 7. This relationship inevitably leads to the study of four-dimensional rotations which is not as well known as three-dimensional rotations. To make the material accessible to junior researchers, a detailed discussion of four-dimensional rotations is presented.

Chapter 8 contains more properties of the coupled fractional Fourier transform, such as its extension, convolution and uncertainty relations. It also introduces the short-time coupled fractional Fourier transform and its basic properties.

Chapter 9 is an introduction to the class of linear canonical transforms which contains the fractional Fourier transform as a special case. Several properties of the linear canonical transform, such as its convolution, sampling theorems and metaplectic group representation are presented. The metaplectic representation of the linear canonical transform facilitates the extension of the transform to higher dimensions and makes the derivation of its properties managable.

Chapter 10, which is the last chapter of the book, introduces the Special Affine Fourier Transform and another variant of it, known as the *offset linear canonical transform.* The special affine Fourier transform is a generalization of the linear canonical transform and is the most general inhomogeneous, lossless linear mapping in phase space. Elementary properties, Poisson summation formula, Zak transform and sampling theorems for the special affine Fourier transform are presented in this chapter.

The book ends with a bibliography that contains a wholesome list of references. Although there are thousands of articles written on these topics, it was impractical to include most of them. I have limited my choices to those references that are closely related to the topics presented in this volume. I apologize to the authors whose work I have missed. The majority of the references are published in engineering and optics research journals which attests to the fact that the subject of fractional integral transforms is not only relevant to mathematicians, but to engineers and physicists as well.

I strived to make some chapters self-contained. To this end, I had to restate results presented in previous chapters and provide proofs with significant level of detail. This might have come at the expense of brevity and elegancy. To appreciate the development of any mathematical topic, one ought to know its history and how it started. To help the reader achieve that, I have included a brief historical introduction to each of the main integral transforms presented in this volume. Inspite of concerted efforts of everyone involved in this book project, one may expect to find some typographic errors which I hope will be few and obvious and will not cause any distraction.

Some of the work presented in this volume is a result of the author's own research and publications, either individually or in collaboration over the years with several colleagues, of whom the most recent are Professors Ayush Bhandari, Firdous Shah, Azhar Tantary and Rajakumar Roopkumar. My collaboration with Professor Bhandari began when he was an undergraduate student at Nanyang Technological University, Singapore. But ironically, I have never met either Professors Shah, Tantary or Roopkumar, but thanks to the internet that made our collaboration possible.

This project took more time than I anticipated because most of the work done on it took place while I was chairing a large mathematics department with more than 70 full-time and part-time instructors, having other university administrative duties and dealing with restricted working conditions under Covid-19 lockdown.

I take this opportunity to acknowledge the encouragement and guidance I have received over the years from Professors Paul Butzer, John Benedetto and Gilbert Walter.

Finally, I would like to express my appreciation to the staff of CRC Press, Taylor & Francis Group, in particular, to Mansi Kabra and Kumar Shashi for their support throughout the production process.

CHAPTER 1

Introduction and Preliminaries

In this preliminary chapter, we introduce some of the definitions and notation that will be used throughout the book, and state some results that will be recalled later on in our presentation.

1.1 NOTATION

The notation used in this book is standard, yet it may differ from that used in physics and engineering literature. For example, the imaginary number, $\sqrt{-1}$, is denoted by i, while in most engineering literature, it is denoted by j.

We use $\mathbb{C}$ and $\mathbb{R}$ to denote the sets of complex and real numbers, respectively, $\mathbb{Z}$ to denote the integers and $\mathbb{N}$ to denote the set of natural numbers. The real and complex n-dimensional Euclidean spaces are denoted by $\mathbb{R}^n$ and $\mathbb{C}^n$; that is, $x \in \mathbb{R}^n\,(\mathbb{C}^n)$ means $x = (x_1, \ldots, x_n), x_i \in \mathbb{R}\,(\mathbb{C}), i = 1, \ldots, n$.

The conjugate, $x - iy$, of a complex number $z = x + iy$ is denoted by $\bar{z}$, the real and imaginary parts of z are denoted by Re z and Im z, respectively. The absolute value and the argument of z will be denoted by $|z|$, $\arg z$, respectively, where

$$|z| = \sqrt{z\bar{z}} = \sqrt{x^2 + y^2} \text{ and } \arg z = \tan^{-1}(y/x).$$

Sometimes, we may use the vector notation $\mathbf{x} = (x_1, \ldots, x_n)$ to emphasize that x has n components. The inner product of $\mathbf{x}, \mathbf{y} \in \mathbb{C}^n$, is defined as

$$\mathbf{x} \cdot \mathbf{y} = \sum_{k=1}^{n} x_k \bar{y}_k,$$

and hence the norm of a vector $\mathbf{x}$ is given by

$$\|\mathbf{x}\| = \sqrt{\sum_{k=1}^{n} |x_k|^2}.$$

The transpose of an $m \times n$, matrix M is an $n \times m$ matrix which will be denoted by M^T and the inverse of an $n \times n$ matrix will be denoted by M^{-1} if it exists.

DOI: 10.1201/9781003089353-1

The differential operator

$$D^{\alpha} = \frac{\partial^{|a|}}{\partial x_1^{\alpha_1} \cdots \partial x_n^{\alpha_n}}$$

has the usual meaning: $\alpha = (\alpha_1, \ldots, \alpha_n)$ is a multi-index with $\alpha_i\,(i = 1, \ldots, n)$ being a non-negative integer and $|\,\alpha\,| = \sum_{i=1}^{n} \alpha_i$.

For $n = 1$, the kth derivative of a function $f(x)$ is denoted by $D^k f(x)$ or $f^{(k)}(x)$, where $D = \frac{d}{dx}$.

The *support* of a function $f(x)$ defined on some open set I in $\mathbb{R}^n$, denoted by supp f, is the closure with respect to I of the set of points where $f(x) \neq 0$. The *characteristic function* of a set A is defined as

$$\chi_A(x) = \begin{cases} 1 & \text{if } x \in A \\ 0 & \text{if } x \notin A. \end{cases}$$

The Heaviside function, $H(x)$, is defined as $H(x) = \chi_{[0,\infty)}(x)$. For a real number $x \neq 0$, we define the sign of x, denoted by $\operatorname{sgn} x$, as $x/|x|$, or equivalently

$$\operatorname{sgn} x = \begin{cases} 1 & \text{if } x > 0, \\ -1 & \text{if } x < 0. \end{cases}$$

The *translation*, *modulation*, and *dilation* operators are defined by

$$T_a f(x) = f(x - a),$$

$$E_a f(x) = e^{iax} f(x),$$

$$D_a f(x) = \frac{1}{\sqrt{|\,a\,|}} f(x/a), a \neq 0,$$

respectively.

The *supremum, infimum, maximum* and *minimum* are abbreviated as $\sup, \inf, \max$ and $\min$, respectively. The Kronecker delta is defined as

$$\delta_{m,n} = \begin{cases} 0 & \text{if } m \neq n \\ 1 & \text{if } m = n. \end{cases}$$

Let I denote an open set in $\mathbb{R}^n (n \geq 1)$ and $d\mu$ be an arbitrary Borel measure on I. We define $L^p(I,\mu)$, for $0 < p < \infty$, as the set of all measurable functions f on I such that $|f|^p$ is integrable on I with respect to $d\mu$, i.e.,

$$L^p(I,\mu) = \left\{ f : f \text{ is measurable on } I \text{ and } \int_I |f(\mathbf{x})|^p \, d\mu(\mathbf{x}) < \infty \right\}.$$

When $I = \mathbb{R}^n$ and $d\mu(\mathbf{x}) = d\mathbf{x}$ (the Lebesgue measure), the integral $\int_I f(\mathbf{x})\, d\mathbf{x}$ will mean

$$\int_{-\infty}^{\infty} \cdots \int_{-\infty}^{\infty} f(x_1, \ldots, x_n)\, dx_1 \cdots dx_n.$$

For $p=\infty$, we define $L^{\infty}(I,\mu)$ as the set of all functions that are essentially bounded on I with respect to μ. A function f is said to be *essentially bounded* on I with respect to μ if there exists a positive real number M such that $\mu\{x\in I:|f(x)|>M\}=0$. In this case, we define the *essential supremum*, abbreviated ess. sup, by

$$\text{ess.}\sup f=\inf\{M:\mu\{x\in I:|f(x)|>M\}=0\}.$$

If $\text{ess.}\sup f=M$, then $|f(x)|\leq M$ for almost all $x\in I$; the exception is a subset E of I with $\mu(E)=0$.

We define $L^p_{loc}(I,\mu)(0<p<\infty)$ as

$$L^p_{loc}(I,\mu)=\{f: f \text{ is measurable on } I \text{ and } \int_J|f(\mathbf{x})|^p d\mu(\mathbf{x})<\infty,$$

$$\text{for every open set } J \text{ whose closure } \bar{J} \text{ is a compact subset of } I\}.$$

For $p\geq 1$ and positive $W(x)$, $L^p(I,W)$ is a Banach space with norm

$$\|f\|_{L^p(I,W)}=\begin{cases}\left(\int_I|f(x)|^pW(x)\,dx\right)^{1/p}, & 1\leq p<\infty\\ \text{ess.sup}_{x\in I}|f(x)|, & p=\infty\end{cases}.$$

For $1\leq p<\infty$, and $f,g\in L^p(I)$, we have Minkowski's inequality :

$$\|f+g\|_{L^p(I)}\leq\|f\|_{L^p(I)}+\|g\|_{L^p(I)}.$$

When $I=\mathbb{R}^n$ and $W=1$, we may write $\|f\|_p$ instead of $\|f\|_{L^p}$.

We call two positive real numbers p and q conjugate numbers if $(1/p)+(1/q)=1$. Let I be a measurable set and f, g be measurable functions on I. Then

$$\int_I|f(x)g(x)|\,d\mu(x)\leq\left(\int_I|f(x)|^p\,d\mu(x)\right)^{1/p}\left(\int_I|g(x)|^q\,d\mu(x)\right)^{1/q}.$$

This inequality is known as *Hölder's inequality*. The case in which $p=q=2$ is known as *Cauchy–Schwarz inequality*.

If $f:[a,b]\to\mathbb{R}$ is a nonnegative Lebesgue integrable function and ϕ is a convex function, then the following Jensen's inequality holds

$$\phi\left(\frac{1}{b-a}\int_a^b f(x)dx\right)\leq\frac{1}{b-a}\int_a^b\phi(f(x))\,dx.$$

The standard convolution of two functions f,g, which is associated with the Fourier transform and is denoted by $*$, is defined as

$$(f*g)(x)=\frac{1}{\sqrt{2\pi}}\int_{\mathbb{R}}f(t)g(x-t)dt.$$

The set of all continuous functions on an open set I is denoted by $C(I)$, and the set of all functions having $k(k\geq 0)$ continuous derivatives on I is denoted by $C^k(I)$. We adopt the notation $C^0(I) = C(I)$. If I is an interval in $\mathbb{R}^1$, it is sometimes convenient to extend the definition of $C^k(I)$ to intervals that are not open. For example, if $I = [a,b)$, we say that $f \in C^k(I)$ if it is in $C^k(J)$, where $J = (a,b)$, and f has k right-hand derivatives at a, and they are continuous from the right at a. Similarly, one can extend the definition of $C^k(I)$ to the intervals $(a,b]$ and $[a,b]$.

The collection of all infinitely differentiable functions on some open set I in $\mathbb{R}^n$ will be denoted by $C^\infty(I)$, while $\mathcal{D}(I)$ will stand for the collection of all functions $f(x)$ that are infinitely differentiable on I such that the support of f is a compact subset of I.

Finally, the set of all infinitely differentiable functions on $\mathbb{R}^n$ that, together with their derivatives of all orders, are rapidly decreasing on $\mathbb{R}^n$, which is known as the Schwartz space of functions, will be denoted by S.

1.2 SPECIAL FUNCTIONS AND ORTHOGONAL POLYNOMIALS

In this section, we introduce some special functions and orthogonal polynomials that will be used in the sequel. For more details, the reader may consult [24, 96, 97, 110, 117, 121, 175, 284, 293, 307, 308, 352].

1.2.1 The Gamma Function

The Γ-function is defined as

$$\Gamma(z) = \int_0^\infty t^{z-1}e^{-t}dt, \qquad \operatorname{Re} z > 0.$$

Hence,

$$\Gamma(z) = \int_0^1 t^{z-1}e^{-t}dt + \int_1^\infty t^{z-1}e^{-t}dt = P(z) + F(z),$$

where

$$P(z) = \int_0^1 t^{z-1}e^{-t}dt = \int_0^1 t^{z-1}\left(\sum_{n=0}^\infty \frac{(-t)^n}{n!}\right)dt = \sum_{n=0}^\infty \frac{(-1)^n}{n!(z+n)},$$

and $F(z) = \int_1^\infty t^{z-1}e^{-t}dt$ is an entire function.

Therefore, the Γ-function is a meromorphic function with simple poles at $z = -n$, $n = 0, 1, 2, \ldots$. The residue at the pole $z = -n$ is $(-1)^n/n!$. It follows from the definition of the Γ-function that $\Gamma(1) = 1$ and

$$\Gamma(z) = \frac{1}{z}\int_0^\infty t^z e^{-t}dt = \frac{1}{z}\Gamma(z+1)$$

or

$$\Gamma(z+1) = z\Gamma(z). \tag{1.1}$$

Thus, if n is a positive integer, we have

$$\Gamma(z+n) = z(z+1)(z+2)\cdots(z+n-1)\Gamma(z),$$

which implies that

$$\Gamma(n+1) = 1.2.3.\cdots n = n!.$$

Since $1/\Gamma(z)$ is an entire function, it can be written in terms of the canonical product of its zeros (the poles of Γ) as suggested by the Hadamard factorization theorem for entire functions. In fact, we have

$$\frac{1}{\Gamma(z)} = ze^{\gamma z}\prod_{n=1}^{\infty}\left[\left(1+\frac{z}{n}\right)e^{-z/n}\right], \tag{1.2}$$

where γ is *Euler's constant* defined by

$$\gamma = \lim_{n\to\infty}\left(\sum_{k=1}^{n}\frac{1}{k} - \ln n\right) = 0.577215\cdots. \tag{1.3}$$

From (1.2), it follows easily that

$$\frac{1}{\Gamma(z)\Gamma(-z)} = -z^2\prod_{n=1}^{\infty}\left(1-\frac{z^2}{n^2}\right) = -\pi^{-1}z\sin(\pi z),$$

or

$$\Gamma(z)\Gamma(-z) = -\frac{\pi}{z\sin(\pi z)} \qquad (z \text{ non-integer}),$$

which gives

$$\Gamma(z)\Gamma(1-z) = \frac{\pi}{\sin(\pi z)} \qquad (z \text{ non-integer}). \tag{1.4}$$

Upon replacing z by $z+1/2$, we obtain

$$\Gamma\left(\frac{1}{2}+z\right)\Gamma\left(\frac{1}{2}-z\right) = \pi\sec(\pi z),$$

in particular, $\Gamma\left(\frac{1}{2}\right) = \sqrt{\pi}$.

1.2.2 The Beta Function

The Beta function $B(x,y)$ is defined as

$$B(x,y) = \int_0^1 t^{x-1}(1-t)^{y-1}dt; \ \text{Re}\, x,\ \text{Re}\, y > 0.$$

In the above integral, if we make the change of variable, $t = 1-u$, we find that

$$B(x,y) = B(y,x),$$

and if we set $t = u/(u+1)$, we obtain

$$B(x,y) = \int_0^{\infty}\frac{u^{x-1}}{(1+u)^{x+y}}du; \qquad \text{Re}\, x,\ \text{Re}\, y > 0.$$

By multiplying the relation

$$\int_0^\infty t^{x+y-1}e^{-(1+v)t}dt = \frac{\Gamma(x+y)}{(1+v)^{x+y}},$$

by v^{x-1}, integrating with respect to v between 0 and ∞, and interchanging the integrals, we obtain the main relation between the gamma and the beta functions

$$B(x,y) = \frac{\Gamma(x)\Gamma(y)}{\Gamma(x+y)}.$$

1.2.3 The Hermite Polynomials $H_n(x)$

1. **Differential Equation:** The Hermite polynomials H_n of degree n satisfy the differential equation

$$y'' - 2xy' + 2ny = 0.$$

2. **Orthogonality Relation:** The Hermite polynomials are orthogonal on the real line with respect to the weight function e^{-x^2}

$$\int_{-\infty}^{\infty} e^{-x^2} H_n(x)H_m(x)dx = \sqrt{\pi}2^n n!\,\delta_{m,n}.$$

3. **Recurrence Relations:**

   (a) **Three-Term Recurrence Relation:**

$$H_n(x) = 2xH_{n-1}(x) - 2(n-1)H_{n-2}(x), \quad n = 2,3,4,\ldots$$

$$H_0(x) = 1 \text{ and } H_1(x) = 2x.$$

   Moreover, $H_n(-x) = (-1)^n H_n(x)$.

   (b) **Recurrence Relation for the Derivative:**

$$\frac{d}{dx}\{H_n(x)\} = 2nH_{n-1}(x),$$

   and

$$\frac{d}{dx}\{H_n(x)\} = 2xH_n(x) - H_{n+1}(x).$$

4. **Generating Function:**

$$\sum_{n=0}^{\infty} \frac{H_n(x)}{n!}w^n = \exp(2xw - w^2).$$

5. **Mehler's Formula** [175, p. 61]: A generating function for the product of Hermite polynomials is given by the formula

$$\sum_{n=0}^{\infty} \frac{H_n(x)H_n(t)}{2^n n!}z^n = \frac{1}{\sqrt{1-z^2}} \exp\left(\frac{2xtz - (x^2+y^2)z^2}{1-z^2}\right).$$

6. **Rodrigues' Formula:**

$$H_n(x) = (-1)^n e^{x^2} \frac{d^n}{dx^n}\left(e^{-x^2}\right).$$

7. **The Hermite Functions:** The Hermite functions are defined as

$$\phi_n(x) = e^{-x^2/2} H_n(x)$$

and they are the eigenfunctions of the Sturm-Liouville problem

$$y'' - x^2 y = \lambda y, \quad -\infty < x < \infty,$$

$$y(-\infty) \text{ and } y(\infty) \text{ are finite,}$$

with $-(2n+1)$ being the corresponding eigenvalues.

One of the most important properties of the Hermite functions for our forthcoming discussion is that they are also eigenfunctions of the Fourier transformation $\mathcal{F}$ with $\exp(in\pi/2)$ being the corresponding eigenvalues, i.e.,

$$\mathcal{F}[\phi_n(x)](w) = e^{in\pi/2} \phi_n(w).$$

8. **Orthogonality Relation:** The Hermite functions are orthogonal in the sense

$$\int_{-\infty}^{\infty} \phi_n(x)\phi_m(x)dx = \sqrt{\pi}2^n n!\, \delta_{m,n}.$$

Moreover, they form an orthogonal basis of $L^2(\mathbb{R})$; hence, if $f \in L^2(\mathbb{R})$, then the expansion

$$f(x) = \sum_{n=0}^{\infty} a_n \phi_n(x), \tag{1.5}$$

is valid in the sense of $L^2(\mathbb{R})$, where

$$a_n = \frac{1}{2^n n!\,\sqrt{\pi}} \int_{-\infty}^{\infty} f(x)\phi_n(x)dx. \tag{1.6}$$

1.2.4 The Laguerre Polynomials $L_n^{\alpha}(x)(\alpha > -1)$

1. **Differential Equation:** The generalized Laguerre polynomials are solutions to the differential equation

$$xy'' + (\alpha + 1 - x)y' + ny = 0.$$

2. **Orthogonality Relation:** The generalized Laguerre polynomials are orthogonal on the half line $[0, \infty)$ with respect to the weight function $x^{\alpha}e^{-x}$, i.e.,

$$\int_0^{\infty} e^{-x} x^{\alpha} L_n^{\alpha}(x) L_m^{\alpha}(x) dx = \frac{\Gamma(n+\alpha+1)}{\Gamma(n+1)} \delta_{m,n}.$$

3. **Generating Functions:** We have

$$\sum_{n=0}^{\infty} L_n^{\alpha}(x) w^n = (1-w)^{-\alpha-1} e^{-xw/(1-w)},$$

and

$$\sum_{n=0}^{\infty} \frac{L_n^{\alpha}(x)}{\Gamma(n+\alpha+1)} w^n = e^w (xw)^{-\alpha/2} J_{\alpha}\left[2\sqrt{xw}\right],$$

where J_ν is the Bessel function.

4. **Generating Function for the Product** [293, 5.1.15]:

$$\sum_{n=0}^{\infty} \frac{n! e^{in\alpha}}{\Gamma(\nu+n+1)} L_n^{\nu}(x^2) L_n^{\nu}(y^2)$$
$$= \frac{(xy)^{-\nu} e^{(i\nu/2)(\pi/2-\alpha)}}{1-e^{i\alpha}} \exp\left[-\frac{e^{i\alpha}(x^2+y^2)}{1-e^{i\alpha}}\right] J_\nu\left(\frac{xy}{\sin(\alpha/2)}\right),$$

5. **Rodrigues' Formula:**

$$L_n^{\alpha}(x) = \frac{x^{-\alpha} e^x}{n!} \frac{d^n}{dx^n}\left(e^{-x} x^{n+\alpha}\right).$$

1.2.5 The Jacobi Polynomials $P_n^{(\alpha,\beta)}(x)(\alpha,\beta > -1)$

1. **Differential Equation:** The Jacobi polynomials are solutions to the differential equation

$$(1-x^2)y'' + [\beta - \alpha - (\alpha+\beta+2)x]\,y' + n(n+\alpha+\beta+1)y = 0.$$

2. **Orthogonality Relation:** They are orthogonal on $[-1,1]$ with respect to the weight function $(1-x)^\alpha(1+x)^\beta$,

$$\int_{-1}^{1} (1-x)^\alpha (1+x)^\beta \, P_n^{(\alpha,\beta)}(x)\, P_m^{(\alpha,\beta)}(x)\, dx =$$
$$\frac{2^{\alpha+\beta+1}}{2n+\alpha+\beta+1} \frac{\Gamma(n+\alpha+1)\Gamma(n+\beta+1)}{\Gamma(n+1)\Gamma(n+\alpha+\beta+1)} \delta_{m,n}.$$

3. **Generating Function:**

$$\sum_{n=0}^{\infty} P(x) w^n = \frac{2^{\alpha+\beta}}{\sqrt{1-2xw+w^2}} \left[1-w+\sqrt{1-2xw+w^2}\right]^{-\alpha}$$
$$\times \left[1+w+\sqrt{1-2xw+w^2}\right]^{-\beta},$$

for w sufficiently small.

4. **Rodrigues' Formula:**

$$P_n^{(\alpha,\beta)}(x) = \frac{(-1)^n}{2^n n!} (1-x)^{-\alpha} (1+x)^{-\beta} \frac{d^n}{dx^n}\left[(1-x)^{n+\alpha}(1+x)^{n+\beta}\right].$$

1.2.6 The Bessel Functions

Using Frobenius' method to solve the differential equation

$$z^2y'' + zy' + (z^2 - \nu^2)y = 0, \qquad \nu \in \mathbb{R}$$

in a neighborhood of the regular singularity $z = 0$, leads to the solution

$$J_\nu(z) = \sum_{k=0}^{\infty} \frac{(-1)^k z^{\nu+2k}}{2^{\nu+2k} k! \Gamma(\nu + k + 1)},$$

and $J_{-\nu}$.

1. **Definition:** The first solution $J_\nu(z)$ is called the Bessel function of the first kind and order ν.

   If ν is not an integer, J_ν and $J_{-\nu}$ are independent. If ν is a negative integer, the factor $1/\Gamma(\nu + k + 1)$ must be replaced by 0 whenever $\nu + \alpha + 1 \leq 0$. For all values of $\nu, z^{-\nu} J_\nu(z)$ is an even entire function in z. It is easy to see that

   $$J_{-\frac{1}{2}}(z) = \sqrt{\frac{2}{\pi z}} \cos z \quad \text{and } J_{\frac{1}{2}}(z) = \sqrt{\frac{2}{\pi z}} \sin z.$$

   Other types of Bessel functions are:

   $$Y_\nu(z) = \frac{1}{(\sin \nu\pi)} \{J_\nu(z) \cos \nu\pi - J_{-\nu}(z)\}, \quad \nu \neq -1, -2, \ldots,$$

   and

   $$K_\nu(z) = \frac{\pi}{(2 \sin \pi\nu)} \{I_{-\nu}(z) - I_\nu(z)\},$$

   where

   $$I_\nu(z) = \sum_{k=0}^{\infty} \frac{(z/2)^{\nu+2k}}{\Gamma(k+1)\Gamma(k+\nu+1)}, \qquad |z|, \infty.$$

   The function Y_ν is called the *Bessel function of the second kind* or *Neumann's function*, K_ν is the *modified Bessel function of the third kind* or *MacDonald's function*, and I_ν is the modified Bessel function of the first kind.

2. **Differential Equation:**

   The functions $y_1 = x^a J_\nu(bx^c)$ and $y_2 = x^a Y_\nu(bx^c)$ are solutions of the differential equation

   $$y'' - (\frac{2a-1}{x})y' + \left(b^2c^2x^{2c-2} + \frac{a^2 - \nu^2c^2}{x^2}\right) y = 0, \; bx^c > 0.$$

3. **Differentiation and Recurrence Relations:**
   We list some differentiation formulas and recurrence relations for the Bessel function

of the first kind; similar formulas and relations for the other types of the Bessel functions can be obtained:

$$\frac{d}{dz}\left[z^{\nu}J_{\nu}(z)\right] = z^{\nu}J_{\nu-1}(z), \tag{1.7}$$

$$\frac{d}{dz}\left[z^{-\nu}J_{\nu}(z)\right] = -z^{-\nu}J_{\nu+1}(z). \tag{1.8}$$

Moreover, we have

$$2J'_{\nu}(z) = J_{\nu-1}(z) - J_{\nu+1}(z), \tag{1.9}$$

and

$$J'_{\nu}(z)J_{-\nu}(z) - J'_{-\nu}(z)J_{\nu}(z) = \frac{2\sin\nu\pi}{\pi z}, \ (\nu \text{ is non-integer}).$$

The three-term recurrence relation is

$$2\nu J_{\nu}(z) = z\left[J_{\nu-1}(z) + J_{\nu+1}(z)\right]. \tag{1.10}$$

4. **Eigenfunctions:** The eigenfunctions of the Sturm-Liouville problem

$$y'' - \left(\frac{\nu^2 - \frac{1}{4}}{x^2}\right)y = \lambda y, \quad 0 \le x \le b, \quad \nu > -\frac{1}{2},$$

$$y(0) = 0 = y(b),$$

are $\phi_n(x) = \sqrt{x}J_{\nu}(\sqrt{\lambda_n}x)$ and the eigenvalues, λ_n, are the zeros of $J_{\nu}(b\sqrt{\lambda})$. The orthogonality relation of the eigenfunctions takes the form

$$\int_0^b xJ_{\nu}\left(\sqrt{\lambda_m}x\right)J_{\nu}\left(\sqrt{\lambda_n}x\right)dx = 0 \text{ if } m \neq n.$$

5. **Integral Representations (Poisson's Integral Representation):**

$$\begin{aligned} J_{\nu}(z) &= \frac{z^{\nu}}{2^{\nu}\sqrt{\pi}\Gamma\left(\nu+\frac{1}{2}\right)}\int_{-1}^{1}(1-t^2)^{\nu-1/2}e^{izt}dt \\ &= \frac{z^{\nu}}{2^{\nu-1}\sqrt{\pi}\Gamma(\nu+1/2)}\int_0^{\pi/2}\cos(z\sin\theta)(\cos\theta)^{2\nu}d\theta, \end{aligned} \tag{1.11}$$

for $\operatorname{Re} v > -\frac{1}{2}$. Moreover, we have

$$J_n(z) = \frac{1}{2\pi}\int_{\alpha}^{\alpha+2\pi}e^{i(n\theta - z\sin\theta)}d\theta, \quad [307, \text{p. } 20] \tag{1.12}$$

and

$$K_{\mu}(z) = \frac{\sqrt{\pi}(z/2)^{\mu}}{\Gamma(\mu+1/2)}\int_0^{\infty}e^{-z\cosh t}(\sinh t)^{2\mu}dt. \quad [307, \text{p. } 172] \tag{1.13}$$

6. **Generating Function:**
A generating function of the Bessel functions of the first kind and integer order is given by

$$\sum_{k=-\infty}^{\infty} J_k(z)t^k = \exp\left[\frac{z}{2}\left(t - \frac{1}{t}\right)\right].$$

7. **Results Related to the Zeros of the Bessel Function:**
The zeros of the Bessel function $J_\nu(z)$ (ν real), i.e., the solutions of the equation $J_\nu(z) = 0$, are all simple, except possibly the zero at the origin. If ν is an integer n, J_n has no complex zeros and has an infinite number of real zeros which are symmetrically located about the origin. The zero at the origin is of order n.

Since $z^{-\nu}J_\nu(z)$ is an entire function in z, it can be expressed in terms of the canonical product of its zeros $\alpha_{\nu,k}$ as suggested by the Hadamard factorization theorem for entire functions. In fact, we have

$$J_\nu(z) = \frac{z^\nu}{2^\nu\Gamma(\nu+1)}\prod_{k=1}^{\infty}\left(1 - \frac{z^2}{\alpha_{\nu,k}^2}\right).$$

8. **Orthogonality Relation:** A very useful property of the Bessel functions is the orthogonality relation

$$\int_0^b xJ_\nu(\alpha_{\nu,m}x/b)J_\nu(\alpha_{\nu,n}x/b)dx = \frac{b^2}{2}\left[J_{\nu+1}(\alpha_{\nu,n})\right]^2 \delta_{m,n};\ m,n = 1,2,3,\ldots, \tag{1.14}$$

where $b > 0,\ \nu > -1$.

1.2.7 The Mittag–Leffler function

The Mittag-Leffler function $E_a(z)$ is defined as

$$E_a(z) = \sum_{k=0}^{\infty}\frac{z^k}{\Gamma(ak+1)}, \quad a > 0.$$

It is an entire function of order $1/a$ that reduces to e^z when $a = 1$. In fact,

$$E_1(z) = e^z, \qquad E_2(z^2) = \cosh z.$$

The *generalized Mittag–Leffler function* $E_{a,b}(z)$ is defined as

$$E_{a,b}(z) = \sum_{k=0}^{\infty}\frac{z^k}{\Gamma(ak+b)}, \quad a,b > 0,$$

and it reduces to the Mittag–Leffler function when $b = 1$. It can also be defined by the integral

$$E_{a,b}(z) = \frac{1}{2\pi i}\int_C \frac{t^{a-b}e^t}{t^a - z}dt,$$

where the integration path, c, is a loop that starts and ends at $-\infty$ and encircles the disc $|t| \leq |z|^{1/a}$ in the positive sense. Moreover, it satisfies the relations

$$\int_0^\infty e^{-t} t^{b-1} E_{a,b}(t^a z)\, dt = \frac{1}{1-z},$$

and

$$\left(\frac{d}{dz}\right)^m [z^{b-1} E_{a,b}(z^a)] = z^{b-m-1} E_{a,b-m}(z^a).$$

1.2.8 The Hypergeometric and q-Hypergeometric Functions

Recall the Pochhammer symbol

$$(a)_0 = 1,\ (a)_n = a(a+1)(a+2)\cdots(a+n-1),\ n = 1,2,3,\ldots.$$

Or

$$(a)_n = \frac{\Gamma(a+n)}{\Gamma(a)},\ n = 0,1,2,\ldots.$$

The hypergeometric function [107]

$${}_pF_q(a_r; b_t; z) = \sum_{n=0}^{\infty} \frac{(a_1)_n \cdots (a_p)_n}{(b_1)_n \cdots (b_q)_n} \frac{z^n}{n!},$$

which converges for all z if $p \leq q$, diverges (except for $z = 0$) if $p > q+1$, and converges for $|z| < 1$ if $p = q+1$.

We use the notation

$$\begin{aligned}
(a;q)_0 &= 1,\quad (a;q)_n = \prod_{k=1}^{n} (1 - aq^{k-1}),\ |q| < 1,\ n = 1,2,\ldots,\infty, \\
(a_1,\ldots,a_m;q)_n &= \prod_{l=1}^{m} (a_l;q)_n, \\
(a;q)_\alpha &= \frac{(a;q)_\infty}{(aq^\alpha,q)_\infty}.
\end{aligned}$$

Note that

$$\lim_{q \to 1^-} \frac{(q^a;q)_n}{(1-q)^n} = (a)_n = a(a+1)\cdots(a+n-1);$$

hence for $a = 1$ the limit is $n!$.

The symbol ${}_{r+1}\phi_r$ stands for the q-hypergeometric function

$${}_{r+1}\phi_r(a_1,\ldots,a_{r+1}; b_1,\ldots,b_r; q,z) = \sum_{n=0}^{\infty} \frac{(a_1,\ldots,a_{r+1};q)_n}{(q,b_1,\ldots,b_r;q)_n} z^n.$$

1.3 NON-ORTHOGONAL BASES AND FRAMES IN A HILBERT SPACE

1.3.1 Non-orthogonal Bases and Frames

In this section, we assume that the reader is familiar with the rudiments of the theory of Hilbert spaces. By a *non-orthogonal expansion* in a Hilbert space $\mathcal{H}$ we mean a series expansion of any $x \in \mathcal{H}$ in the form $x = \sum_n a_n x_n$, where $\{x_n\}$ is a complete set of vectors that is not necessarily orthogonal. Although non-orthogonal expansions are less convenient to work with than orthogonal expansions, they provide a more general frame work; see [31, 67, 73, 74, 127, 357]. This section is devoted to the study of such expansions.

Let $\mathcal{H}$ be a Hilbert space with inner product $\langle \cdot, \cdot \rangle$ and norm $\|x\| = \sqrt{\langle x, x \rangle}$, for $x \in \mathcal{H}$.

Definition 1. *A sequence of vectors $\{x_n\}_{n=1}^{\infty}$ in a Hilbert space $\mathcal{H}$ is said to be a basis (Schauder basis) of $\mathcal{H}$ if to each $x \in \mathcal{H}$, there corresponds a unique sequence of scalars $\{c_n\}_{n=1}^{\infty}$ such that*

$$x = \sum_{n=1}^{\infty} c_n x_n, \tag{1.15}$$

where the convergence is understood to be in the norm, that is

$$\left\| x - \sum_{n=1}^{N} c_n x_n \right\| \to 0 \text{ as } N \to \infty.$$

Recall that a basis $\{x_n\}_{n=1}^{\infty}$ of $\mathcal{H}$ is said to be orthogonal if $\langle x_m, x_n \rangle = 0$, whenever $m \neq n$, and an orthogonal basis is said to be orthonormal if, in addition, $\langle x_n, x_n \rangle = 1$ for all n.

An orthogonal basis is complete in the sense that if $\langle x, x_n \rangle = 0$ for all n, then $x = 0$. Every orthogonal set in a separable Hilbert space is countable and every separable Hilbert space has an orthonormal basis. For orthonormal basis, the expansion (1.15) is given by

$$x = \sum_{n=1}^{\infty} \langle x, x_n \rangle x_n, \tag{1.16}$$

with

$$\|x\|^2 = \sum_{n=1}^{\infty} |\langle x, x_n \rangle|^2. \tag{1.17}$$

More generally, for any $x, y \in \mathcal{H}$

$$\langle x, y \rangle = \sum_{n=1}^{\infty} \langle x, x_n \rangle \overline{\langle y, x_n \rangle}. \tag{1.18}$$

It can be shown [328, p. 29] that every basis $\{x_n\}_{n=1}^{\infty}$ of a Hilbert space possesses a unique *biorthonormal basis* $\{x_n^*\}_{n=1}^{\infty}$, which means that

$$\langle x_m, x_n^* \rangle = \delta_{m,n},$$

and $\{x_n^*\}_{n=1}^{\infty}$, is also a basis of $\mathcal{H}$. Moreover, for every $x \in \mathcal{H}$ we have

$$x = \sum_{n=1}^{\infty} \langle x, x_n^* \rangle x_n = \sum_{n=1}^{\infty} \langle x, x_n \rangle x_n^*.$$

Such a non-orthogonal expansion is called a *biorthogonal expansion*. If $\langle x_m, x_n^* \rangle = 0$ whenever $m \neq n$, but $\langle x_n, x_n^* \rangle$ is not necessarily equal to one, $\{x_n^*\}_{n=1}^{\infty}$ will be called a biorthogonal basis of $\{x_n\}_{n=1}^{\infty}$. In this case, we have for any $x \in \mathcal{H}$

$$x = \sum_{n=1}^{\infty} \left(\bar{d}_n\right)^{-1} \langle x, x_n^* \rangle x_n = \sum_{n=1}^{\infty} (d_n)^{-1} \langle x, x_n \rangle x_n^*, \tag{1.19}$$

where $d_n = \langle x_n^*, x_n \rangle \neq 0$. It can be verified that for all $n, d_n \neq 0$.

Definition 2. *Let $\mathcal{G} = \{g_n\}$ be a basis in a separable Hilbert space $\mathcal{H}$. Then*

1. *$\mathcal{G}$ is called unconditional if*

$$\sum c_n g_n \in \mathcal{H} \text{ implies that } \sum |c_n| g_n \in \mathcal{H}.$$

2. *$\mathcal{G}$ is said to be bounded if there exist two non-negative numbers A and B such that for all n*

$$A \leq \|g_n\| \leq B.$$

3. *$\mathcal{G}$ is said to be a Riesz basis if there exist a topological isomorphism $T : \mathcal{H} \to \mathcal{H}$ and an orthonormal basis $\{u_n\}$ of $\mathcal{H}$ such that $T g_n = u_n$ for each n.*

Definition 3. *Let $\mathcal{G} = \{g_n\}$ be a sequence in $\mathcal{H}$ (not necessarily a basis of $\mathcal{H}$). Then $\mathcal{G}$ is called a frame if there exist two numbers $A, B > 0$ such that for any $f \in \mathcal{H}$, we have*

$$A \|f\|^2 \leq \sum_n |\langle f, g_n \rangle|^2 \leq B \|f\|^2.$$

The numbers A and B are called the frame bounds. The frame is said to be tight if $A = B$ and is exact if it ceases to be a frame whenever any single element is deleted from the frame. To every frame $\mathcal{G}$ there corresponds an operator S, known as the frame operator, which maps $\mathcal{H}$ into itself and is defined by

$$Sf = \sum_n \langle f, g_n \rangle g_n, \text{ for all } f \in \mathcal{H}.$$

In view of the definition, frames are complete, since if $\langle f, g_n \rangle = 0$ for all n, then

$$A \|f\|^2 \leq \sum_n |\langle f, g_n \rangle|^2 = 0,$$

which implies that $f = 0$.

For the proof of the following theorem, see [357, pp. 261].

Theorem 1. *In a separable Hilbert space $\mathcal{H}$ the following conditions are equivalent:*

1. *$\{g_n\}$ is an exact frame ,*
2. *$\{g_n\}$ is a bounded unconditional basis,*
3. *$\{g_n\}$ is a Riesz basis.*

If $\{g_n\}$ is an orthonormal basis of $\mathcal{H}$, then the Parseval equality

$$\|f\|^2 = \sum_n |\langle f, g_n \rangle|^2, \qquad f \in \mathcal{H}$$

shows that an orthonormal basis is a tight exact frame with frame bounds $A = B = 1$.

If $\{g_n\}$ is a tight frame with frame bound $A = 1 = B$, and if $\|g_n\| = 1$ for all n, then $\{g_n\}$ is an orthonormal basis. For, if $\{g_n\}$ is a tight frame with frame bound $= 1$, then

$$\|g_k\|^2 = \sum_n |\langle g_k, g_n \rangle|^2 = \|g_k\|^4 + \sum_{n \neq k} |\langle g_k, g_n \rangle|^2.$$

But since $\|g_k\| = 1$, it follows that $\langle g_k, g_n \rangle = 0$ for all $n \neq k$. The completeness of $\{g_n\}$ is a consequence of the fact that frames are complete.

The following theorem summarizes some of the main properties of frames and frame operators that will be needed later. For a proof, see [357, pp. 263-265].

Theorem 2. *Let $\mathcal{H}$ be a separable Hilbert space and $\{g_n\}_{n=-\infty}^{\infty} \subset \mathcal{H}$ be a frame with frame bounds A and B. Then*

1. *The frame operator $Sf = \sum_n \langle f, g_n \rangle g_n$ is a bounded linear operator on $\mathcal{H}$ with $AI \leq S \leq BI$, where I is the identity operator.*
2. *S is invertible with $B^{-1}I \leq S^{-1} \leq A^{-1}I$. Moreover, S^{-1} is a positive operator; hence it is self-adjoint.*
3. *$\{S^{-1}g_n\}$ is a frame with frame bounds B^{-1}, A^{-1}.*
4. *Every $f \in \mathcal{H}$ can be written in the form*

$$f = \sum_n \left\langle f, S^{-1}g_n \right\rangle g_n = \sum_n \langle f, g_n \rangle S^{-1} g_n. \tag{1.20}$$

5. *If there exists a sequence of scalars $\{c_n\}$ such that $f = \sum_n c_n g_n$, then*

$$\sum_n |c_n|^2 = \sum_n |a_n|^2 + \sum_n |a_n - c_n|^2,$$

where $a_n = \langle f, S^{-1} g_n \rangle$.

6. *In addition, if $\{g_n\}$ is an exact frame, then $\{g_n\}$ and $\{S^{-1}g_n\}$ are biorthonormal, i.e., $\langle g_m, S^{-1}g_n \rangle = \delta_{m,n}$.*

The frame $\{S^{-1}g_n\}$ is called the *dual frame* of the frame $\{g_n\}$. Equation (1.20) is the analogue of (1.19) for frames and the two coincide if any of the conditions of Theorem 1 is satisfied, and in this case $S^{-1}g_n = g_n^*$.

Part (5) shows that although the expansion of f in terms of the frame $\{g_n\}$ is not necessarily unique, using the expansion with coefficients $a_n = \langle f, S^{-1}g_n \rangle$ is the most economical in the sense that for any other coefficients $\{c_n\}$ with $f = \sum_n c_n g_n$, we have $\sum_n |c_n|^2 \geq \sum_n |a_n|^2$.

1.3.2 Reproducing-Kernel Hilbert Spaces

Let $\mathcal{H}$ be a Hilbert space consisting of complex-valued functions defined on some set X. A function $K(x,t)$, $x,t \in X$ is called a reproducing kernel of $\mathcal{H}$ if

1. For every t, $K(x,t)$ as a function of x belongs to $\mathcal{H}$,
2. For every $x,t \in X$ and $f \in \mathcal{H}$
$$f(t) = \langle f(x), K(x,t) \rangle.$$

A Hilbert space $\mathcal{H}$ is called a reproducing-kernel Hilbert space (RKHS) if it possesses a reproducing kernel. A Hilbert space $\mathcal{H}$ has a reproducing kernel if and only if for every $x \in X$, the evaluation functional $\mathcal{F}(f) = f(x)$, $f \in \mathcal{H}$ is continuous for every $x \in X$, i.e. there exists a constant $C > 0$ such that

$$|f(x)| \leq C\|f\|_{\mathcal{H}},$$

for all $f \in \mathcal{H}$, [21].

By the Riesz representation theorem, for each $x \in X$ there exists a unique element $K_x \in \mathcal{H}$ such that for all $f \in \mathcal{H}$, $f(x) = \langle f, K_x \rangle_{\mathcal{H}}$. The reproducing kernel is given by

$$K(x,t) = \langle K_x, K_t \rangle_{\mathcal{H}}.$$

The reproducing kernel if it exists, it is unique. Below is a short list of some of the properties of reproducing kernels that will be needed in the sequel:

1. $K(x,t) = \overline{K(t,x)}$
2. $K(x,x) \geq 0$.
3. $|K(x,t)| \leq \sqrt{K(x,x)}\,\sqrt{K(t,t)}, \quad x,t \in X$
4. Every finite-dimensional function space $\mathcal{H}$ is a RKHS. In fact, if $\{e_k(x)\}_{k=0}^{m}$ is an orthonormal basis for $\mathcal{H}$, then
$$K(x,t) = \sum_{k=0}^{m} e_k(x)\overline{e_k}(t)$$
is a reproducing kernel. More generally, we have
5. If $\mathcal{H}$ is a separable reproducing-kernel Hilbert space with orthonormal basis $\{e_k(x)\}_{k=0}^{\infty}$, then
$$K(x,t) = \sum_{k=0}^{\infty} e_k(x)\overline{e_k}(t)$$
is the reproducing kernel. Not every separable Hilbert space is a RKHS.

Definition 4. *A basis* $\{S_n(x)\}_{n=0}^{\infty}$ *of a reproducing-kernel Hilbert space* $\mathcal{H}$ *is called a sampling basis if there exists a sequence* $\{t_n\}$ *in* X *such that for every* $f \in \mathcal{H}$

$$f(x) = \sum_n f(t_n)S_n(x),$$

where the series converges in the norm.

Let $\{S_n(x)\}_{n=0}^{\infty}$ be an orthonormal basis of a reproducing-kernel Hilbert space $\mathcal{H}$. Then $\{S_n\}_{n=0}^{\infty}$ is a sampling basis if and only if $K(x,t_n) = S_n(x)$, where $K(x,t)$ is the reproducing kernel. For, Since $\{S_n\}_{n=0}^{\infty}$ is an orthonormal basis, we have

$$f(x) = \sum_{n=0}^{\infty} \langle f, S_n \rangle S_n(x).$$

Therefore, if $\{S_n(x)\}_{n=0}^{\infty}$ is a sampling basis, we have

$$f(t_n) = \langle f(x), S_n(x) \rangle = \langle f(x), K(x,t_n) \rangle.$$

Hence $S_n(x) = K(x,t_n)$. Conversely, Since

$$f(x) = \sum_{n=0}^{\infty} \langle f(t), S_n(t) \rangle S_n(x),$$

it follows that

$$f(x) = \sum_{n=0}^{\infty} \langle f(t), K(t,t_n) \rangle S_n(x) = \sum_{n=0}^{\infty} f(t_n)S_n(x),$$

hence, $\{S_n(x)\}$ is a sampling basis.

Because in a separable Hilbert space every basis possesses a biorthogonal basis, the above results can be extended to the case in which $\{S_n(x)\}$ is not necessarily an orthogonal basis. For details, see [357, Section 10.1].

1.4 SHIFT-INVARIANT SPACES

Shift-invariant spaces have been the focus of many research papers in recent years because of their close connection with sampling theory [5, 95, 299], wavelets and multiresolution analysis [84, 302, 303, 304]. They have many applications in signal and image processing [34, 36]. For example, in many signal processing applications, it is of interest to represent a signal as a linear combination of shifted versions of some basis functions, $\varphi_1, \ldots, \varphi_n$, called the generators of the space, that generate a stable basis for the space.

Shift-invariant spaces came to light with the introduction of wavelets and multiresolution analysis. Recall [84, 191] that the scaling function ϕ of a multiresolution analysis is assumed to satisfy the following conditions:

1. The translates of ϕ, i.e., $\{\phi(x-n)\}_{n\in\mathbb{Z}}$ are orthonormal on the real line, i.e.

$$\int_{\mathbb{R}} \phi(x-m)\phi(x-n)dx = 1. \tag{1.21}$$

2. ϕ satisfies the two-scale relation

$$\phi(x) = \sqrt{2} \sum_{n=-\infty}^{\infty} \alpha_n \phi(2x-n), \quad \text{with} \quad \sum_{n=-\infty}^{\infty} |\alpha_n|^2 < \infty \tag{1.22}$$

where $\alpha_n = \langle \phi, \phi_{-1,n} \rangle$ and $\phi_{-1,n}(x) = \sqrt{2}\phi(2x-n)$.

By taking the Fourier transform of (1.21) and (1.22), one can show that the equivalence of these two conditions in the Fourier transform domain are:

$$\sum_{k=-\infty}^{\infty} |\hat{\phi}(w+2\pi k)|^2 = 1 \quad \text{a.e.}$$

and

$$\hat{\phi}(w) = h\left(\frac{w}{2}\right)\hat{\phi}\left(\frac{w}{2}\right),$$

where

$$h\left(\frac{w}{2}\right) = \frac{1}{\sqrt{2}} \sum_{n=-\infty}^{\infty} \alpha_n e^{inw/2}.$$

The function $h(w)$ is a periodic function with period 2π that belongs to $L^2[0,2\pi]$. Some of these ideas permeated to the general theory of shift-invariant and sampling spaces.

We consider only subspaces of $L^2(\mathbb{R})$ with one generator in the form

$$\mathcal{V}(\varphi) = \left\{ f(t) = \sum_{n=-\infty}^{+\infty} c[n]\varphi(t-n), \varphi \in L^2(\mathbb{R}), c = \{c[n]\} \in \ell_2 \right\}, \tag{1.23}$$

where ℓ_2 is the space of all square-summable sequences and $\|c\|_{\ell_2}^2$ is the squared ℓ_2-norm of the sequence.

The closure of $\mathcal{V}(\varphi)$ is a closed subspace of $L^2(\mathbb{R})$. Furthermore, it is shift-invariant in the sense that for all $f \in \mathcal{V}(\varphi)$, its shifted version, $f(\cdot - k) \in \mathcal{V}(\varphi)$, $k \in \mathbb{Z}$, where $\mathbb{Z}$ denotes the set of integers.

For the basis functions to be stable, it is required that the family of functions $\{\varphi(t-n)\}_{n=-\infty}^{\infty}$ forms a Riesz basis or equivalently, there exists two positive constants $0 < \eta_1, \eta_2 < +\infty$, such that

$$\forall c \in \ell_2, \quad \eta_1 \|c\|_{\ell_2}^2 \leqslant \left\| \sum_{n=-\infty}^{\infty} c[k]\varphi(t-k) \right\|_{L^2}^2 \leqslant \eta_2 \|c\|_{\ell_2}^2 \tag{1.24}$$

If we denote the Fourier transform of a function $h(t)$ by $\widehat{h}(\omega) = \frac{1}{\sqrt{2\pi}} \int_{-\infty}^{+\infty} h(t)e^{-j\omega t}dt$, we obtain the Fourier domain equivalent of (1.24) as

$$\eta_1 \leqslant \sum\nolimits_{n=-\infty}^{+\infty} |\widehat{\varphi}(\omega + 2\pi n)|^2 \leqslant \eta_2. \tag{1.25}$$

The ratio $\rho = \eta_2/\eta_1$ is called the *condition number* of the Riesz basis. The basis is shift-orthonormal [1] or a tight frame if $\rho = 1$.

---

[1]Shift-orthonormality means that $\langle \varphi, \varphi(\cdot - k) \rangle = \delta_k$ where $\langle x, y \rangle = \int_{-\infty}^{+\infty} x(t)y^*(t)dt$ is the L^2-inner product and $\delta_k = \begin{cases} 1, & \text{if } k = 0 \\ 0, & \text{if } k \neq 0 \end{cases}$ denotes the Kronecker delta.

1.5 GENERALIZED FUNCTIONS AND DISTRIBUTIONS

In this section, we give a brief introduction to the theory of generalized functions (distributions). The reader may consult the following references for more details [43, 52, 108, 204, 207, 208, 209, 258, 352, 361].

1.5.1 Testing-Function Spaces and Their Duals

Generalized functions are defined as elements of the dual space of some testing-function space. Therefore, to introduce generalized functions, we first need to introduce the concept of a testing-function space.

Definition 5. *Let I be an open subset of $\mathbb{R}^n$. A set of functions $V(I)$ is said to be a testing-function space (on I) if:*

1. *all the elements of $V(I)$ are infinitely differentiable functions on I, i.e., they belong to $C^\infty(I)$,*
2. *$V(I)$ is either a complete countably normed or a complete countable-union space,*
3. *the convergence of a sequence $\{\phi_m\}_{m=1}^{\infty}$ to ϕ_0 in $V(I)$ implies that $\lim_{m\to\infty} D^k\phi_m = D^k\phi_0$ uniformly on every compact subset of I, for all $|k| = 0, 1, 2, \ldots$.*

Examples of testing-function spaces:

1. The space $\mathcal{E}(I)$ of all infinitely differentiable functions on I. When $I = R^n$, we write $\mathcal{E}$.
2. The space $\mathcal{P}_T \subset \mathcal{E}$, of all infinitely differentiable periodic functions with period T.
3. The space $\mathcal{D}(I)$, of all infinitely differentiable functions with compact support in I and when $I = \mathbb{R}^n$, we write $\mathcal{D}$.
4. The Schwartz space of functions S which is the space of all rapidly decreasing infinitely differentiable functions, i.e., the space of infinitely differentiable functions in $\mathbb{R}^n$ that satisfy the relations

$$\gamma_{m,|k|}(\phi) = \sup_{\mathbf{x}\in\mathbb{R}^n} \left|\left(1+|\mathbf{x}|^2\right)^m D^k\phi(\mathbf{x})\right| < \infty, \quad , m, |k| = 0, 1, 2, \ldots,$$

where $\gamma_{m,|k|}(\phi)$ is a semi-norm.

Definition 6. *The dual space $V^*(I)$ of $V(I)$, which is the space of all continuous linear functionals acting on $V(I)$, is called a space of generalized functions on I and any element f of $V^*(I)$ is called a generalized function on I. The action of f on $\phi \in V(I)$ is denoted by $\langle f, \phi\rangle$.*

The spaces of generalized functions are complete. If f is a conventional function that is Lebesgue measurable on I such that for any $\phi \in V(I)$, the integral

$$\int_I f(\mathbf{x})\phi(\mathbf{x})d\mathbf{x}$$

exists in the Lebesgue sense, and if for any sequence $\{\phi_m\}_{m=1}^{\infty}$ converging to ϕ_0 in $V(I)$, we have

$$\int_I f(\mathbf{x})\phi_m(\mathbf{x})d\mathbf{x} \to \int_I f(\mathbf{x})\phi_0(\mathbf{x})d\mathbf{x}$$

as $m \to \infty$, we say that f is a regular generalized function belonging to $V^*(I)$. More precisely, f can be identified with an element $\tilde{f}$ in $V^*(I)$ where

$$\left\langle \tilde{f}, \phi \right\rangle = \int_I f(\mathbf{x})\phi(\mathbf{x})d\mathbf{x},$$

for every $\phi \in V(I)$. In view of the linearity and the continuity of the integral, it readily follows that $\tilde{f} \in V^*(I)$. We identify f with $\tilde{f}$ and write

$$\langle f, \phi \rangle = \int_I f(\mathbf{x})\phi(\mathbf{x})d\mathbf{x};$$

hence $f \in V^*(I)$. A generalized function $f \in V^*(I)$ that is not regular is called singular.

Definition 7. *A generalized function $f \in V^*(I)$ is said to vanish in an open set $J \subset I$ if for any $\phi \in V(I)$ with supp $\phi \subset J$, we have $\langle f, \phi \rangle = 0$. The null set of f is the union of all open subsets J of I on which f vanishes. The complement of the null set of f with respect to I is called the support of f. f is said to be concentrated in a subset A of I if A contains the support of f.*

Thus, the support of f, written supp$f$, is the smallest closed set B (with respect to I) such that f is zero on $I - B$.

Definition 8. *Two generalized functions f and g in $V^*(I)$ are said to be equal on an open set $J \subset I$ if and only if $f - g$ vanishes on J. A generalized function is said to be regular on J if it coincides with a regular generalized function concentrated in J.*

The limit of a sequence of regular generalized functions may not be regular. For example, consider $f_\epsilon(x) = (1/x)\,\Phi_\epsilon(x), \epsilon > 0$, where $\Phi_\epsilon(x)$ is the characteristic function of the set $(-\infty, -\epsilon) \cup (\epsilon, \infty)$. Then, for any $\phi \in \mathcal{D}$

$$\langle f_\epsilon, \phi \rangle = \int\limits_{|x|>\epsilon} \frac{\phi(x)}{x} dx,$$

and it follows that $\lim_{\epsilon \to 0} f_\epsilon = f$, where $f(x) = 1/x$ if $x \neq 0$, yet f is not locally integrable in any neighborhood of the origin; hence it is not regular.

1.5.2 Spaces of Generalized Functions

Now we give examples of spaces of generalized functions, but first let us observe that the spaces $\mathcal{E}, \mathcal{P}_T, \mathcal{D}, \mathcal{S}, \mathcal{E}(I)$ and $\mathcal{D}(I)$ are all testing-function spaces; therefore, their dual spaces are spaces of generalized functions. Moreover, since $\mathcal{D} \subset \mathcal{S} \subset \mathcal{E}$ and $\mathcal{D}$ is dense in $\mathcal{E}$, it follows that $\mathcal{D}^* \supseteq \mathcal{S}^* \supset \mathcal{E}^*$, and similarly $\mathcal{E}^*(I) \subset \mathcal{D}^*(I)$.

1. The space $\mathcal{D}^*$ is known as the space of *Schwartz distributions* and $\mathcal{D}^*(I)$ as the space of *Schwartz distributions* on I.

2. The space $\mathcal{E}^*$ is the space of all Schwartz distributions with compact support, i.e., $f \in \mathcal{E}^*$ if and only if $f \in \mathcal{D}^*$ and f has compact support.

3. The space $\mathcal{P}_T^*$ is called the space of periodic distributions with period T.

4. The space $\mathcal{S}^*$ is the space of *tempered distributions*. It contains all locally integrable functions with polynomial growth at $\pm\infty$ (cf. Section 1.1), which are sometimes called *distributions with slow growth*.

1.5.3 A Special Type of Generalized Functions

In this section, we introduce the δ-function and some of its generalizations because of the important role they play in function transformations.

1. The *Dirac-delta function* , $\delta(\mathbf{x})$, is defined by

$$\langle \delta, \phi \rangle = \phi(\mathbf{0}), \ \phi \in \mathcal{D}.$$

It is a prototype of singular generalized functions. Showing that it is indeed singular is easy. Assume to the contrary that it is regular. Hence, there exists a locally integrable function f such that

$$\int_{\mathbb{R}^n} f(\mathbf{x})\phi(\mathbf{x})\, d\mathbf{x} = \phi(\mathbf{0}), \text{ for all } \phi \in \mathcal{D}.$$

For $a > 0$, let

$$\phi_a(\mathbf{x}) = \begin{cases} \exp(a^2/\left(|\mathbf{x}|^2 - a^2\right), & |\mathbf{x}| < a \\ 0, & |\mathbf{x}| \geq a \end{cases}.$$

Then $\phi_a \in \mathcal{D}$ and

$$\int\limits_{\mathbb{R}^n} f(\mathbf{x})\phi_a(\mathbf{x})d\mathbf{x} = \int\limits_{\bar{B}_a(\mathbf{0})} f(\mathbf{x})\phi_a(\mathbf{x})d\mathbf{x} = \frac{1}{e},$$

where $\bar{B}_a(\mathbf{0})$ is the closed ball with center at $\mathbf{0} = (0,\ldots,0)$ and radius a, i.e., $\bar{B}_a(\mathbf{0}) = \{\mathbf{x} \in \mathbb{R}^n, |\mathbf{x}| \leq a\}$. But if f were a locally integrable function, Lebesgue's theorem would imply that

$$\lim_{a\to 0} \int\limits_{\bar{B}_a(\mathbf{0})} f(\mathbf{x})\phi_a(\mathbf{x})d\mathbf{x} = 0,$$

which is a contradiction.

2. It can be shown that

$$\lim_{w\to 0} \frac{e^{-x^2/w^2}}{\sqrt{\pi}w} = \delta(x). \tag{1.26}$$

3. From the inversion formula of the Fourier transform of the delta function, we obtain

$$\frac{1}{2\pi}\int_{-\infty}^{\infty} e^{iw(x-t)}dw = \delta(x-t). \tag{1.27}$$

4. Let S be a manifold in $\mathbb{R}^n$of dimension less than $n, d\sigma$ be the induced Lebesgue measure on S and f be a locally integrable function. For any $\phi \in \mathcal{D}$, let

$$\langle f, \phi \rangle = \int\limits_S f(\mathbf{x})\phi(\mathbf{x})d\sigma.$$

This equation defines a continuous linear functional or a generalized function that is concentrated in S.

5. If S is the hyperplane whose equation is $x_1 = 0$, we have

$$\langle \delta(x_1), \phi(\mathbf{x}) \rangle = \int \delta(x_1)\phi(\mathbf{x})d\mathbf{x} = \int \left(\int \delta(x_1)\phi(x_1, \ldots, x_n)dx_1 \right) dx_2 \cdots dx_n$$

$$= \int \phi(0, x_2, \ldots, x_n)dx_2 \cdots dx_n.$$

Similarly,

$$\left\langle \delta^{(k)}(x_1), \phi(\mathbf{x}) \right\rangle = \int \delta^{(k)}(x_1)\phi(\mathbf{x})d\mathbf{x} = (-1)^k \int \left. \frac{\partial^k \phi}{\partial x_1^k} \right|_{x_1=0} dx_2 \cdots dx_n.$$

1.6 SAMPLING AND THE PALEY-WIENER SPACE

In this section, we discuss how functions, under certain conditions, can be reconstructed from their values (samples) at a discrete set of points. One of the fundamental theorems in this setting is the Whittaker-Shannon-Kotel'nikov (WSK) sampling Theorem [167, 264, 315] which shows how a function $f \in L^2(\mathbb{R})$ whose Fourier transform $\hat{f}$ has support on an interval $[-\sigma, \sigma]$ can be reconstructed from its samples.

Such a function is called bandlimited to $[-\sigma, \sigma]$, and σ is called the bandwidth. The term bandlimited functions (signals) came from electric engineering where it means that the frequency content of a time signal $f(t)$ is limited by certain bounds from below and above [342]. In communication engineering when f represents a signal, its energy is represented by its L^2 norm and, hence; the WSK theorem is a tool to reconstruct finite energy, bandlimited signals from their samples.

The WSK theorem, which has numerous practical applications [95, 149, 299, 357], has been extended to integral transforms other than the Fourier transform, such as the Fresnel, Bessel-Hankel, Hartley, Sturm-Liouville transforms, and many more integral transforms [113, 130, 349, 359, 360]. Sampling theorems for fractional integral transforms will be discussed in details in later chapters, but for more information on sampling theory, the reader may consult the following references [19, 30, 128, 129, 168, 176, 229, 231, 351, 356, 357, 358].

Before introducing the WSK theorem, we recall the following result by Paley and Wiener, which gives a description of the class of functions bandlimited to $[-\sigma, \sigma]$ and which is denoted by PW_σ.

Theorem 3 (Paley-Wiener, [228]). *A function f is band-limited to $[-\sigma, \sigma]$ if and only if*

$$f(t) = \int_{-\sigma}^{\sigma} e^{-i\omega t} g(\omega)\, d\omega \qquad (t \in \mathbb{R}), \tag{1.28}$$

for some function $g \in L^2(-\sigma,\sigma)$ and if and only if f is an entire function of exponential type σ that is square integrable on the real line, i.e., f is an entire function such that

$$|f(z)| \le \sup_{x\in\mathbb{R}} |f(x)| \exp(\sigma\,|y|), \quad z = x+iy,$$

and

$$\int_{\mathbb{R}} |f(x)|^2\,dx < \infty.$$

For an extension of the Paley-Wiener Theorem , see [242]

The Whittaker-Shannon-Kotel'nikov (WSK) sampling theorem states:

Theorem 4. *Let f be a function band-limited to $[-\sigma,\sigma]$, then f can be reconstructed from its samples, $f(k\pi/\sigma)$, that are taken at the equally spaced nodes $k\pi/\sigma$ on the time axis $\mathbb{R}$ via the construction formula*

$$f(t) = \sum_{k=-\infty}^{\infty} f\left(\frac{k\pi}{\sigma}\right)\frac{\sin(\sigma t - k\pi)}{(\sigma t - k\pi)} \qquad (t\in\mathbb{R}), \tag{1.29}$$

the series being absolutely and uniformly convergent on $\mathbb{R}$. See, e.g., [357, p. 16].

The points $\{t_k = k\pi/\sigma\}$ are called the sample points and the functions

$$S_k(t) = \frac{\sin\sigma(t-t_k)}{\sigma(t-t_k)} = \text{sinc}\,(\sigma(t-t_k)/\pi), \tag{1.30}$$

where

$$\text{sinc}\,(z) = \begin{cases} \sin\pi z/(\pi z), & z \neq 0 \\ 1, & z = 0, \end{cases}$$

are called the sampling functions. It should be noted that the sampling functions in this case are shifts of one function, namely the sinc function. The space of functions bandlimited to $[-\sigma,\sigma]$ may also be denoted by B_σ, because it is a special case of more general spaces of functions known as the Bernstein spaces. The space of all functions bandlimited to a compact set $S\subset\mathbb{R}$ may be denoted by $B(S)$.

The sampling theorem may be viewed from a different perspective as a by-product of the theory of reproducing-kernel Hilbert spaces which was discussed in a previous section. To this end, let us note that the space $L^2(\mathbb{R})$ is not a reproducing-kernel Hilbert space, but one of its subspaces, namely the Paley-Wiener space PW_π of all functions bandlimited to $[-\pi,\pi]$ is a RKHS with reproducing kernel

$$K(x,t) = \frac{\sin\pi(x-t)}{\pi(x-t)}.$$

Moreover,

$$\left\{K(x,t_n) = \frac{\sin\pi(x-t_n)}{\pi(x-t_n)}, \quad t_n = n, n = 0,\pm1,\pm2,\dots\right\}$$

is an orthonormal sampling basis for the space PW_π. Consequently, we have

$$\frac{\sin\pi(x-t)}{\pi(x-t)} = \sum_{n=-\infty}^{\infty}\frac{\sin\pi(x-n)}{\pi(x-n)}\frac{\sin\pi(t-n)}{\pi(t-n)},$$

and for any $f \in PW_\pi$, the following formula holds

$$f(x) = \sum_{n=-\infty}^{\infty} f(n) \frac{\sin \pi(x-n)}{\pi(x-n)}.$$

The series in Equation (1.29) can be put in the Lagrange-type interpolation form

$$f(t) = \sum_{k=-\infty}^{\infty} f(t_k) \frac{G(t)}{(t-t_k)G'(t_k)}, \tag{1.31}$$

where $t_k = k\pi/\sigma$ and $G(t) = \sin \sigma t = \sigma t \prod_{k=1}^{\infty} (1 - t^2/t_k^2)$.

We have pointed out that the sampling theorem has been extended to integral transforms other than the Fourier transform. The key link to that extension lies in Kramer's sampling theorem which may be stated as follows:

Theorem 5 (Kramer [168]). *Let there exist a function $K(x,t)$ continuous in t such that $K(x,t) \in L^2(I)$ as a function of x for every real number t, where I is a finite interval $I = [a,b]$. Assume that there exists a sequence of real numbers $\{t_n\}_{n\in\mathbb{Z}}$ such that $\{K(x,t_n)\}_{n\in\mathbb{Z}}$ is a complete orthogonal family in $L^2(I)$. Then for any function of the form*

$$F(t) = \int_a^b f(x)\overline{K}(x,t)\, dx = \langle f, K \rangle, \tag{1.32}$$

with $f \in L^2(I)$, we have

$$F(t) = \sum_{n=-\infty}^{\infty} F(t_n) S_n^*(t), \tag{1.33}$$

where

$$S_n^*(t) = \frac{\int_a^b \overline{K}(x,t)K(x,t_n)dx}{\int_a^b |K(x,t_n)|^2 dx}. \tag{1.34}$$

If there exists a non-negative function $g(x) \in L^1(I)$ such that $|K(x,t)|^2 \le g(x)$, for almost all $x \in I$ and all t, then the series converges uniformly on compact subsets of $\mathbb{R}$

The WSK Sampling Theorem is a special case of Kramer's sampling theorem because if we take $I = [-\sigma,\sigma]$, $K(x,t) = e^{ixt}$ and $t_n = \frac{n\pi}{\sigma}$, it is easy to see that $\{e^{it_n t}\}_{n=-\infty}^{\infty}$ is a complete orthogonal set in $L^2(I)$, and in addition

$$S_n^*(t) = \frac{\sin \sigma(t-t_n)}{\sigma(t-t_n)} = \text{sinc}\,(\sigma(t-t_n)/\pi),$$

where

$$\text{sinc}\,(z) = \begin{cases} \sin \pi z/(\pi z), & z \neq 0 \\ 1, & z = 0. \end{cases}$$

Hence Equations (1.32), (1.33) and (1.34) reduce to (1.28), (1.29) and (1.30).

Kramer [168] noted that the kernel function $K(x,t)$ and the sampling points $\{t_n\}_{n\in\mathbb{Z}}$ may be obtained from certain boundary-value problems. For example, consider the differential operator L defined as

$$L = p_0(x)\frac{d^n}{dx^n} + \cdots + p_{n-1}(x)\frac{d}{dx} + p_n(x), \quad x \in I,$$

where $p_k(x)$ is a complex-valued function with $n-k$ continuous derivatives, $k = 0,1,\ldots,n$, for any $x \in I = (a,b)$, and $p_0(x) \neq 0$ for any $x \in [a,b]$, with $-\infty \leq a < b \leq \infty$. The adjoint operator L^* is defined as

$$L^*g = (-1)^n \frac{d^n}{dx^n}(\overline{p}_0 g) + (-1)^{n-1}\frac{d^{n-1}}{dx^{n-1}}(\overline{p}_1 g) + \cdots + \overline{p}_n g.$$

The operator L is said to be formally *self-adjoint* if $L = L^*$. If the coefficient functions $p_k\,(k = 0,1,\ldots,n)$ are real-valued, then it is easy to see that if L is formally self-adjoint, then n is even.

Let $U_j(y) = 0$, $j = 1,\ldots,n$, be linearly independent homogeneous boundary conditions of the form

$$U_j(y) = \sum_{k=1}^{n}\left(\alpha_{j,k}y^{(k-1)}(a) + \beta_{j,k}y^{(k-1)}(b)\right), \quad j = 1,2,\ldots,n.$$

To any such system of boundary conditions, there exists an associated system of boundary conditions, known as the *adjoint system*, and if the two systems are equivalent, we say that they are self-adjoint [72, p. 287], [297].

The boundary-value problem:

$$Ly = -ty, \qquad x \in I, \tag{1.35}$$

$$U_j(y) = 0, \qquad j = 1,\ldots,n, \tag{1.36}$$

is said to be *self-adjoint* if the differential operator and the boundary conditions are self-adjoint. It is known that the eigenfunctions of a regular, self-adjoint boundary-value problem are complete in $L^2(I)$.

Kramer observed that if the regular, self-adjoint boundary-value problem (1.35) and (1.36) possesses a function $\phi(x,t)$ that generates the eigenfunctions of the problem $\{\phi_n(x)\}$ when the eigenvalue parameter t is replaced by the eigenvalues $\{t_n\}$, i.e., $\phi(x,t_n) = \phi_n(x)$, then one can take the sampling points to be $\{t_n\}$ and the kernel function $K(x;t)$ to be $\phi(x,t)$.

As a special case of Kramer's result, Campbell [60] (see also [45]) considered the regular second order Sturm- Liouville boundary-value problem:

$$L(\varphi) := -\varphi'' + q(x)\varphi = \mu^2\varphi, \qquad x \in [0,\pi] \tag{1.37}$$

$$a\varphi(0,\mu) + b\varphi'(0,\mu) = 0 \tag{1.38}$$

$$c\varphi(\pi,\mu) + d\varphi'(\pi,\mu) = 0 \tag{1.39}$$

where $q \in L^1(0,\pi)$, $a,b,\ c,\ d \in \mathbb{R}$, $a^2+b^2 \neq 0$ and $c^2+d^2 \neq 0$. Since the problem is regular, the spectrum is discrete, bounded from below and simple. If we denote by μ_n^2 the eigenvalues and $\varphi(x,\mu_n)$ the normalized eigenfunctions, Kramer's theorem would read: If

$$F(\mu) = \int_0^\pi f(x)\varphi(x,\mu)dx\,, \tag{1.40}$$

where $f \in L^2(0,\pi)$ and $\varphi(x,\mu)$ is a solution of the initial-value problem (1.37) and (1.38), then

$$F(\mu) = \sum_{n\geq 0} F(\mu_n)S_n^*(\mu)\,, \tag{1.41}$$

where $S_n^*(\mu)$ is given by (1.34).

The importance of Campbell's theorem is reflected in the fact that the kernels of many integral transformations arise from such Sturm- Liouville boundary-value problems. We will see several examples of that in forthcoming chapters. For related references, see [19, 60, 128, 129, 168, 176, 177, 214, 229, 231, 336, 349, 351, 356, 357, 358].

1.7 POISSON SUMMATION FORMULA

Another formula related to the sampling theorem is the *Poisson Summation Formula* [42, 57] which relates the Fourier series of a function to its Fourier transform. It is also closely related to the Zak transform which will be introduced in a later chapter [144, 143]. There are different versions of this formula depending on one's definition of the Fourier transform. Following our definition of the Fourier transform, the formula may be stated as follows: Suppose that $f \in L^1(\mathbb{R})$ and $\sigma > 0$. If we denote the Fourier transform of f by $\hat{f}$ and assume that $\sum_{k=-\infty}^{\infty} |\hat{f}(2k\sigma)| < \infty$, then

$$\frac{1}{\sqrt{2\pi}} \sum_{k=-\infty}^{\infty} f(t+t_k) = \frac{\sigma}{\pi} \sum_{k=-\infty}^{\infty} \hat{f}(2k\sigma)e^{2ik\sigma t}, \tag{1.42}$$

where $t_k = k\pi/\sigma$. In particular, if $\sigma = \pi$, we have

$$\frac{1}{\sqrt{2\pi}} \sum_{k=-\infty}^{\infty} f(t+k) = \sum_{k=-\infty}^{\infty} \hat{f}(2k\pi)e^{2ik\pi t},$$

from which we have

$$\frac{1}{\sqrt{2\pi}} \sum_{k=-\infty}^{\infty} f(k) = \sum_{k=-\infty}^{\infty} \hat{f}(2k\pi).$$

The last equation in turn implies that if f is bandlimited to $[-\pi,\pi]$, we have

$$\frac{1}{\sqrt{2\pi}} \sum_{k=-\infty}^{\infty} f(k) = \hat{f}(0).$$

A sufficient condition for the absolute convergence of both sides of the formula is that

$$|f(x)| \leq C\frac{1}{(1+|x|)^{1+\epsilon}}, \qquad \left|\hat{f}(w)\right| \leq D\frac{1}{(1+|w|)^{1+\epsilon}}; \quad \epsilon > 0,$$

for some positive constants, C and D. The formula has extensions to higher dimensions and to other lattice points; see [114].

1.8 UNCRTAINTY PRINCIPLE

The uncertainty principle was first introduced in quantum mechanics by the German physicist Werner Heisenberg in 1927 who was awarded the Nobel Prize in Physics in 1932 for his work on the creation of quantum mechanics. The uncertainty principle states that the position and the momentum of a particle cannot simultaneously be determined precisely, the more precisely the position is determined, the less precisely the momentum can be determined. Mathematically, this may be stated as follows: if we denote the standard deviation of the position by σ_x and the standard deviation of the momentum by σ_p, then $\sigma_x\sigma_p \geq \hbar/2$, where $\hbar$ is the reduced Planck constant, $h/(2\pi)$.

In mathematics there is a number of different analogues of the uncertainty principle, but the most basic one involves a function and its Fourier transform. It says that if a signal $f(t)$ is concentrated in the region $|t-t_0| \leq \epsilon_t$, it cannot also have its Fourier transform $\hat{f}(\omega)$ be concentrated in $|\omega-\omega_0| \leq \epsilon_\omega$, unless $1 \leq \epsilon_t\epsilon_\omega$. Another formulation of the uncertainty principle for the Fourier transform may be stated as follows: if $f \in L^2(\mathbb{R})$, then

$$\frac{1}{2}\int_{\mathbb{R}} |f(t)|^2\,dt \leq \left(\int_{\mathbb{R}} t^2\,|f(t)|^2\,dt\right)^{1/2}\left(\int_{\mathbb{R}} \omega^2\left|\hat{f}(\omega)\right|^2 d\omega\right)^{1/2}, \tag{1.43}$$

with the equality holds only if $f(t) = Ce^{-\pi t^2}$. For a proof, see [357, P. 34]. A slightly more general version of (1.43) is obtained by replacing $f(t)$ by $e^{it\omega_0}f(t+t_0)$ which is clearly in $L^2(\mathbb{R})$. With some calculations, this yields

$$\begin{aligned}\frac{1}{2}\|f(t)\|^2 &= \frac{1}{2}\int_{\mathbb{R}} |f(t)|^2\,dt \\ &\leq \left(\int_{\mathbb{R}} (t-t_0)^2\,|f(t)|^2\,dt\right)^{1/2}\left(\int_{\mathbb{R}} (\omega-\omega_0)^2\left|\hat{f}(\omega)\right|^2 d\omega\right)^{1/2},\end{aligned}$$

with the equality holds only if

$$f(t) = Ce^{-\pi(t-t_0)^2 - i\omega_0(t-t_0)}.$$

It should be noted that there are different versions of the uncertainty principle depending on one's definition of the Fourier transform; [29, 68, 319]. For more information on the uncertainty priciple, see [81, 23, 262, 263, 286].

CHAPTER 2

Integral Transformations

2.1 INTRODUCTION AND BRIEF HISTORY

Integral transforms have a long history that goes back to the eighteenth century. One of the earliest integral transforms is the Laplace transform which was named after the French mathematician Pierre Simon Laplace (1749–1827) who put probability theory on solid foundation in his work *Théorie analytique des probabilités* (Analytical Theory of Probabilities) [173]. The Laplace transform still plays an important role in both theory and applications of mathematics.

Another classical integral transform that has ubiquitous applications is the Fourier transform which was an extension of the Fourier series introduced by the French mathematician Jean Baptiste Fourier (1768–1830) in his work *Théorie analytique de la chaleur* (The Analytical Theory of Heat) [103].

Some integral transforms were introduced by engineers, such as the Gabor transform which was named after the Hungarian electrical engineer and physicist Dennis Gabor who won the Nobel Prize in 1971 for inventing holography [104]. The Gabor transform is a special case of the windowed (sliding window) Fourier transform. Some integral transforms were introduced by physicists, such as the wavelets transform introduced by the French geophysicists A. Grossmann and J. Morlet [118], J. Morlet et al. [210, 211]. Other transforms were introduced by mathematicians, such as the Laplace, Fourier, Hilbert and the Mellin transforms.

A number of transforms were discovered by scientists in one discipline and later were rediscovered by other scientists in a different discipline. The Radon transform is a good example of that. The Radon transform was introduced by the Austrian mathematician, Johann Radon in 1917 in his seminal paper [247] "*Über die Bestimmung Von Funcktionen durch ihre Integralwerte länge gewisser Manningfaltigkeiten*," or "On the determination of functions from their integrals along certain manifolds." in which he showed how to reconstruct a function from its integrals along certain manifolds. In particular, he showed how to reconstruct a function of two variables from its integrals along straight lines.

This mathematical idea, which went unnoticed for more than forty years, turned out to be the foundation of one of the revolutionary developments in medical imaging, namely computerized tomography, or CT Scan. It was Allen Cormack, a South African physicist working at the Radiology Department at the University of Cape Town's medical school, who in his quest for a solution to a radiology problem, realized the significance of Radon's

DOI: 10.1201/9781003089353-2

paper, which he, together with Godfrey Hounsfield, a British engineer, used to develop a new X-ray machine that revolutionized the field of medical imaging; it was the first CT scan machine ever invented. For more on the Radon transform, see [89, 352, 354].

Therefore, it is generally fair to say that major contributions to the field of integral transforms are shared equally by mathematicians, physicists and engineers.

2.2 WHAT IS AN INTEGRAL TRANSFORM?

A general form of integral transforms may be written as

$$\mathcal{F}[f(\mathbf{x})](\mathbf{y}) = F_{\mathbf{a}}(\mathbf{y}) = \int_{\Omega} K_{\mathbf{a}}(\mathbf{x}, \mathbf{y}, f) d\mu(\mathbf{x}),$$

where $\mathbf{x} \in \mathbb{R}^n(\mathbb{C}^n), \mathbf{y} \in \mathbb{R}^m(\mathbb{C}^m)$ and $\mathbf{a} \in \mathbb{R}^k$ is a set of k parameters, Ω is a measurable subset of $\mathbb{R}^n(\mathbb{C}^n)$ and $d\mu$ is an appropriate measure. The function $K_{\mathbf{a}}$ is called the kernel of the transform, f is the transformed function and F is its integral transform. In other words, an integral transform is a transformation $\mathcal{F}$ that maps f into F

Although the subject of non-linear integral transforms deserves considerable attention, throughout this book our focus will be on linear integral transforms, and from now on the term integral transform will mean linear integral transform.

Integral transforms differ from one another by the choice of the kernel K, the set Ω and the measure $d\mu$.

A linear integral transform generally takes the form

$$\mathcal{F}[f(\mathbf{x})](\mathbf{y}) = \mathcal{F}[f](\mathbf{y}) = F_{\mathbf{a}}(\mathbf{y}) = \int_{\Omega} K_{\mathbf{a}}(\mathbf{x}, \mathbf{y}) f(\mathbf{x}) d\mu(\mathbf{x}).$$

The above integral transform is clearly linear, i.e.,

$$\mathcal{F}[af + bg] = a\mathcal{F}[f] + b\mathcal{F}[g], \qquad a, b \in \mathbb{R}(\mathbb{C}).$$

Because a number of integral transforms have physical applications in which f and F may represent physical quantities, we may visualize an integral transform as a device in which the input is represented by f, the output by F, and the core of the device is the kernel of the transform K. The action of the kernel on the input is represented by an integral over some measurable set Ω.

2.3 EXAMPLES OF INTEGRAL TRANSFORMS

In this section, we give examples of integral transforms in one and higher dimensions.

2.3.1 One-Dimensional Integral Transforms

Below are examples of some classical one-dimensional integral transforms that follow the pattern described above:

1. *The Laplace Transform* [40]: $K(x,t) = e^{-xt}$, $\Omega = [0, \infty)$, $d\mu = dx$,

$$F(t) = \int_0^{\infty} f(x) e^{-xt} dx.$$

2. *The Fourier Transform* [49, 228, 281]: $K(x,t) = e^{ixt}/\sqrt{2\pi}$, $\Omega = \mathbb{R}, d\mu = dx$,

$$F(t) = \frac{1}{\sqrt{2\pi}} \int_{\mathbb{R}} f(x) e^{ixt} dx.$$

Another variant of the Fourier transform is given by

$$F(t) = \int_{\mathbb{R}} f(x) e^{2\pi ixt} dx.$$

3. *The Cosine and Sine Transforms:*

$$F(t) = \sqrt{2/\pi} \int_0^\infty f(x) \cos(tx) dx, \qquad F(t) = \sqrt{2/\pi} \int_0^\infty f(x) \sin(tx) dx.$$

An intimately related transform to the cosine and sine transforms is the Hartley transform defined as

$$H[f](t) = \frac{1}{\sqrt{2\pi}} \int_{\mathbb{R}} f(x) Cas\,(tx) dx,$$

where the kernel is $Cas(z) = \cos z + \sin z$. The Hartley transform was introduced by an engineer R. Hartley [123] in 1942 to overcome what he considered a drawback of the Fourier transform which transforms a real-valued function into a complex-valued function [47, 48].

4. *The Mellin Transform* [196]: $K(x,t) = x^{t-1}, \Omega = [0, \infty), d\mu = dx$,

$$F(t) = \int_0^\infty f(x) x^{t-1} dx.$$

5. *The Hankel Transform* [70, 71, 80]: $K(x,t) = \sqrt{xt} J_\nu(xt)$, $\Omega = [0, \infty), d\mu = dx$, where J_ν is the Bessel function of order $\nu > -1/2$,

$$F(t) = \int_0^\infty f(x) \sqrt{xt} J_\nu(xt) dx.$$

Another version of the Hankel transform is given by

$$F(t) = \int_0^\infty x f(x) J_\nu(xt) dx.$$

6. *The Legendre transform* [58]: $K(x,t) = (1/2) P_t(x)$, $\Omega = [-1, 1]$, $d\mu = dx$,

$$F(t) = \frac{1}{2} \int_{-1}^{1} f(x) P_t(x) dx, \quad t \in \mathbb{R},$$

where

$$P_t(x) = F\left(-t, t+1; 1, (1-x)/2\right),$$

is the Legendre function and

$$F(a, b; c, x) = {}_2F_1(a, b; c, x)$$

is the hypergeometric function. The Legendre transform is a special case of the Jacobi transform [90, 306].

7. *The Sturm-Liouville Trannsform* [359, 352, 305]: Consider the boundary-value problem (BVP)

$$y'' - q(x)y = -\lambda y\,, \quad -\infty < a \le x \le b < \infty\,, \tag{2.1}$$

$$y(a)\cos\alpha + y'(a)\sin\alpha = 0 \tag{2.2}$$

$$y(b)\cos\beta + y'(b)\sin\beta = 0\,, \tag{2.3}$$

and let $\phi(x,\lambda), \psi(x,\lambda)$ be the solutions of (2.1) such that

$$\begin{aligned} \phi(a,\lambda) &= \sin\alpha\,, \quad \phi'(a,\lambda) = -\cos\alpha \\ \psi(b,\lambda) &= \sin\beta\,, \quad \psi'(b,\lambda) = -\cos\beta\,. \end{aligned}$$

There exists a linear integral transformation T from $L^2[a,b]$ into the space of all entire functions of order *one* and type at most $(b-a)$ defined by

$$(Tf)(\lambda) = F(\lambda) = \int_a^b f(x)\phi(x,\lambda)dx, \quad f \in L^2[a,b]\,, \lambda \in \mathbb{C} \tag{2.4}$$

Moreover, F is completely determined by its values at the eigenvalues of the problem $\{\lambda_n\}_{n=0}^{\infty}$ and can be reconstructed from these values according to the formula

$$F(\lambda) = \sum_{n=0}^{\infty} F(\lambda_n)\frac{P(\lambda)}{(\lambda-\lambda_n)P'(\lambda_n)}$$

where $P(\lambda)$ is the Wronskian of ϕ and ψ.

8. *The Stieltjes Transform* [289]: $K(x,t) = \frac{1}{x+t}$, $\Omega = [0,\infty)$, $d\mu$ is a Lebesque measure

$$F(t) = \int_0^{\infty} \frac{f(x)}{x+t} d\mu(x).$$

9. *The Hilbert Transform* [122]: $K(x,t) = \frac{1}{\pi(x-t)}$, $\Omega = \mathbb{R}, d\mu = dx$,

$$F(t) = \frac{1}{\pi}\int_{\mathbb{R}} \frac{f(x)}{x-t} dx.$$

10. *The Weierstrass Transform* [71, 131]: $K(x,t) = \frac{1}{\sqrt{2\pi\sigma}} e^{-(x-t)^2/2\sigma}$, $\sigma > 0$, $\Omega = \mathbb{R}$, $d\mu = dx$,

$$F(t) = \frac{1}{\sqrt{2\pi\sigma}} \int_{\mathbb{R}} f(x) e^{-(x-t)^2/2\sigma} dx.$$

11. *The Bargmann Transform* [25, 26, 145]: $K(z,t) = \frac{1}{\sqrt[4]{\pi}} e^{-\frac{1}{2}(z^2+t^2)+\sqrt{2}zt}$, $\Omega = \mathbb{R}$, $d\mu = dt$,

$$F(z) = \frac{1}{\sqrt[4]{\pi}} \int_{\mathbb{R}} f(t) e^{-\frac{1}{2}(z^2+t^2)+\sqrt{2}zt} dt,$$

where $z = x + iy$.

12. *The Fresnel Transform* [113, 137]: $K(x,t) = \frac{1}{\sqrt{i\lambda z}} e^{i\pi(x-t)^2/\lambda z}$, $\Omega = \mathbb{R}, d\mu = dt$,

$$F(t) = \frac{1}{\sqrt{i\lambda z}} \int_{\mathbb{R}} f(x) e^{i\pi(t-x)^2/\lambda z} dx, \quad z \neq 0, \lambda > 0.$$

13. *The Wavelet Transform* [84, 85, 86, 314]: $K(x,a,b) = \frac{1}{\sqrt{|a|}} \psi\left(\frac{x-b}{a}\right)$, $\Omega = \mathbb{R}, d\mu = dx$,

$$F(a,b) = \int_{\mathbb{R}} f(x) \psi_{a,b}(x) dx,$$

where $\psi_{a,b}(x) = \frac{1}{\sqrt{|a|}} \psi\left(\frac{x-b}{a}\right)$, $a > 0$, $b \in \mathbb{R}$, and ψ satisfies the admissibility condition

$$\int_{\mathbb{R}} \frac{|\hat{\psi}(w)|^2}{|w|} dw < \infty, \tag{2.5}$$

where $\hat{\psi}$ is the Fourier transform of ψ.

14. *The Windowed-Fourier Transform, or Short-time Fourier Transform, or Sliding Window Fourier Transform* [84, 294]: $K(x,t,w) = g(x-t)e^{iwx}/\sqrt{2\pi}$, $\Omega = \mathbb{R}$, $d\mu = dx$, where $g \in L^2(\mathbb{R})$ is the window function,

$$F(t,w) = \frac{1}{\sqrt{2\pi}} \int_{\mathbb{R}} f(x) g(x-t) e^{iwx} dx, \ w,t \in \mathbb{R}.$$

Closely related to the windowed-Fourier transform is the convolution transform, which is defined for a function f with respect to a kernel function G as

$$F(t) = \int_{\mathbb{R}} G(t-x) f(x) dx,$$

where G is assumed to satisfy certain conditions; see [133, 134] and [352, Ch.17].

15. *The Gabor Transform* [104, 146]: The Gabor transform is a special case of the windowed Fourier transform where the window function is the Gaussian function $K(x,t,w) = e^{-\pi(t-x)^2} e^{iwx}/\sqrt{2\pi}$, $\Omega = \mathbb{R}$, $d\mu = dx$,

16. *The Stockwell Transform* [93, 290]: $K(x,t,w) = \frac{|w|}{\sqrt{2\pi}} e^{(x-t)^2 w^2/2} e^{-2\pi i w x}$, $\Omega = \mathbb{R}, d\mu = dx$,

$$F(t,w) = \frac{|w|}{\sqrt{2\pi}} \int_{\mathbb{R}} f(x) e^{(x-t)^2 w^2/2} e^{-2\pi i w x} dx, \ w \neq 0.$$

17. *The Linear Canonical Transform* [213, 212]:

$$K_{a,b,c,d}(x,t) = \frac{1}{\sqrt{2\pi i b}} \exp\left\{\frac{i}{2b}\left[ax^2 - 2xt + dt^2\right]\right\},$$

where $ad - bc = 1$, $b \neq 0$, $\Omega = \mathbb{R}, d\mu = dx$,

$$F(t) = \frac{1}{\sqrt{2\pi i b}} \int_{\mathbb{R}} f(x) \exp\left\{\frac{i}{2b}\left[ax^2 - 2xt + dt^2\right]\right\} dx.$$

18. *The Special Affine Fourier Transform* [1, 2]:

$$K_{a,b,c,d;p,q}(x,t) = \frac{1}{\sqrt{2\pi|b|}} \exp\left\{\frac{i}{2b}\left[ax^2 + dt^2 + 2x(p-\omega) + 2t(bq - dp)\right]\right\},$$

where $ad - bc = 1$, $b \neq 0$, $p, q \in \mathbb{R}$, $\Omega = \mathbb{R}$, $d\mu = dx$. Hence

$$F(t) = \frac{1}{\sqrt{2\pi|b|}} \int_{\mathbb{R}} f(x) \exp\left\{\frac{i}{2b}\left[ax^2 + dt^2 + 2x(p-\omega) + 2t(bq - dp)\right]\right\} dx.$$

19. *The Offset Linear Canonical Transform* [234, 236]:

$$K(x,t)_{a,b,c,d;p,q} = \frac{e^{iqt}}{\sqrt{2i\pi b}} \exp\left\{\frac{i}{2b}\left[ax^2 + d(t-p)^2 + 2x(p-t)\right]\right\},$$

where $ad - bc = 1$, $b \neq 0$, $p, q \in \mathbb{R}$, $\Omega = \mathbb{R}$, $d\mu = dx$,

$$F(t) = \frac{e^{iqt}}{\sqrt{2i\pi b}} \int_{\mathbb{R}} f(x) \exp\left\{\frac{i}{2b}\left[ax^2 + d(t-p)^2 + 2x(p-t)\right]\right\} dx.$$

For more examples, we refer the reader to [46, 53, 109, 111, 112, 120, 165]. The first few examples may give the impression that most one-dimensional integral transforms are taken over the full real line or the half line $[0, \infty)$. But examples 6 and 7 show that this is not always the case.

The integral transforms in Examples 13-16 deserve special attention because they map functions of one variable into functions of two variables and in higher dimensions, they map functions of n variables into functions of $2n$ variables. This type of transformations has interesting applications. For example, in signal processing when f represents a signal in the time domain, $F(t, w)$ represents a time-frequency representation of the signal. The Hilbert transform deviates slightly from the pattern because its kernel is a singular function.

2.3.2 Higher Dimensional Transforms

The generalization of most of these integral transforms to higher dimensions is usually straightforward as can be seen from the next few examples.

- **The Fourier Transform:**

$$F(\mathbf{t}) = \frac{1}{(2\pi)^{n/2}} \int_{\mathbb{R}^n} f(\mathbf{x}) e^{i(\mathbf{t}\cdot\mathbf{x})} d\mathbf{x}, \qquad t \in \mathbb{R}^n,$$

where $\mathbf{x} \cdot \mathbf{t} = \sum_{k=1}^{n} x_k t_k$.

- **Windowed-Fourier Transform:**

$$F(\mathbf{t}, \mathbf{w}) = \frac{1}{(2\pi)^{n/2}} \int_{\mathbb{R}^n} f(\mathbf{x}) g(\mathbf{t} - \mathbf{x}) e^{i(\mathbf{w}\cdot\mathbf{x})} d\mathbf{x}, \quad \mathbf{t}, \mathbf{w} \in \mathbb{R}^n.$$

- **The Two-dimensional Fresnel Transform:**

$$F(u,v) = \frac{e^{ikz}}{i\lambda z}\int_{\mathbb{R}}\int_{\mathbb{R}} f(x,y)\exp\left\{\frac{ik}{2z}\left[(u-x)^2+(v-y)^2\right]\right\}dxdy,$$

where $k = 2\pi/\lambda$ is the wavenumber and λ is the wave length. This transform arises in optics to model Fresnel diffraction.

- **The Continuous Wavelet Transform**:

$$F(a,\mathbf{b}) = \int_{\mathbb{R}^n} f(\mathbf{x})\psi_{a,\mathbf{b}}(\mathbf{x})d\mathbf{x},$$

where $\psi \in L^2(\mathbb{R}^n)$ with

$$\int_{\mathbb{R}^n} \frac{|\hat{\psi}(\mathbf{w})|^n}{|\mathbf{w}|^n}d\mathbf{w} < \infty. \tag{2.6}$$

and

$$\psi_{a,\mathbf{b}}(\mathbf{x}) = a^{-n/2}\psi\left(\frac{\mathbf{x}-\mathbf{b}}{a}\right), \quad a > 0, \mathbf{b} \in \mathbb{R}^n.$$

2.3.3 Special Cases of Higher Dimensional Transforms

In this subsection, we give examples of integral transforms that exist in dimensions $n \geq 2$ and have no analogue in dimension one.

- **The Directional Wavelet Transform** [20]: In $\mathbb{R}^2$, one can define a continuous wavelet transform with rotation given by the matrix

$$R_\theta = \begin{pmatrix} \cos\theta & -\sin\theta \\ \sin\theta & \cos\theta \end{pmatrix},$$

so that the kernel of the transformation is

$$\psi^{a,b,\theta}(\mathbf{x}) = a^{-1}\psi\left(\frac{R_\theta(\mathbf{x}-\mathbf{b})}{a}\right).$$

- **The Continuous Ridgelet Transform** [63]: The two-dimensional continuous ridgelet transform of $f \in L^2(\mathbb{R}^2)$ is defined as

$$\mathcal{R}_\psi f(a,b,\theta) = \int_{\mathbb{R}^2} f(\mathbf{x})\psi_{a,b,\theta}(\mathbf{x})d\mathbf{x},$$

where

$$\psi_{a,b,\theta}(x_1,x_2) = a^{-1/2}\overline{\psi}\left(\frac{(x_1\cos\theta + x_2\sin\theta) - b}{a}\right),$$

$\psi \in L^2(\mathbb{R})$ and satisfies condition (2.6) with $a > 0, b \in \mathbb{R}, \theta \in [0, 2\pi)$. The ridgelet transform is related to the Radon transform. The definition of the Radon transform is given at the end of this section for a reason explained there. Let us denote the Radon transform of f by

$$(Rf)(p,\mathbf{u}) = f\dagger(p,\mathbf{u}),$$

where $\mathbf{u}$ is a unit vector. We use the notation

$$(R_{\mathbf{u}}f)(p) = f^{\dagger}_{\mathbf{u}}(p).$$

Then

$$\mathcal{R}_{\psi}f = a^{-1/2}\int_{\mathbb{R}} \psi\left(\frac{p-b}{a}\right) f^{\dagger}_{\mathbf{u}}(p)dp.$$

Thus, the Ridgelet transform may be viewed as an application of a one-dimensional wavelet transform to a slice of the Radon transform; see [91, 92].

- **The Continuous Shearlet Transform** [171]: To introduce the shearlet transform , let us first introduce the dilation and translation operators D_A and T_t in $\mathbb{R}^n$ as follows:

$$D_A\psi(\mathbf{x}) = |\text{det A}|^{-1/2}\psi(A^{-1}\mathbf{x}),$$

and

$$T_{\mathbf{t}}\psi(\mathbf{x}) = \psi(\mathbf{x}-\mathbf{t}),$$

where $\mathbf{x},\mathbf{t} \in \mathbb{R}^n$ and A is an $n \times n$ non-singular matrix.

Consider the parabolic scaling matrix

$$A = \begin{pmatrix} a & 0 \\ 0 & a^{1/2} \end{pmatrix}, \quad a > 0,$$

and the shearing matrix

$$S = \begin{pmatrix} 1 & s \\ 0 & s \end{pmatrix}, \quad s \in \mathbb{R}.$$

The kernel of the *Continuous Shearlet Transform* is given by

$$\psi_{a,s,t}(\mathbf{x}) = T_{\mathbf{t}}D_A D_S\psi(\mathbf{x}),$$

and the continuous shearlet transform is defined as

$$\mathcal{SH}f(a,s,\mathbf{t}) = F(a,s,\mathbf{t}) = \int_{\mathbb{R}^2} f(\mathbf{x})\psi_{a,s,t}(\mathbf{x})d\mathbf{x}, \quad \mathbf{x} \in \mathbb{R}^2,$$

where $\psi \in L^2(\mathbb{R}^2)$ satisfying

$$\int_{\mathbb{R}^2} \frac{|\hat{\psi}(\omega_1,\omega_2)|^2}{|\omega_1|^2} d\omega_1 d\omega_2 < \infty.$$

- **The Gyrator Transform** [251, 152]: The Gyrator transform is an important but less known two-dimensional integral transform that is closely related to the two-dimensional Fourier and the two-dimensional fractional Fourier transforms. The Gyrator transform of order α is defined as

$$(R_{\alpha}f)(\omega_1,\omega_2) = \int_{\mathbb{R}}\int_{\mathbb{R}} f(t_1,t_2)G_{\alpha}(t_1,t_2,\omega_1,\omega_2)\,dt_1dt_2,$$

where

$$G_\alpha(t_1,t_2,\omega_1,\omega_2) = \frac{1}{2\pi|\sin\alpha|} \times \exp\left\{\frac{i}{\sin\alpha}\left[(t_1t_2+\omega_1\omega_2)\cos\alpha-(t_2\omega_1+t_1\omega_2)\right]\right\}.$$

This transform originated in optics and was implemented in paraxial optics. For more on this transform, see Chapter 6.

- **The Radon Transform** [89]: Although the Radon transform does not exactly fit in the standard pattern, it is mentioned here because of its importance and connection to the ridgelet transform. The Radon transform of $f \in L^2(\mathbb{R}^n)$, is defined as

  $$Rf(\zeta,t) = \int_{\mathbb{R}^n} f(\mathbf{x})\delta(t-\zeta\cdot\mathbf{x})\,d\mathbf{x},$$

  where ζ is a unit vector in $\mathbb{R}^n$ and $t \in \mathbb{R}^n$. It does not exactly fit in the standard pattern because the kernel is not a function but is the Dirac delta function which is a *generalized function,* [205, 206].

2.4 GENERAL PROPERTIES OF INTEGRAL TRANSFORMATIONS

In the study of linear integral transformations, the following questions are of particular interest:

- What are the domain and range of the transformation? That is what are the function spaces to which f and F belong?
- More importantly, does the transform have an inverse? That is given F, can one find f? The answer to this question is of paramount importance in applications. A transform that has no inverse is practically useless.
- It would be interesting if the transform preserves the product formula, i.e., $\mathcal{F}[fg] = \mathcal{F}[f]\mathcal{F}[g]$; however, this is not always feasible. Alternatively, one may ask: can we find an operation, called convolution, denoted by $*$ so that $\mathcal{F}[f*g] = \mathcal{F}[f]\mathcal{F}[g]$?

The answer to the last question is sometimes yes, but finding the appropriate convolution can be very challenging.

The inversion formulas for many integral transforms take the form

$$f(x) = \int_D F(t)H(t,x)d\eta(t),$$

for some domain D and Borel measure $d\eta$. If both the transform and its inverse are given by

$$F(t) = \int_{\mathbb{R}} f(x)K(t,x)dx, \quad \text{and } f(x) = \int_{\mathbb{R}} F(t)H(x,t)dt,$$

we formally have

$$\int_{\mathbb{R}} K(t,x)H(x,y)dx = \delta(t-y),$$

as can be seen for the Fourier transform, where

$$\frac{1}{2\pi}\int_{\mathbb{R}} e^{itx}e^{-iyx}dx = \delta(t-y).$$

With some calculations, one can show that a similar formula holds for the fractional Fourier transform; see Chapter 4.

In some cases where the integral transform is given by

$$F(t) = \int_0^\infty f(x)K(tx)dx,$$

the inversion formula may also take the form

$$f(x) = \int_0^\infty F(t)H(tx)dt.$$

In this case, we have the following relation between K and H, [281]

$$\tilde{K}(s)\tilde{H}(1-s) = 1, \tag{2.7}$$

where the tilde stands for the Mellin transform. For, if we take the Mellin transform of F and interchange the integrals, we formally obtain

$$\begin{aligned}\tilde{F}(s) = \int_0^\infty t^{s-1}F(t)dt &= \int_0^\infty f(x)dx\int_0^\infty t^{s-1}K(tx)dt \\ &= \int_0^\infty f(x)dx\int_0^\infty (\frac{\eta}{x})^{s-1}K(\eta)\frac{d\eta}{x} \\ &= \int_0^\infty x^{-s}f(x)dx\int_0^\infty \eta^{s-1}K(\eta)d\eta = \tilde{f}(1-s)\tilde{K}(s).\end{aligned}$$

Repeating the same process for f, we obtain $\tilde{f}(s) = \tilde{F}(1-s)\tilde{H}(s)$, which when combined with the previous equation, yields Eq. (2.7)

$$\tilde{K}(s)\tilde{H}(1-s) = 1.$$

In the special case where the forward and inverse transforms are the same, as for the cosine. sine and Hankel transforms, we have

$$\tilde{H}(s)\tilde{H}(1-s) = 1, \quad \text{or} \quad \tilde{H}(1-s) = \frac{1}{\tilde{H}(s)}. \tag{2.8}$$

For the cosine transform

$$F(t) = \sqrt{2/\pi}\int_0^\infty f(x)\cos tx dx,$$

the Mellin transform of the kernel, $K(y) = \sqrt{2/\pi}\cos y$, is

$$\tilde{K}(s) = \sqrt{2/\pi}\Gamma(s)\cos(\pi s/2)$$

and hence

$$\tilde{K}(1-s) = \sqrt{2/\pi}\Gamma(1-s)\sin(\pi s/2).$$

In view of Eq. (1.4),

$$\Gamma(s)\Gamma(1-s) = \frac{\pi}{\sin\pi s},$$

we have $\tilde{K}(s)\tilde{K}(1-s) = 1$, which is consistent with Eq. (2.8). Similar argument applies to the sine transform.

As for the Hankel transform given by

$$F(t) = \int_0^\infty f(x)\sqrt{tx}J_\nu(tx)dx,$$

we have

$$\begin{aligned}
\tilde{F}(s) = \int_0^\infty t^{s-1}F(t)dt &= \int_0^\infty \sqrt{x}f(x)dx \int_0^\infty t^{s-1}\sqrt{t}J_\nu(tx)dt \\
&= \int_0^\infty \sqrt{x}f(x)dx \int_0^\infty (\frac{\eta}{x})^{s-1/2}J_\nu(\eta)\frac{d\eta}{x} \\
&= \int_0^\infty x^{-s}f(x)dx \int_0^\infty \eta^{s-1/2}J_\nu(\eta)d\eta \\
&= \tilde{f}(1-s)\tilde{K}(s),
\end{aligned}$$

where $\tilde{K}(s)$ is the Mellin transform of the kernel $K(y) = \sqrt{y}J_\nu(y)$. But it can be verified that the Mellin transform of the kernel is given by

$$\tilde{K}(s) = 2^{s-1/2}\frac{\Gamma\left(\frac{\nu+s+1/2}{2}\right)}{\Gamma\left(\frac{\nu-s+3/2}{2}\right)},$$

which yields

$$\tilde{K}(1-s) = 2^{1/2-s}\frac{\Gamma\left(\frac{\nu-s+3/2}{2}\right)}{\Gamma\left(\frac{\nu+s+1/2}{2}\right)}.$$

Therefore, $\tilde{K}(s)\tilde{K}(1-s) = 1$.

It should be noted that the Hankel transform pair may also be defined as

$$F(t) = \int_0^\infty xf(x)J_\nu(tx)dx, \text{ and } f(x) = \int_0^\infty tF(t)J_\nu(tx)dt.$$

Repeating the same process as above, we obtain

$$\tilde{F}(s) = \tilde{f}(2-s)\tilde{K}(s) \quad \text{and } \tilde{f}(s) = \tilde{F}(2-s)\tilde{K}(s),$$

which yields $\tilde{K}(s)\tilde{K}(2-s) = 1$, where $\tilde{K}$ is the Mellin transform of the kernel $K(\eta) = J_\nu(\eta)$. The last relation looks slightly different from Eq. (2.7) because the input function $f(x)$ is replaced by $xf(x)$.

In view of the fact that the Mellin transform of the Bessel function of order ν is given by

$$\tilde{K}(s) = \frac{2^{s-1}\Gamma\left(\frac{\nu+s}{2}\right)}{\Gamma\left(\frac{\nu-s+2}{2}\right)},$$

we have

$$\tilde{K}(2-s) = \frac{2^{1-s}\Gamma\left(\frac{\nu-s+2}{2}\right)}{\Gamma\left(\frac{\nu+2}{2}\right)},$$

and; hence, $\tilde{K}(s)\tilde{K}(2-s) = 1$.

2.5 WHY INTEGRAL TRANSFORMS?

One may wonder what integral transforms are good for. In this section, we will try to answer this question.

1. In solving some difficult mathematical problems, it may be useful to transform these problems by means of some function transformation to a new setting where the problems can be easily solved. Having solved the problems in the new setting, one has to transform the solution back to the original setting, and this is where the inverse transformation comes into play.

   A prototype situation in which this method has been used is in solving certain initial-value and boundary-value problems involving ordinary or partial differential equations. These problems may be converted into a system of algebraic equations that can be solved and then the solution is converted back to the original setting using the inverse transformation. Several classical linear integral transforms, such as the Laplace, Fourier, Hankel and Mellin transforms, are good examples of that.

2. Sometimes the transform function F itself has some physical meaning and ought to be studied in its own right. For example, in electrical engineering the original function $f(t)$ may represent a signal that is a function of time, e.g., a radar or radio signal. The Fourier transform $F(\omega)$ of $f(t)$ in this case represents the frequency content of the signal f. Thus, both f and F give two different representations of the same object.

   Another example is the windowed Fourier transform in which $f(t)$ may represent a signal in the time domain and its transform function $F(t,w)$ represents its time-frequency distribution. A black and white image may be represented by a function of two variables $f(x,y)$ and its transform may represent the output of an operation on the image, such as rotation, cropping, or magnification.

3. In some applications, the transform function F, is a set of data that can be measured experimentally and from which the original function f could be reconstructed. The original function f may represent the internal distribution, perhaps density of some physical quantity, of a hard-to-reach object like a human brain.

   An example of such a transformation is the Radon transform, which enables one to reconstruct a function f of $n(n \geq 2)$ variables from the values of its integrals over all hyperplanes of dimension $n-1$. As a special case, if $n=2$, the Radon transform F, can be measured physicall, and then one can use the Radon transform to reconstruct the function $f(x,y)$ from the values of its integrals over all straight lines in the x-y plane. This was the main mathematical idea behind the invention of computerized tomography or CT scan.

For more information on integral transforms and their applications, see [51, 52, 87, 88, 96, 117, 142, 243, 255, 256, 257, 298, 321, 341, 352, 361].

CHAPTER 3

Fractional Integral Transforms

3.1 INTRODUCTION

Fractional integral transforms have a relatively short history compared to the conventional integral transforms. But before we talk about fractional integral transforms, let us examine the word *fraction*. According to the Webster Dictionary: fraction means a piece broken off, or fragment, or portion of something. So, how does this apply to integral transforms?

In general, a fractional integral transform is a variation of a known integral transform. It depends on some parameters and reduces to the known integral transform for special values of the parameters. For simplicity, if the fractional transform depends on one parameter only, say $0 \leq \alpha \leq 1$, it ought to reduce to the original transform when $\alpha = 1$. In most cases, the fractional transform reduces to the identity transformation when $\alpha = 0$.

The term *Fractional Integral Transform*, to the best of my knowledge, started to appear in the literature in the early 1980s in the work of Victor Namias [217] who introduced an integral transform that generalized the Fourier transform and which he called *the fractional order Fourier transform*.

Nowadays, there are several fractional integral transforms, such as fractional Hankel, fractional wavelets, fractional Radon, fractional Stockwell, fractional windowed Fourier transforms and so on. Most of them are related directly or indirectly to the fractional Fourier transform and because of this relationship we will begin our discussion by the latter.

Namias introduced the concept of Fourier transform of fractional order in which the ordinary Fourier transform is regarded as a transform of order one and the identity transformation as a transform of order zero, and what is in between is called fractional Fourier transform. Moreover, he sought to maintain the additive property so that two successive applications of the transform of order one half would yield the ordinary Fourier transform.

More explicitly, the fractional Fourier transform may be viewed as a family of transformations, $\{\mathcal{F}_\alpha\}$ indexed by a parameter α, with $0 \leq \alpha \leq 1$, such that $\mathcal{F}_0$ is the identity transformation and $\mathcal{F}_1$, is the standard Fourier transformation. That is

$$\mathcal{F}_0[f] = f, \quad \mathcal{F}_1[f] = \hat{f},$$

where $\hat{f}$ is the Fourier transform of f, and, in addition, $\mathcal{F}_\alpha \mathcal{F}_\beta = \mathcal{F}_{\alpha+\beta}$.

DOI: 10.1201/9781003089353-3

The range of the parameter does not have to be the interval $[0,1]$ because this interval can be mapped by a simple substitution into the interval $[a,b]$, $a<b$. In fact, because of its periodicity, the fractional Fourier transform is parameterized by an angle $0 \leq \theta \leq 2\pi$, where $\mathcal{F}_0$, is the identity transformation and the conventional Fourier transform is obtained when $\theta = \pi/2$. This is one of the reasons that some authors call the fractional Fourier transform the angular Fourier transform. It is easy to show that $\mathcal{F}_\theta = \mathcal{F}_{\theta+2\pi}$, i.e.,

$$\mathcal{F}_0[f] = f, \quad \mathcal{F}_{\pi/2}[f] = \hat{f}, \quad \mathcal{F}_\pi[f(x)] = f(-x), \quad \mathcal{F}_{2\pi}[f] = f. \tag{3.1}$$

Eq. (3.1) has a physical meaning in optics as we will see in a later chapter.

Recall that the kernel $K(t,x)$ of the Fourier transform is the exponential function e^{itx}. The kernel of the fractional Fourier transform, $K_\alpha(t,x)$, $0 \leq \alpha \leq 1$, is a function that generalizes the exponential function but reduces to it when the parameter is equal to one. But there are a number of generalizations of the exponential function that depend on a parameter $0 \leq \alpha \leq 1$, and yield the exponential function when $\alpha = 1$. As a result any integral transform with such a function as a kernel may be called *fractional Fourier transform*.

The question is: What other properties do these functions share with the exponential function? and, What properties do these so-called fractional Fourier transforms share with the standard Fourier transform?

In what follows we shall give examples of such functions.

The idea of generalizing the sine, cosine and Fourier transforms, can be traced back to the work of W. H. Young in 1912 [329] who introduced infinite integrals involving a generalization of the sine and cosine functions, and hence, the exponential function. Young's approach was to introduce the function

$$C_p(t) = \sum_{k=0}^{\infty} \frac{t^{p+2k}(-1)^k}{\Gamma(p+2k+1)}; \quad 0 \leq p \leq 2,$$

which yields $C_0(t) = \cos t$, $C_1(t) = \sin t$. He used this function to construct the following generalizations of the cosine, sine transforms and; hence, of the Fourier transform:

$$\hat{f}(w) = \int_0^\infty f(t) C_p(wt)\, dt.$$

The function C_p has other similar properties to the cosine and sine functions:

$$C_p(t) = \frac{d}{dt} C_{p+1}(w),$$

and

$$\int_0^\infty e^{-t} C_p(wt)\, dt = \frac{w^p}{(1+w^2)},$$

which is a generalization of the relations

$$\int_0^\infty e^{-t} \sin(wt)\, dt = \frac{w}{(1+w^2)},$$

and

$$\int_0^\infty e^{-t}\cos(wt)dt = \frac{1}{(1+w^2)}.$$

To present another example of a fractional cosine, sine and Fourier transforms, let us recall from Chapter 1 that the Mittag-Leffler function is defined as

$$E_{\alpha,A}(x) = \sum_{k=0}^{\infty}\frac{(Ax)^k}{\Gamma(1+\alpha k)}.$$

It is an entire function of order $1/\alpha$ and type A. In particular, for $\alpha = 1$, we have $E_{1,A}(x) = e^{Ax}$. The *Mittag-Leffler Transform* of $f(x)$ is defined as

$$F_{\alpha,A}(t) = \int_{\mathbb{R}} f(x)E_{\alpha,A}(tx)dx,$$

whenever the integral exists [352].

For $\alpha = 1, A = 1$, and $f(x) = 0, x < 0$, the Mittag-Leffler Transform reduces to the Laplace transform and for $\alpha = 1, A = i$, we have the Fourier transform.

Let us denote $E_{\alpha,1}$ by E_α, for short and define

$$E_\alpha(ix^\alpha) = \cos_\alpha x^\alpha + i\sin_\alpha x^\alpha, \qquad 0 < \alpha \leq 1,$$

where

$$\cos_\alpha x^\alpha = \sum_{k=1}^{\infty}(-1)^k\frac{x^{2k\alpha}}{\Gamma(2k\alpha+1)},$$

and

$$\sin_\alpha x^\alpha = \sum_{k=1}^{\infty}(-1)^k\frac{x^{(2k+1)\alpha}}{\Gamma(2k\alpha+\alpha+1)}.$$

It is evident that for $\alpha = 1$, these functions reduce to the cosine and sine functions. Using these functions as kernels, the author in [150, 151] obtained fractional cosine, sine and Fourier transforms; see also [192].

Another form of generalization of the cosine, sine and exponential functions are given by their q-analogue as follows. Consider the functions defined by

$$C_q(\cos\theta;\omega) = \frac{(-\omega^2;q^2)_\infty}{(-q\omega^2;q^2)_\infty}\,{}_2\phi_1(-qe^{2i\theta},-qe^{-2i\theta};q;q^2,-\omega^2)$$

and

$$S_q(\cos\theta;\omega) = \frac{(-\omega^2;q^2)_\infty}{(-q\omega^2;q^2)_\infty}\left(\frac{2q^{1/4}\omega}{1-q}\right)\cos\theta$$
$$\times\,{}_2\phi_1(-q^2e^{2i\theta},-q^2e^{-2i\theta};q^3;q^2,-\omega^2),$$

where $0 < q \leq 1$. It can be shown that these functions reduce to the cosine and sine functions as $q \to 1$.

The symbol ${}_{r+1}\phi_r$ stands for the function (see Section 1.2.8)

$$ {}_{r+1}\phi_r(a_1,\ldots,a_{r+1};b_1,\ldots,b_r;q,z)=\sum_{n=0}^{\infty}\frac{(a_1,\ldots,a_{r+1};q)_n}{(q,b_1,\ldots,b_r;q)_n}z^n. $$

To see how these functions resembles the cosine and sine functions, first let us set

$$ w(\cos\theta)=\frac{(e^{2i\theta},e^{-2i\theta};q)_\infty}{\sin\theta\,(q^{1/2}e^{2i\theta},q^{1/2}e^{-2i\theta};q)_\infty},\qquad W(x)=\sqrt{1-x^2}\,w(x). \tag{3.2} $$

One interesting feature of these functions is that they form an orthogonal system on $(-1,1)$ like the system $\{1,\cos n\pi x,\sin m\pi x\}$, $m,n=1,2,\ldots$. It has been shown [56, 292] that,

$$ \int_{-1}^{1}C_q(x;\omega)C_q(x;\omega')w(x)\,dx=\int_{-1}^{1}S_q(x;\omega)S_q(x;\omega')w(x)\,dx=0, $$
$$ \int_{-1}^{1}C_q(x;\omega)S_q(x;\omega')w(x)\,dx=0, $$
$$ \int_{-1}^{1}C_q^2(x;\omega)w(x)\,dx=\int_{-1}^{1}S_q^2(x;\omega)w(x)\,dx $$
$$ =\pi\frac{(q^{1/2},-q^{1/2}\omega^2;q)_\infty}{(q,-\omega^2;q)_\infty}\frac{(-\omega^2;q^2)_\infty}{(-q\omega^2;q^2)_\infty}\,{}_2\phi_1(q^{1/2},-\omega^2;-q^{1/2}\omega^2;q,q) $$

where ω and ω' are different solutions of the equation

$$ S_q\left(\frac{1}{2}(q^{1/4}+q^{1/4});\omega\right)=\frac{(-i\omega;\sqrt{q})_\infty-(i\omega;\sqrt{q})_\infty}{2i(-q\omega^2;q^2)_\infty}=0. \tag{3.3} $$

Let us denote the non-negative zeros of the above equation by ω_n, where $\omega_0=0$, $\omega_1<\omega_2<\ldots$.

The q-exponential function is defined as in the standard way (see Ismail and Zhang [141]) as

$$ \mathcal{E}_q(x;i\omega)=C_q(x;\omega)+iS_q(x;\omega). \tag{3.4} $$

Bustoz and Suslov [56, 292] proved that $\{\mathcal{E}_q(x;i\omega_n)\}_{n=-\infty}^{\infty}$ is a complete orthogonal system in $L^2(-1,1)$ with respect to the weight function $w(x)$ and satisfies the orthogonality relation

$$ \int_{-1}^{1}\mathcal{E}_q(x;i\omega_m)\,\mathcal{E}_q(x;-i\omega_n)w(x)dx=2k(\omega_n)\delta_{m,n}, \tag{3.5} $$

where

$$ k(\omega)=\pi\frac{(q^{1/2},-q^{1/2}\omega^2;q)_\infty(-\omega^2;q^2)_\infty}{(q,-\omega^2;q)_\infty(-q\omega^2;q^2)_\infty}\,{}_2\phi_1(q^{1/2},-\omega^2;-q^{1/2}\omega^2;q,q). \tag{3.6} $$

Using the function $\mathcal{E}_q(x;i\omega)$ one may define a fractional Fourier transform that reduces to the Fourier transform as $q\to 1$. For more details and generalizations of those functions, we refer the reader to the work of Ismail in [139, 140]. For other types of q-analogue of the Fourier and Hankel transforms, we refer the reader to the work of T. Koornwinder [164].

We close this section by another form of a fractional Fourier transform which was introduced in [283] and is given by

$$(\mathcal{F}_\alpha f)(w) = \int_{\mathbb{R}} e^{i(sign\, w)|w|^{1/\alpha} t} f(t)dt, \quad w \in \mathbb{R} \text{ and } 0 < \alpha \leq 1,$$

For $w > 0$ and $\alpha = 1$, we get the standard Fourier transform.

3.2 PRELUDE TO FRACTIONAL INTEGRAL TRANSFORMS

In the previous section we saw examples of generalizations of the exponential function that depend on a parameter, say $0 \leq \alpha, q \leq 1$, that yield the exponential function when the parameter is equal to one. Therefore, one may argue that any integral transform with such a function as a kernel, may be called a fractional Fourier transform.

However, it is widely regarded by researchers in the field that the only authentic fractional Fourier transform is the one introduced by Namias because it generalizes in a natural way almost all the important properties of the Fourier transform and it extends its applications to new domains. Thus, from now on and throughout this book, the term fractional Fourier transform will refer to Namias's fractional Fourier transform.

In this section, we will briefly introduce Namias's approach, which was used to obtain other fractional integral transforms, such as the fractional Hankel transform. A more extensive study of the fractional Fourier transform will be presented in subsequent chapters.

3.2.1 The Fractional Fourier Transform

Let's recall that the eigenfunctions of the Fourier transformation as an operator $\mathcal{F}$ are the Hermite functions (see Section 1.2.3) defined as

$$\phi_n(x) = e^{-x^2/2} H_n(x), \quad n = 0, 1, 2, ...$$

where $H_n(x)$ is the Hermite polynomial of degree n. The corresponding eigenvalues are $\exp(in\pi/2)$. That is

$$\mathcal{F}[\phi_n(x)](w) = e^{in\pi/2} \phi_n(w).$$

It is also known that the Hermite functions are orthogonal basis of $L^2(\mathbb{R})$; hence, for any $f \in L^2(\mathbb{R})$, we have

$$f(x) = \sum_{n=0}^{\infty} a_n \phi_n(x), \tag{3.7}$$

where the convergence is (at least) in the sense of $L^2(\mathbb{R})$, and the expansion coefficients are given by

$$a_n = \frac{1}{2^n n! \sqrt{\pi}} \int_{-\infty}^{\infty} f(x) \phi_n(x) dx. \tag{3.8}$$

Namias introduced his fractional Fourier transform as follows. Consider the operator $\mathcal{F}_\alpha$ satisfying the eigenvalue equation

$$\mathcal{F}_\alpha[\phi_n(x)] = e^{in\alpha} \phi_n(x).$$

The operator $\mathcal{F}_\alpha$ is called the fractional Fourier transformation. It can be shown that using the differential equation for the Hermite functions

$$y'' - x^2 y = -(2n+1)y,$$

the transformation $\mathcal{F}_\alpha$ can be written in the form

$$\mathcal{F}_\alpha = e^{i\alpha A},$$

where

$$A = -\frac{1}{2}\frac{d^2}{dx^2} + \frac{1}{2}x^2 - 1/2,$$

so that

$$e^{i\alpha A}[\phi_n(x)] = e^{in\alpha}\phi_n(x). \tag{3.9}$$

Equation (3.9) shows that the Hermite functions are also eigenfunctions of the operator $e^{i\alpha A}$, but with eigenvalues $e^{in\alpha}$.

On account of equation (3.7), the effect of $\mathcal{F}_\alpha$ on $f(x)$ is given by

$$\mathcal{F}_\alpha[f](x) = \sum_{n=0}^{\infty} a_n e^{in\alpha}\phi_n(x). \tag{3.10}$$

The following properties of the fractional Fourier transform are evident from Eq. (3.10):

1. $\mathcal{F}_0 = I$, the identity operator,
2. $\mathcal{F}_{\pi/2} = F$, where F is the Fourier transform,
3. $\mathcal{F}_\alpha \mathcal{F}_\beta = \mathcal{F}_{\alpha+\beta}$
4. $\mathcal{F}_\pi = \tilde{I}$, where $\tilde{I}[f(x)] = f(-x)$,
5. $\mathcal{F}_{\alpha+2\pi} = \mathcal{F}_\alpha$.

Property (4) follows from the fact that the Hermite functions $\phi_n(x)$ are even for n even and odd for n odd, i.e., $\phi_n(-x) = (-1)^n\phi_n(x)$.

From Eqs. (3.7) and (3.8), we have

$$\mathcal{F}_\alpha[f](x) = \int_{-\infty}^{\infty} f(t)K_\alpha(x,t)dt,$$

where

$$K_\alpha(x,t) = \frac{1}{\sqrt{\pi}}\sum_{n=0}^{\infty}\frac{1}{2^n n!}\phi_n(t)\phi_n(x)e^{in\alpha} = \frac{e^{-(x^2+t^2)/2}}{\sqrt{\pi}}\sum_{n=0}^{\infty}\frac{e^{in\alpha}}{2^n n!}H_n(t)H_n(x).$$

Using Mehler's formula [175, p. 61] (see Section 1.2.3)

$$\sum_{n=0}^{\infty}\frac{H_n(x)H_n(t)}{2^n n!}z^n = \frac{1}{\sqrt{1-z^2}}\exp\left(\frac{2xtz-(x^2+y^2)z^2}{1-z^2}\right).$$

we obtain

$$\begin{aligned}
K_\alpha(x,t) &= \frac{e^{-(x^2+t^2)/2}}{\sqrt{\pi(1-e^{2i\alpha})}} \exp\left\{\frac{2xte^{i\alpha}-(x^2+t^2)e^{2i\alpha}}{1-e^{2i\alpha}}\right\} \\
&= \frac{e^{-(x^2+t^2)/2}}{\sqrt{\pi(1-e^{2i\alpha})}} \exp\left\{\frac{2xt-(x^2+t^2)e^{i\alpha}}{e^{-i\alpha}-e^{i\alpha}}\right\} \\
&= \frac{1}{\sqrt{\pi(1-e^{2i\alpha})}} \exp\left\{\frac{ixt}{\sin\alpha} - \frac{(x^2+t^2)}{2}\left[\frac{ie^{i\alpha}}{\sin\alpha}+1\right]\right\} \\
&= \frac{1}{\sqrt{\pi(1-e^{2i\alpha})}} \exp\left\{\frac{ixt}{\sin\alpha} - i\frac{(x^2+t^2)}{2}\cot\alpha\right\}.
\end{aligned}$$

The constant $\frac{1}{\sqrt{\pi(1-e^{2i\alpha})}}$ may be simplified to

$$\frac{1}{\sqrt{\pi(1-e^{2i\alpha})}} = \frac{1}{\sqrt{2\pi}}\frac{e^{i(\pi/4-\alpha/2)}}{\sqrt{\sin\alpha}} = \frac{\sqrt{1+i\cot\alpha}}{\sqrt{2\pi}}.$$

After some calculations, the kernel $K_\alpha(t,x)$ of the fractional Fourier transform is reduced to

$$\begin{aligned}
K_\alpha(t,x) &= \frac{e^{i(\pi/4-\alpha/2)}}{\sqrt{2\pi\sin\alpha}} \exp\left(\frac{ixt}{\sin\alpha} - \frac{i(x^2+t^2)\cot\alpha}{2}\right) \\
&= \frac{d(\alpha)}{\sqrt{2\pi}} \exp\left[ib(\alpha)xt - ia(\alpha)(x^2+t^2)\right] \\
&= c(\alpha)\exp\left\{i\left[b(\alpha)xt - a(\alpha)(x^2+t^2)\right]\right\}
\end{aligned}$$

where

$$a(\alpha) = \frac{\cot\alpha}{2}, \quad b(\alpha) = \csc\alpha, \quad d(\alpha) = \sqrt{1+i\cot\alpha} \quad \text{and, } c(\alpha) = \frac{d(\alpha)}{\sqrt{2\pi}}.$$

It is evident that $K_\alpha(t,x)$ is symmetric in the variables t and x so that $K_\alpha(t,x) = K_\alpha(x,t)$ and $K_{\pi/2}(t,x) = (1/\sqrt{2\pi})e^{ixt}$. Therefore, for all $0 \le \alpha \le 2\pi$, the one-dimensional fractional Fourier transform of a function f is given by

$$\mathcal{F}_\alpha[f](t) = F_\alpha(t) = \int_{-\infty}^{\infty} f(x)K_\alpha(t,x)dx \tag{3.11}$$

where

$$K_\alpha(t,x) = \begin{cases} c(\alpha)\cdot e^{-i[a(\alpha)(t^2+x^2)-b(\alpha)xt]}, & \alpha \ne p\pi \\ \delta(t-x), & \alpha = 2p\pi \\ \delta(t+x), & \alpha = (2p-1)\pi \end{cases} \tag{3.12}$$

where $a(\alpha), b(\alpha)$ are given as above, $c(\alpha) = \sqrt{\frac{1+i\cot\alpha}{2\pi}}$, $\alpha \in \mathbb{R}$, p is an integer and δ is the Dirac delta function.

In the case where $\alpha = 2p\pi$, we rewrite K_α in the form

$$\begin{aligned} K_\alpha(x,t) &= c(\alpha)\exp\left\{-i\frac{(x^2+t^2)}{2\tan\alpha}+i\frac{tx}{\sin\alpha}\right\} \\ &= c(\alpha)\exp\left\{-i\frac{(x-t)^2}{2\tan\alpha}+ixt\left(\frac{1}{\sin\alpha}-\frac{1}{\tan\alpha}\right)\right\} \\ &= c(\alpha)\exp\left\{-i\frac{(x-t)^2}{2\tan\alpha}+ixt\tan(\alpha/2)\right\}, \end{aligned}$$

where

$$c(\alpha) = \sqrt{\frac{i+\tan\alpha}{2\pi\tan\alpha}}.$$

Hence, with some easy calculations, one can show that the limit of $K_\alpha(x,t)$ may be put in the form

$$\begin{aligned} \lim_{\alpha\to 0} K_\alpha(x,t) &= \lim_{\alpha\to 0}\sqrt{\frac{i}{2\pi\tan\alpha}}\exp\left\{\frac{-i}{2\tan\alpha}(x-t)^2\right\} \\ &= \lim_{\alpha\to 0}\frac{1}{\sqrt{\pi}}\frac{1}{\lambda}\exp\left\{-(x-t)^2/\lambda^2\right\} \\ &= \delta(t-x), \end{aligned}$$

where $1/\lambda = \sqrt{i/(2\tan\alpha)}$. The last step follows from the known relation

$$\lim_{w\to 0}\frac{-e^{x^2/w^2}}{\sqrt{\pi}w} = \delta(x);$$

see Eq. (1.26). The proof for the case $\alpha = (2p+1)\pi$ is similar.

3.2.2 The Fractional Hankel Transform

The second fractional integral transform introduced in the literature was the fractional Hankel transform, also introduced by Namias in [218, 362]. Namias used the method he developed to derive the fractional Fourier transform to obtain a fractional Hankel transform.

Recall (Chapter 2) that the Hankel transform pair of order ν is given by

$$\mathcal{H}_\pi[f] = F(t) = \int_0^\infty xf(x)J_\nu(tx)dx, \qquad f(x) = \int_0^\infty tF(t)J_\nu(tx)dt,$$

where J_ν is the Bessel function; see Section 1.2.6. Namias denoted the Hankel transform by $\mathcal{H}_\pi$ for reasons to be clear later. Observing that the functions

$$\phi_n^\nu(x) = x^\nu e^{-x^2/2}L_n^\nu(x^2),$$

are the eigenfunctions of the Hankel transform with eigenvalues $(-1)^n$, where $L_n^\nu(x)$ are the generalized Laguerre polynomials (see Section 1.2.4), i.e.

$$\mathcal{H}_\pi\left[\phi_n^\nu(x)\right] = e^{in\pi}\phi_n^\nu(x).$$

Namias sought a fractional transform, as he did with the Fourier transform, such that

$$\mathcal{H}_\alpha\left[\phi_n^\nu(x)\right] = e^{in\alpha}\phi_n^\nu(x). \tag{3.13}$$

The fractional operator $\mathcal{H}_\alpha$ may also be represented in the form $e^{i\alpha A}$, where

$$A = -\frac{1}{4}\frac{d^2}{dx^2} - \frac{1}{4x}\frac{d}{dx} + \frac{x^2}{4} + \frac{\nu^2}{4x^2} - \frac{(\nu+1)}{2}.$$

Using the fact that the generalized Laguerre polynomials form a complete orthogonal system in $L^2[0,\infty)$, with the orthogonality relation

$$\int_0^\infty e^{-y}y^\nu L_n^\nu(y)L_m^\nu(y)dy = \frac{\Gamma(\nu+n+1)}{n!}\delta_{m,n}, \quad m,n = 0,1,2,\ldots,$$

we have for an appropriate function f, the expansion

$$f(x) = \sum_{n=0}^\infty a_n\phi_n^\nu(x), \tag{3.14}$$

where the coefficients a_n are given by

$$a_n = \frac{2n!}{\Gamma(\nu+n+1)}\int_0^\infty f(y)e^{-y^2/2}y^{\nu+1}L_n^\nu(y^2)dy.$$

Thus, it follows in view of Eqs. (3.13) and (3.14) that

$$\mathcal{H}_\alpha[f(x)] = \sum_{n=0}^\infty e^{in\alpha}a_n\phi_n^\nu(x).$$

Taking account of the formula (see Section 1.2.4)

$$\begin{aligned}&\sum_{n=0}^\infty \frac{n!e^{in\alpha}}{\Gamma(\nu+n+1)}L_n^\nu(x^2)L_n^\nu(y^2)\\ &= \frac{(xy)^{-\nu}e^{(i\nu/2)(\pi/2-\alpha)}}{1-e^{i\alpha}}\exp\left[-\frac{e^{i\alpha}(x^2+y^2)}{1-e^{i\alpha}}\right]J_\nu\left(\frac{xy}{\sin(\alpha/2)}\right),\end{aligned}$$

we obtain, with some calculations, the fractional Hankel transform formula

$$\begin{aligned}\mathcal{H}_\alpha[f](x) &= \frac{e^{i(\nu+1)(\pi/2-\alpha/2)}}{\sin(\alpha/2)}\exp\left(-\frac{ix^2}{2}\cot(\alpha/2)\right)\\ &\times \int_0^\infty yf(y)\exp\left(-\frac{iy^2}{2}\cot(\alpha/2)\right)J_\nu\left(\frac{xy}{\sin(\alpha/2)}\right)dy,\end{aligned}$$

which reduces to the Hankel transform when $\alpha = \pi$.

3.3 GENERAL CONSTRUCTION OF FRACTIONAL INTEGRAL TRANSFORMS

In this section, we present a general approach to constructing fractional integral transforms that yields Namias's fractional Fourier and Hankel transforms as special cases. For more details, we refer the reader to [343].

Let H be a separable Hilbert space with inner product $\langle\,,\rangle$, and A be a continuous linear transformation on H with a complete set of normalized eigenvectors $\{\phi_n\}_{n=0}^{\infty}$ such that $\|\phi_n\| = 1$. Let the corresponding eigenvalues be denoted by $\{\lambda_n\}_{n=0}^{\infty}$ so that $A[\phi_n] = \lambda_n \phi_n$ for all $n = 0, 1, 2, \ldots$.

A prototype of the operator A is any self-adjoint, compact operator. If $f \in H$, we have

$$f = \sum_{n=0}^{\infty} \langle f, \phi_n \rangle \phi_n = \sum_{n=0}^{\infty} \hat{f}_n \phi_n,$$

where

$$\hat{f}_n = \langle f, \phi_n \rangle.$$

It follows that

$$A[f] = \sum_{n=0}^{\infty} \lambda_n \langle f, \phi_n \rangle \phi_n,$$

provided that the series converges. A sufficient condition for this to hold is that $\{\lambda_n\}_{n=0}^{\infty}$ be uniformly bounded. For,

$$\begin{aligned}\left\| \sum_{k=m}^{n} \lambda_k \hat{f}_k \phi_k \right\|^2 &= \sum_{k=m}^{n} |\lambda_k|^2 \, |\hat{f}_k|^2 \\ &\leq \sup_k |\lambda_k|^2 \sum_{k=m}^{n} |\hat{f}_k|^2 \to 0 \quad \text{as} \quad m, n \to \infty.\end{aligned}$$

We now define the fractional transformation of order $0 \leq \alpha$ of the operator A, denoted by A_α, as

$$A_\alpha f = \sum_{n=0}^{\infty} \lambda_n^{\alpha} \langle f, \phi_n \rangle \phi_n, \tag{3.15}$$

whenever the series converges. If $\{\lambda_n\}_{n=0}^{\infty}$ is uniformly bounded, so is $\{\lambda_n^{\alpha}\}_{n=0}^{\infty}$ and the operator A_α is well defined, but multi-valued because the power function, $f(z) = z^{\alpha}$, is a multi-valued function. The operator is unique once the branch of the power function is determined. In most cases the principal branch is chosen. For more details, see [220, p. 139-142].

Let $I = (a, b)$, where $-\infty \leq a < b \leq \infty$ and $L_\rho^2(I)$ be the Hilbert space of all square integrable functions on I, with respect to a measure $d\rho$. Let A, ϕ_n and λ_n have the same meaning as above and assume that $\phi_n(x) \in L_\rho^2(I)$. If

$$\sum_{n=0}^{\infty} |\lambda_n|^{2\alpha} \, |\phi_n(x)|^2 < \infty, \tag{3.16}$$

it is easy to see that for each fixed x, the function $k_\alpha(x,t)$, defined by

$$k_\alpha(x,t) = \sum_{n=0}^{\infty} \overline{\lambda_n^\alpha \phi_n}(x)\phi_n(t).$$

is in $L^2_\rho(I)$ as a function of t. For,

$$\|k_\alpha\|_t^2 = \int_I |k_\alpha(x,t)|^2 \, d\rho(t) = \sum_{n=0}^{\infty} |\lambda_n|^{2\alpha} |\phi_n(x)|^2 < \infty.$$

Since for each fixed x, the series defining $k_\alpha(x,t)$ converges to it in the sense of $L^2_\rho(I)$, it also converges to it weakly. Therefore, we can interchange the inner product and summation to obtain an integral representation of the operator A_α given by (3.15) as

$$A_\alpha[f](x) = \int_I f(t)\overline{k}_\alpha(x,t)\, d\rho(t) = \langle f(t), k_\alpha(x,t)\rangle.$$

It is evident that the transformation A_α satisfies the following analogous properties to those of the fractional Fourier transform F_α:

1. $A_0 = I$
2. $A_1 = A$
3. $A_\alpha A_\beta = A_{\alpha+\beta}$

If, in addition $\lambda_n = e^{i\beta n}$ for some β, then the transformation is periodic with period $2\pi/\beta$, i.e., $A_{\alpha+2\pi/\beta} = A_\alpha$.

Let us consider the case where the eigenvalues of A are of the form $\lambda_n = \lambda^n$, where $\lambda = |\lambda| e^{i\beta}$ so that $A\phi_n = \lambda^n \phi_n$, and require that the series

$$\sum_{n=0}^{\infty} \lambda^n |\phi_n(x)|^2 < \infty, \quad \text{for } |\lambda| < 1. \tag{3.17}$$

This series converges if, for example, $\phi_n(x) = O(n^p)$ for some p and all x as $n \to \infty$.

We now define the fractional (angular) transform, A_α, of order $0 \leq \alpha \leq 1$ as

$$A_\alpha[f](x) = \lim_{|\lambda| \to 1^-} \sum_{n=0}^{\infty} |\lambda|^{n\alpha} e^{in\alpha\beta} \langle f, \phi_n\rangle \phi_n(x) = \sum_{n=0}^{\infty} e^{in\alpha\beta} \langle f, \phi_n\rangle \phi_n(x),$$

whenever the last series converges. This definition yields the integral representation

$$\begin{aligned} A_\alpha[f](x) &= \lim_{|\lambda| \to 1^-} \int_I f(t)\overline{k}_\alpha(x,t,\lambda)\, d\rho \\ &= \int_I f(t)\overline{k}_\alpha(x,t,1)\, d\rho = \int_I f(t)\overline{k}_\alpha(x,t)\, d\rho \end{aligned} \tag{3.18}$$

or

$$A_\alpha[f](x) = \lim_{|\lambda| \to 1^-} \langle f(t), k_\alpha(x,t,\lambda)\rangle = \langle f(t), k_\alpha(x,t)\rangle,$$

where

$$k_\alpha(x,t,\lambda) = \sum_{n=0}^{\infty} |\lambda|^{n\alpha} e^{-in\alpha\beta} \overline{\phi}_n(x)\phi_n(t),$$

and

$$k_\alpha(x,t) = \sum_{n=0}^{\infty} e^{-in\alpha\beta} \overline{\phi}_n(x)\phi_n(t). \tag{3.19}$$

If we use the polar coordinates, (r,θ), the kernel $k_\alpha(x,t)$ can be viewed as the radial limit of the function $k_\alpha(x,t,\lambda)$ as $|\lambda| \to 1$ along the ray $\theta = \alpha$. This gives more meaning to the phrase *angular transform* as it has been used in connection with the fractional Fourier transform.

3.3.1 Examples of the General Construction

We now demonstrate our definition of a fractional integral transform with the following examples, which include Namias's fractional Fourier and Hankel transforms as special cases.

Example 1: (The Fractional Fourier Transform)

Let $I = \mathbb{R}$, $H = L^2(\mathbb{R})$, and ϕ_n be the Hermite functions

$$\phi_n(x) = \frac{1}{\sqrt{2^n n! \sqrt{\pi}}} e^{-x^2/2} H_n(x),$$

and the transformation A be the Fourier transformation so that

$$A[\phi_n(x)] = \lambda_n \phi_n(x), \quad \text{where} \quad \lambda_n = e^{in\beta}, \text{ where } \beta = \pi/2.$$

To obtain an integral representation of A_α, we combine (3.18) and (3.19)

$$\begin{aligned} A_\alpha[f](x) &= \sum_{n=0}^{\infty} e^{in\alpha\beta} \left(\int_{-\infty}^{\infty} f(t)\phi_n(t)dt \right) \phi_n(x) \\ &= \int_{-\infty}^{\infty} f(t) k_\alpha(x,t)\, dt, \quad \alpha \neq n\pi; n = 0,1,2,3,\ldots, \end{aligned}$$

where

$$k_\alpha(x,t) = \sum_{n=0}^{\infty} e^{in\alpha\pi/2} \phi_n(t)\phi_n(x).$$

On account of Mehler's formula, we obtain after some calculations

$$k_\alpha(x,t) = \frac{1}{\sqrt{\pi}\sqrt{1-e^{2i\alpha\beta}}} e^{-(x^2+t^2)/2} \exp\left(\frac{2xte^{i\alpha\beta} - (x^2+t^2)e^{2i\alpha\beta}}{1-e^{2i\alpha\beta}} \right),$$

which can be simplified to

$$\begin{aligned} k_\alpha(x,t) &= \frac{e^{i(\pi/4 - \alpha\pi/4)}}{\sqrt{2\pi \sin(\alpha\pi/2)}} \exp\left(\frac{ixt}{\sin(\alpha\pi/2)} - \frac{i(x^2+t^2)\cot(\alpha\pi/2)}{2} \right) \\ &= \frac{c(\alpha)}{\sqrt{2\pi}} \exp\left[ib(\alpha)xt - ia(\alpha)(x^2+t^2) \right], \end{aligned}$$

where

$$a(\alpha) = \frac{\cot(\alpha\pi/2)}{2}, \quad b(\alpha) = \csc(\alpha\pi/2) \quad \text{and} \quad c(\alpha) = \sqrt{1+i\cot(\alpha\pi/2)}.$$

The discrepancy between our formula and Namias's formula is that in Namias's formula the standard Fourier transform is obtained when $\alpha = \pi/2$ while in ours it is obtained when $\alpha = 1$.

Example 2: (The Fractional Mellin Transform) The Mellin transform of a function $f(t)$ is defined as

$$F(\omega) = \int_0^\infty t^{\omega-1} f(t)\,dt.$$

It can be obtained from the Fourier or the two-sided Laplace transforms by a change of variables. Likewise, we define the fractional Mellin transform of a function $\tilde{f}(x)$ as

$$\tilde{F}(\omega) = \frac{c(\alpha)}{\sqrt{2\pi}} \int_0^\infty x^{i\omega(\csc\alpha)-1} \tilde{f}(x)\,dx.$$

This is obtained from the fractional Fourier transform with kernel $k_\alpha(\omega,t)$ by setting

$$x = e^t, \quad \tilde{f}(x) = f(\ln x)\exp[-ia(\alpha)(\ln x)^2] \text{ and } \quad \tilde{F}(\omega) = e^{ia(\alpha)\omega^2} F(\omega).$$

Example 3: (The Fractional Hankel Transform)

Let $I = R^+ = (0,\infty)$, $H = L^2(0,\infty)$ and

$$\phi_n^\nu(x) = \sqrt{\frac{n!}{\Gamma(n+\nu+1)}}\, e^{-x/2} x^{\nu/2} L_n^\nu(x), \quad \nu > -1,$$

where $L_n^\nu(x)$ is the generalized Laguerre polynomial of degree n and index ν. We choose the transformation A to be the Hankel transformation

$$A^\nu[f](x) = \int_0^\infty f(t) J_\nu(\sqrt{xt})dt,$$

where $J_\nu(z)$ is the Bessel function of the first kind and order ν.

In Formula 7.4.21 (4) in [117], which states that

$$\int_0^\infty x^{\nu+1} e^{-\beta x^2} L_n^\nu(ax^2) J_\nu(xy)dx = \frac{(\beta-a)^n}{2^{\nu+1}\beta^{\nu+n+1}}\, y^\nu e^{-y^2/(4\beta)} L_n^\nu\left[\frac{ay^2}{4\beta(a-\beta)}\right],$$

where a, Re $\beta > 0$, Re $\nu > -1$, if we put $x = \sqrt{z}$, $a = 1$, $\beta = \frac{1}{2}$ and $y = \sqrt{u}$, we obtain

$$A^\nu[\phi_n^\nu](u) = \int_0^\infty \phi_n^\nu(z) J_\nu(\sqrt{uz})dz = \lambda_n \phi_n^\nu(u), \quad \lambda_n = 2(-1)^n = 2e^{in\pi}.$$

Therefore, the fractional Hankel transform of order α is defined as

$$A_\alpha^\nu[f](x) = \int_0^\infty f(t)k_\alpha^\nu(x,t)dt,$$

where

$$k_\alpha^\nu(x,t) = \sum_{n=0}^{\infty} 2^\alpha e^{in\pi\alpha}\phi_n^\nu(x)\phi_n^\nu(t).$$

To calculate $k_\alpha^\nu(x,t)$ explicitly, recall that

$$k_\alpha^\nu(x,t) = \lim_{|\lambda|\to 1^-} k_\alpha^\nu(x,t,\lambda) = 2^\alpha \sum_{n=0}^{\infty} \lambda^{\alpha n}\phi_n^\nu(x)\phi_n^\nu(t),$$

where $\lambda = |\lambda|e^{i\pi}$.

We now appeal to formula (4.17.6) in [175, p. 79]

$$\sum_{n=0}^{\infty} \frac{n!}{\Gamma(n+\nu+1)} L_n^\nu(x)L_n^\nu(t)z^n$$
$$= \frac{1}{(1-z)}\frac{1}{(xtz)^{\nu/2}} \exp\left(-\frac{(x+t)z}{1-z}\right) I_\nu\left(\frac{2\sqrt{xyz}}{1-z}\right); \quad |z|<1,$$

where $I_\nu(z)$ is the modified Bessel function of the first kind and order ν, and set $z=\lambda^\alpha$ to obtain

$$k_\alpha^\nu(x,t,\lambda) = 2^\alpha e^{-(x+t)/2}(xt)^{\frac{\nu}{2}} \sum_{n=0}^{\infty} \frac{n!}{\Gamma(n+\nu+1)} L_n^\nu(x)L_n^\nu(t)\lambda^{\alpha n}$$
$$= \frac{2^\alpha}{(1-\lambda^\alpha)\lambda^{\alpha\nu/2}} \exp\left[-\frac{(x+t)(1+\lambda^\alpha)}{2(1-\lambda^\alpha)}\right] I_\nu\left(\frac{2\sqrt{xt\lambda^\alpha}}{1-\lambda^\alpha}\right),$$

where $\lambda^\alpha = |\lambda|^\alpha e^{i\pi\alpha}$. Taking the limit as $|\lambda|\to 1$, we obtain after some calculations

$$k_\alpha^\nu(x,t) = \frac{i2^{\alpha-1}e^{-i\alpha\frac{\pi}{2}(1+\nu)}}{\sin\frac{\alpha\pi}{2}} \exp\left[-\frac{i(x+t)}{2}\cot\frac{\alpha\pi}{2}\right] I_\nu\left(\frac{i\sqrt{xt}}{\sin\frac{\alpha\pi}{2}}\right). \tag{3.20}$$

Because $I_\nu(z) = (1/i)^\nu J_\nu(iz)$ and $J_\nu(-z) = (-1)^\nu J_\nu(z)$, we have the following definition of the fractional Hankel transform.

Definition 9. *The fractional Hankel transform $A_\alpha^\nu(\nu \geq 0, 0<\alpha\leq 1)$ of a function $f(t)$ is defined as*

$$A_\alpha^\nu[f](x) = \frac{e^{i\frac{\pi}{2}(1+\nu)(1-\alpha)}}{2^{1-\alpha}\sin\left(\frac{\alpha\pi}{2}\right)} \int_0^\infty f(t)\exp\left[-\frac{i(x+t)}{2}\cot\frac{\alpha\pi}{2}\right] J_\nu\left(\frac{\sqrt{xt}}{\sin\frac{\alpha\pi}{2}}\right) dt,$$

whenever the integral exists.

Clearly, $A_1^\nu[f] = A^\nu[f]$ and $A_0^\nu[f] = f$ as can be seen from (3.15). One can also verify that the above kernel tends to the delta function as $\alpha \to 0$. Furthermore, we can obtain another version of the sine and cosine transforms in view of the relations

$$J_{1/2}(x) = \sqrt{\left(\frac{2}{\pi x}\right)}\sin x, \text{ and } J_{-1/2}(x) = \sqrt{\left(\frac{2}{\pi x}\right)}\cos x, \quad x > 0.$$

Our definition of the fractional Hankel transform agrees with that given in [9, 189, 218]. But our derivation is not only more general than theirs, but also provides a systematic and unified way to treat all these fractional transforms.

The fractional Hankel transform is particularly useful when axial symmetry is present. To demonstrate that, we start by defining the fractional Fourier transform in two dimensions, then show how it can be reduced to the fractional Hankel transform given above.

Let $\mathbf{x} = \langle x_1, x_2\rangle$ and $\mathbf{t} = \langle t_1, t_2\rangle$ be two-dimensional vectors. We use the standard notation

$$\mathbf{x}\cdot\mathbf{t} = x_1t_1 + x_2t_2 \quad \text{and} \quad \|\mathbf{x}\| = \mathbf{x}\cdot\mathbf{x} = x_1^2 + x_2^2.$$

Definition 10. *We define the two-dimensional fractional Fourier transform (of order α) of a function $f(t_1,t_2)$ as*

$$\begin{aligned} F_\alpha(x_1,x_2) &= F_\alpha[f](x_1,x_2) \\ &= \frac{c(\alpha)}{\sqrt{2\pi}}\int_{-\infty}^{\infty}\int_{-\infty}^{\infty} f(t_1,t_2)\exp\left\{-ia(\alpha)\left[\|\mathbf{x}\|^2 + \|\mathbf{t}\|^2 - 2(\mathbf{x}\cdot\mathbf{t})b(\alpha)\right]\right\}dt_1dt_2 \end{aligned}$$

where $a(\alpha)$ and $c(\alpha)$ are as defined before and $b(\alpha) = \sec\alpha$.

Let

$$x_1 = r\cos\theta, \quad x_2 = r\sin\theta, \quad t_1 = \rho\cos\phi, \quad t_2 = \rho\sin\phi,$$

and assume that f is axially symmetric, i.e., independent of ϕ. Set

$$G_\alpha(r,\theta) = F_\alpha(r\cos\theta, r\sin\theta) \quad \text{and} \quad g(\rho) = f(\rho\cos\phi,\rho\sin\phi).$$

Thus,

$$\begin{aligned} G_\alpha(r,\theta) &= \frac{c(\alpha)}{\sqrt{2\pi}}\times \\ &\int_0^\infty\int_0^{2\pi} g(\rho)\exp\left\{-ia(\alpha)\left[r^2+\rho^2 - 2r\rho b(\alpha)\cos(\phi-\theta)\right]\right\}\rho d\rho d\phi \\ &= \frac{c(\alpha)}{\sqrt{2\pi}}e^{-\frac{i}{2}r^2\cot\alpha}\int_0^\infty\int_0^{2\pi} g(\rho)e^{-\frac{i}{2}\rho^2\cot\alpha}\exp\{ir\rho\csc\alpha\cos(\phi-\theta)\}\rho d\rho d\phi \\ &= \frac{c(\alpha)}{\sqrt{2\pi}}e^{-\frac{i}{2}r^2\cot\alpha}\int_0^\infty g(\rho)e^{-\frac{i}{2}\rho^2\cot\alpha}\rho d\rho\int_0^{2\pi} e^{ir\rho\csc\alpha\cos(\phi-\theta)}d\phi. \end{aligned}$$

Using the relation (see Section 1.2.6)

$$J_p(x) = \frac{\left(\frac{x}{2}\right)^p}{2\sqrt{\pi}\,\Gamma\left(p+\frac{1}{2}\right)}\int_0^{2\pi}(\sin\psi)^{2p}e^{ix\cos\psi}d\psi,$$

we obtain

$$G_\alpha(r,\phi) = \sqrt{2\pi(1+i\cot\alpha)}e^{-\frac{i}{2}r^2\cot\alpha}\int_0^\infty g(\rho)^{-\frac{i}{2}\rho^2\cot\alpha}\rho J_0(r\rho\csc\alpha)d\rho,$$

which means, apart from a multiplicative constant, that G_α is the fractional Hankel transform of g with the Bessel function of order zero as the kernel. To see that, set $x=r^2$, $t=\rho^2$, $g(\rho)=f(\rho^2)$ in Definition 9 and then replace α by $2\alpha/\pi$.

3.3.2 Fractional Integral Transforms Associated With the Jacobi Polynomials

Example 4:

Let $I=(-1,1)$, $H=L^2(I)$, and

$$\psi_n^{(\alpha,\beta)}(x) = \frac{1}{\sqrt{h_n^{(\alpha,\beta)}}}(1-x)^{\alpha/2}(1+x)^{\beta/2}p_n^{(\alpha,\beta)}(x), \quad \alpha,\beta>-1,$$

where $p_n^{(\alpha,\beta)}(x)$ is the Jacobi polynomial of degree n (see Section 1.2.5), and

$$h_n^{(\alpha,\beta)} = \frac{2^{\alpha+\beta+1}}{(2n+\alpha+\beta+1)}\frac{\Gamma(n+\alpha+1)\Gamma(n+\beta+1)}{\Gamma(n+1)\Gamma(n+\alpha+\beta+1)}.$$

It is known that $\left\{\psi_n^{(\alpha,\beta)}(x)\right\}_{n=0}^\infty$ is an orthonormal basis of $L^2(I)$. To obtain the series

$$\sum_{n=0}^\infty \lambda^n\psi_n^{(\alpha,\beta)}(x)\psi_n^{(\alpha,\beta)}(t)$$

in closed form, we first make the change of variable $x=\cos 2\theta$ and $t=\cos 2\phi$, and use formula (2.3) in [24]

$$\begin{aligned}\sum_{n=0}^\infty \lambda^n\psi_n^{(\alpha,\beta)}(x)\psi_n^{(\alpha,\beta)}(t) &= \sum_{n=0}^\infty \lambda^n\psi_n^{(\alpha,\beta)}(\cos 2\theta)\psi_n^{(\alpha,\beta)}(\cos 2\phi)\\ &= 2^{\alpha+\beta}(\sin\theta\sin\phi)^\alpha(\cos\theta\cos\phi)^\beta\sum_{n=0}^\infty c_n p_n^{(\alpha,\beta)}(\cos 2\theta)p_n^{(\alpha,\beta)}(\cos 2\phi)\lambda^n\\ &= \frac{\Gamma(\alpha+\beta+2)(1-\lambda)}{2\Gamma(\alpha+1)\Gamma(\beta+1)(1+\lambda)^{\alpha+\beta+2}}(\sin\theta\sin\phi)^\alpha(\cos\theta\cos\phi)^\beta\\ &\qquad F_4\left(\frac{\alpha+\beta+2}{2},\frac{\alpha+\beta+3}{2};\alpha+1,\beta+1;\frac{a^2}{k^2},\frac{b^2}{k^2}\right), \end{aligned} \tag{3.21}$$

where

$$a=\sin\theta\sin\phi, \quad b=\cos\theta\cos\phi, \quad k=\frac{1}{2}\left(\sqrt{\lambda}+\frac{1}{\sqrt{\lambda}}\right),$$

$$c_n = \frac{(2n+\alpha+\beta+1)}{2^{\alpha+\beta+1}}\frac{\Gamma(n+1)\Gamma(n+\alpha+\beta+1)}{\Gamma(n+\alpha+1)\Gamma(n+\beta+1)},$$

and $F_4(a,b;c,c';x,y)$ is Appell's hypergeometric function of the fourth type and two variables [284, p. 53], defined by the equation

$$F_4(a,b;c,c';x,y) = \sum_{r=0}^{\infty}\sum_{s=0}^{\infty} \frac{(a)_{r+s}(b)_{r+s}}{r!s!(c)_r(c')_s} x^r y^s,$$

with

$$(a)_k = a(a+1)(a+2)\cdots(a+k-1) = \frac{\Gamma(a+k)}{\Gamma(a)}.$$

The series in (3.21) converges for all θ and ϕ since $k \geq 1$. A more convenient form of (3.21) can be obtained by applying a transformation of Watson [307] to reduce it to

$$\begin{aligned}
&\sum_{n=0}^{\infty} \lambda^n \psi_n^{(\alpha,\beta)}(\cos 2\theta)\psi_n^{(\alpha,\beta)}(\cos 2\phi) = 2^{\alpha+\beta}(\sin\theta\sin\phi)^{\alpha}(\cos\theta\cos\phi)^{\beta} \\
\times\ &\sum_{n=0}^{\infty} \lambda^n c_n p_n^{(\alpha,\beta)}(\cos 2\theta) p_n^{(\alpha,\beta)}(\cos 2\phi) \\
=\ &\frac{(1-\lambda)}{8\left(i\sqrt{\lambda}\right)^{\alpha+\beta+2}} \int_0^{\infty} zJ_{\alpha}(z\sin\theta\sin\phi)J_{\beta}(z\cos\theta\cos\phi) \\
\times\ &\exp\left(\frac{iz}{2}\left(\sqrt{\lambda}+\frac{1}{\sqrt{\lambda}}\right)\right)dz \\
=\ &\frac{-1}{2(i\sqrt{\lambda})^{\alpha+\beta-1}}\frac{d}{d\lambda}\int_0^{\infty} J_{\alpha}(z\sin\theta\sin\phi)J_{\beta}(z\cos\theta\cos\phi) \\
\times\ &\exp\left(\frac{iz}{2}\left(\sqrt{\lambda}+\frac{1}{\sqrt{\lambda}}\right)\right)dz.
\end{aligned}$$

Thus, the kernel of the associated transform $A_a^{(\alpha,\beta)}$, $0 \leq a \leq 1$ is given by

$$\begin{aligned}
k_a^{(\alpha,\beta)}(x,t,\lambda) &= \sum_{n=0}^{\infty} \lambda^{an} \psi_n^{(\alpha,\beta)}(x)\psi_n^{(\alpha,\beta)}(t) \\
&= C(\lambda)\int_0^{\infty} zJ_{\alpha}(z\sin\theta\sin\phi)J_{\beta}(z\cos\theta\cos\phi)\exp\left(\frac{iz}{2}\left(\lambda^{a/2}+\lambda^{-a/2}\right)\right)dz,
\end{aligned}$$

where

$$C(\lambda) = \frac{(1-\lambda^a)}{8\left(i\lambda^{a/2}\right)^{\alpha+\beta+2}}.$$

Therefore, the associated angular (fractional) integral transform of order a taken along the ray arg $\lambda = \delta$ in the complex λ-plane is defined as

$$A_a^{(\alpha,\beta)}[f](x) = \int_{-1}^{1} f(t)k_a^{(\alpha,\beta)}(x,t,e^{i\delta})dt.$$

Putting $x = \cos 2\theta$ and $t = \cos 2\phi$, we obtain

$$A_a^{(\alpha,\beta)}[f](\cos 2\theta) = 2\int_0^{\pi/2} f(\cos 2\phi)k_a^{(\alpha,\beta)}(\cos 2\theta,\cos 2\phi,e^{i\delta})\sin 2\phi\, d\phi,$$

or

$$
\begin{aligned}
A_a^{(\alpha,\beta)}[f](\cos 2\theta) &= F_a^{(\alpha,\beta)}(\cos 2\theta) \\
&= \frac{\sin(a\delta/2)}{2(i)^{\alpha+\beta+3}e^{ia\delta(\alpha+\beta+1)/2}} \int_0^{\pi/2} f(\cos 2\phi)\sin 2\phi\, d\phi \\
&\quad \int_0^{\infty} zJ_\alpha(z\sin\theta\sin\phi)J_\beta(z\cos\theta\cos\phi)\exp\left(iz\cos(a\delta/2)\right) dz.
\end{aligned}
$$

3.4 FRACTIONAL DERIVATIVES AND INTEGRALS VERSUS FRACTIONAL INTEGRAL TRANSFORMS

In the next example we will show how the general procedure for constructing fractional integral transforms presented in the previous section can be used to obtain the Riemann-Liouville fractional integral of a function and how it is related to fractional derivatives.

As an introduction to the next example, let us recall the definitions of fractional integrals and derivatives.

Because integration may be viewed as an inverse operation to differentiation, it is natural to associate with it the symbol d^{-1}/dx^{-1}. Thus, we adopt the notation

$$
\frac{d^{-1}f}{dx^{-1}} = \int_a^x f(t)dt. \tag{3.22}
$$

Despite its popularity, this notation is rather ambiguous because it does not display the dependence of the definite integral on the lower limit of integration. Nevertheless, we shall use it because the role that the lower limit of integration plays in our investigation is minor.

From (3.22), we obtain by successive integration

$$
\frac{d^{-n}f}{dx^{-n}} = \frac{1}{(n-1)!}\int_a^x (x-t)^{n-1} f(t)\, dt.
$$

Thus, if we replace n by a non-negative number q, we obtain Riemann's definition of the qth integral of a function f as

$$
\frac{d^{-q}f}{dx^{-q}} = \frac{1}{\Gamma(q)}\int_a^x (x-t)^{q-1} f(t)dt.
$$

For example, if $a = 0$, the qth integral ($q > 0$) of x^p, where $p > -1$, can be evaluated with the aid of the Beta Function to yield

$$
\begin{aligned}
\frac{d^{-q}}{dx^{-q}}x^p &= \frac{1}{\Gamma(q)}\int_0^x (x-t)^{q-1}t^p dt \\
&= \frac{x^{p+q}}{\Gamma(q)}\int_0^1 (1-t)^{q-1}t^p dt = \frac{\Gamma(p+1)}{\Gamma(q+p+1)}x^{q+p}.
\end{aligned} \tag{3.23}
$$

Using Riemann's definition of a fractional integral, we can now define the fractional derivative as follows [352, Ch. 20].

Definition 11. *The Riemann–Liouville fractional derivative operator of order q is defined as*

$$D_x^q f(x) = \frac{d^q}{dx^q} f(x) = \begin{cases} 1/\Gamma(-q)\int_a^x (x-t)^{-q-1} f(t)dt, & q<0 \\ \frac{d^m}{dx^m}(D_x^{q-m} f(x)), & q \geq 0, \end{cases}$$

where

$$\frac{d^m}{dx^m}(D_x^{q-m} f(x)) = \frac{1}{\Gamma(m-q)} \frac{d^m}{dx^m} \int_a^x (x-t)^{m-q-1} f(t)dt,$$

and $m-1 \leq q < m, m = 1,2,3,\ldots$. The Weyl fractional integral of order q is defined as

$$(W_q f)(x) = \frac{1}{\Gamma(q)} \int_x^\infty (t-x)^{q-1} f(t)dt.$$

Example 5: (The Fractional Riemann-Liouville Integral) [344, 350]. Let $I = [0,2\pi]$, $H = L^2(I)$ and $\phi_n(x) = (1/\sqrt{2\pi})e^{inx}$, for $n = 0, \pm 1, \pm 2, \ldots$ Define the transformation (integrator) A by

$$A[\phi_n(x)] = \frac{\phi_n(x)}{in}, \qquad n \neq 0.$$

Hence,

$$A^m[\phi_n(x)] = \frac{\phi_n(x)}{(in)^m}, \qquad n \neq 0 \text{ and } m = 0,1,2,\ldots.$$

For any $f \in L^2(I)$ that is periodic with period 2π and satisfies

$$\int_0^{2\pi} f(t)dt = 0,$$

we have the Fourier series expansion

$$f(x) = \sum_{n \neq 0} \hat{f}_n \phi_n(x),$$

where

$$\hat{f}_n = \int_0^{2\pi} f(t)\overline{\phi}_n(t)\, dt,$$

and $\sum_{n\neq 0}$ means $\sum_{n=-\infty, n\neq 0}^{\infty}$. Consequently,

$$A^m[f(x)] = \sum_{n \neq 0} \hat{f}_n \frac{\phi_n(x)}{(in)^m}, \qquad m = 0,1,2,\ldots,$$

and we define the fractional operator A^α by

$$A^\alpha[f(x)] = \sum_{n \neq 0} \hat{f}_n \frac{\phi_n(x)}{(in)^\alpha}, \qquad 0 \leq \alpha \leq 1. \tag{3.24}$$

Here we shall use Zygmund's notation [372, Ch. XII, p. 134], to show that $A^\alpha[f(x)]$ can be represented by a fractional integral transform in the form

$$A^\alpha[f(x)] = \int_0^{2\pi} f(t) k_\alpha(x,t)\, dt\,, \tag{3.25}$$

where

$$k_\alpha(x,t) = \Psi_\alpha(x) = \sum_{n\neq 0} \frac{1}{(in)^\alpha} \phi_n(x)\overline{\phi}_n(t) = \frac{1}{2\pi} \sum_{n\neq 0} \frac{e^{in(x-t)}}{(in)^\alpha}.$$

It is easy to see that

$$A^\alpha \left[A^\beta [f(x)] \right] = A^{\alpha+\beta}[f(x)].$$

We may write (3.25) in the form

$$A^\alpha[f(x)] = \frac{1}{2\pi} \int_0^{2\pi} f(t)\Psi_\alpha(x-t)\, dt\,, \quad \alpha > 0, \tag{3.26}$$

where

$$\Psi_\alpha(x) = \sum_{n\neq 0} \frac{e^{inx}}{(in)^\alpha} = \sum_{n\neq 0} \gamma_n^{(\alpha)} e^{inx}, \tag{3.27}$$

with

$$\gamma_n^{(\alpha)} = (in)^{-\alpha} = |n|^{-\alpha} \exp\left(-\frac{1}{2} i\pi\alpha \operatorname{sgn} n \right) \qquad \text{for } n \neq 0,\ \gamma_0^{(\alpha)} = 0.$$

Because the sequence $\{1/n^\alpha\}$ tends to zero monotonically, the series in (3.27) converges uniformly in each interval of type $[\epsilon, 2\pi - \epsilon]$, $\epsilon > 0$; see Ch. I, Eq. (2.6) in [372]. And by (1.5) and (1.14) of Ch. V [372] (see also p. 186), the series (3.27) is the Fourier series of an integrable function $\Psi_\alpha(x)$.

Since $\Psi_\alpha(x)$ is integrable and $f \in L^2(I) \subset L^1(I)$, the integral in (3.26), being a convolution of two integrable functions, exists and defines an integrable function, namely $A^\alpha[f(x)]$. Moreover, it follows from Theorem 4.4, Ch. III (see also p. 94) that the series (3.24) converges almost everywhere to $A^\alpha[f(x)]$ and it is the Fourier series of $A^\alpha[f(x)]$.

Now we examine the relationship between fractional derivatives and fractional integral transforms. The fractional derivative $f^{(\alpha)}$ of f of order $0 < \alpha < 1$ is defined as

$$f^{(\alpha)}(x) = \frac{d}{dx} A^{1-\alpha}[f(x)], \quad 0 \leq \alpha \leq 1,$$

and hence if $A^{1-\alpha}[f(x)]$ is absolutely continuous, we have

$$f^{(\alpha)}(x) = \sum_{n\neq 0} (in)^\alpha \hat{f}_n \phi_n(x) = \frac{1}{\sqrt{2\pi}} \sum_{n\neq 0} (in)^\alpha \hat{f}_n e^{inx}.$$

If $\alpha > 0$ is arbitrary, we define

$$f^{(\alpha)}(x) = \frac{d^n}{dx^n} A^{n-\alpha}[f(x)], \quad 0 \leq \alpha,$$

where n is the least integer greater than α.

It can be shown that for $0 \leq \alpha$ and $0 \leq x \leq 2\pi$ [372, Ch. XII, (8.8)]

$$\Psi_\alpha(x) = \frac{2\pi}{\Gamma(\alpha)} \lim_{n\to\infty} \left\{ x^{\alpha-1} + (x+2\pi)^{\alpha-1} + \cdots (x+2n\pi)^{\alpha-1} - (2\pi)^{\alpha-1} \frac{n^\alpha}{\alpha} \right\}.$$

Thus, if f is periodic with period 2π and its integral vanishes over a period, then

$$\frac{1}{2\pi}\int_0^{2\pi} f(x-t)\Psi_\alpha(t)\,dt = \lim_{n\to\infty}\frac{1}{\Gamma(\alpha)}\int_0^{2\pi} f(x-t)\sum_{k=0}^{n}(t+2k\pi)^{\alpha-1}dt$$
$$= \frac{1}{\Gamma(\alpha)}\int_0^{\infty} f(x-t)t^{\alpha-1}dt,$$

or

$$A^\alpha[f(x)] = \frac{1}{\Gamma(\alpha)}\int_{-\infty}^{x} f(t)(x-t)^{\alpha-1}dt,$$

which is related to the Weyl fractional integral of f of order α.

3.5 OTHER FRACTIONAL INTEGRAL TRANSFORMS

Because the Fourier transform is related to other integral transforms and the exponential function is used in other transforms, fractional versions of those transforms have been introduced by replacing the kernel of the Fourier transform by the kernel of the fractional Fourier transform. Examples of such transforms are the fractional Radon transform [333], fractional wavelet [198, 272, 275, 283, 363], fractional shearlets [188] and fractional Gabor transform. Different versions of those transforms have also been introduced. For example, in one version of the fractional wavelet transform $F^\alpha(a,b)$ of f with respect to a mother wavelet ψ, the wavelet transform is defined as

$$F^\alpha(a,b) = \int_{\mathbb{R}} F_\alpha(x)\psi_{a,b}(x)dx,$$

where F_α is the fractional Fourier transform of f. In another version,

$$F^\alpha(a,b) = \int_{\mathbb{R}} f(x)\psi^\alpha_{a,b}(x)dx,$$

where $\psi^\alpha_{a,b} = \psi_{a,b}e^{-i(t^2-b^2)/2\cot\alpha}$.

The fractional Gabor (windowed Fourier transform) is defined as the standard Gabor transform but with the exponential function being replaced by the kernel of the fractional Fourier transform, i.e.,

$$F(t,w) = \int_{\mathbb{R}} f(x)g(t-x)k_\alpha(x,w)dx,$$

where k_α is the kernel of the fractional Fourier transform and g is the Gaussian function. For related work, see [83, 166, 268, 283].

CHAPTER 4

The Fractional Fourier Transform (FrFT)

In Chapter 3 we derived the integral representations of several fractional integral transforms, among them was the fractional Fourier transform. In this chapter, we will discuss in more details properties of the fractional Fourier transform, its existence and operational calculus. But we will begin by a historical introduction to give the reader an overview of the development of the subject over the last four decades and then proceed to discuss the properties of the transform.

4.1 HISTORICAL OVERVIEW

The subject of fractional integral transforms started in 1980 with the introduction of the fractional Fourier transform (FrFT) by Victor Namias in his seminal paper *The fractional order Fourier transform and its application to quantum mechanics* [217]. The fractional Fourier transform, which is a generalization of the Fourier transform, has gained considerable attention in recent years because of its important applications in signal analysis, optics and signal recovery, and also because of its ability to treat some mathematical problems that could not otherwise be handled by the standard Fourier transform.

Although Namias was the first to explicitly use the term *fractional order Fourier transform*, his ideas appeared implicitly in the work of N. Wiener in 1929 [316] in which he sought to bring out the relationship between the expansion of a function in a series of orthogonal Hermite functions and its Fourier transform. He then used this relationship to extend certain results of Hermann Weyl which led to what Wiener called Fourier development of *fractional order*.

Another historical thread to the fractional Fourier transform can be traced to the work of E. Condon in 1937 [79]. To shed light on Condon's work, let us recall that the Fourier transform of a function f is given by

$$\mathcal{F}[f](x) = \hat{f}(x) = \frac{1}{\sqrt{2\pi}} \int_{\mathbb{R}} f(t) e^{ixt} dt, \tag{4.1}$$

and its inverse is given by

$$\mathcal{F}^{-1}[\hat{f}](t) = f(t) = \frac{1}{\sqrt{2\pi}} \int_{\mathbb{R}} \hat{f}(x) e^{-ixt} dx, \tag{4.2}$$

DOI: 10.1201/9781003089353-4

whenever the integrals exist. It is easy to see that the Fourier transformation $\mathcal{F}$ as an operator satisfies the relations

$$\mathcal{F}[f] = \hat{f}, \quad \mathcal{F}[\hat{f}] = \mathcal{F}^2[f] = f(-x), \quad \mathcal{F}^3[f] = \hat{f}(-t), \quad \mathcal{F}^4[f] = f, \tag{4.3}$$

and $f = \mathcal{F}^{-1}[\hat{f}] = \mathcal{F}^{-1}\mathcal{F}[f]$, which shows that the Fourier transformation is of period 4.

Observing that the Fourier transformation generates a cyclic group of order 4 that is isomorphic to the group of rotations of a plane about a fixed point through multiples of a right angle, E. Condon, in his article *Immersion of the Fourier transform in a continuous group of functional transformations* [79] showed that there exists a continuous group of functional transformations containing the ordinary Fourier transforms as a subgroup. He then proceeded to obtain an explicit representation of that group and, hence, obtained a representation of the fractional Fourier transform but without calling it that.

Condon's paper did not receive much attention, but a reference to it as a footnote was cited in V. Bargmann's paper *On a Hilbert Space of Analytic Functions and an Associated Integral Transform* in which he studied a Hilbert space of analytic functions and a transformation acting on it given by an integral transform which is now known as the Bargamnn transform [25, 26].

Approximately two years after the publication of Condon's paper, H. Kober [163] in an article written in German but published in an English journal, Quarterly Journal of Mathematics (Oxford University) obtained what he called *roots* of Hankel, Fourier and other continuous transformations and in which he derived the fractional Hankel and Fourier transforms.

In that article Kober cited a paper by E. Hille [132] on semi-groups of transformations in Hilbert spaces which appeared in the *Proceedings of the National Academy of Science* in 1938, but there was no mention of Condon's paper which appeared in the same journal a year earlier. There is considerable overlap between Condon's and Kober's papers but in Kober's paper the kernels of the fractional Hankel and Fourier transforms were explicitly given.

Another link to the fractional Fourier transform appeared in the work of N. G. De Bruijn [50] and his investigation of the Wigner distribution and Weyl correspondence. This link comes as no surprise since, as we will see later, the Wigner distribution and the fractional Fourier transform are closely related. De Bruijn obtained an integral transform that resembles the fractional Fourier transform but with the cosine and sine functions appearing in the kernel of the fractional Fourier transform were replaced by the cosh and sinh functions.

Unaware of Wiener's and Condon's work, V. Namias in 1980 [217] introduced a transform, which he called the fractional order Fourier transform, to solve ordinary and partial differential equations arising in quantum mechanics from classical quadratic Hamiltonians. In Namias's work the fractional Fourier transforms $\{\mathcal{F}_{\theta \in T}\}$, were indexed by a parameter $\theta \in T = \mathbb{R}/2\pi\mathbb{Z}$ and formed a cyclic group of order 4 with

$$\mathcal{F}_0 = \mathcal{I} = \mathcal{F}_{2\pi}, \; \mathcal{F}_{\pi/2} = \mathcal{F}, \; \mathcal{F}_{-\pi/2} = \mathcal{F}^{-1},$$

where $\mathcal{I}$ is the identity transformation and $\mathcal{F}$ is the conventional Fourier transformation.

McBride and Kerr [193, 194] and Kerr [157, 158, 159, 160], put Namias's work on a more solid foundation by providing it with more rigorous arguments. More than ten years after the publication of Namias's work, D. Mustrad [215, 216] derived some results concerning the fractional Fourier transform, such as a fractional convolution, uncertainty principle, and its relationship with the Wigner distribution, but he referred to the transform as the Condon-Bargmann fractional Fourier transform. It is somewhat intriguing that Mustard used Namias's definition and derivation of the fractional Fourier transform without ever citing Namias's work.

Not much was reported on the fractional Fourier transform since the publication of McBride and Kerr's papers until the early 1990's when A. Lohmann, D. Mendlovic, H. Ozaktas and L. Almeida published a number of papers on the implementation and application of the fractional Fourier transform in optics and signal processing [185, 186, 187, 197, 199, 200, 201, 220, 221, 222, 223, 224, 225, 226, 227]. This was a turning point on the development of the subject that has resulted in hundreds of publications so far. The transform has become the focus of many research papers because of its different applications in many fields [6, 7, 8, 12, 13, 14, 15, 16, 27, 32, 33, 54, 61, 62, 65, 66, 69, 99, 100, 119, 126, 162, 169, 170, 174, 182, 183, 189, 190, 202, 203, 221, 227, 233, 234, 235, 238, 239, 240, 241, 244, 245, 250, 259, 267, 285, 291, 325, 332, 339, 348, 350]

4.2 PRELIMINARIES

In this section, we will discuss some analytic properties of the fractional Fourier transform, its existence, and domain of definition. To begin, let us recall the definition of the transform and some of the notation associated with it.

The fractional Fourier Transform or FrFT of a function $f(t) \in L^1(\mathbb{R})$, is defined as [15, 16, 220],

$$\mathcal{F}_\theta\left[f\right](x) = F_\theta(x) = \int_{-\infty}^{\infty} f(t) K_\theta(x,t)\, dt, \tag{4.4}$$

where

$$K_\theta(x,t) = \begin{cases} c(\theta)\cdot e^{-i\left[a(\theta)(t^2+x^2)-b(\theta)xt\right]}, & \theta \neq p\pi \\ \delta(t-x), & \theta = 2p\pi \\ \delta(t+x), & \theta = (2p-1)\pi \end{cases} \tag{4.5}$$

is the transformation kernel with

$$a(\theta) = \cot\theta/2, \quad b(\theta) = \csc\theta \quad \text{and } c(\theta) = \sqrt{\tfrac{1+i\cot\theta}{2\pi}}.$$

To accommodate for $\sqrt{2\pi}$ which appears in some formulas, we may occasionally, write

$$c(\theta) = d(\theta)/\sqrt{2\pi}, \text{ where } d(\theta) = \sqrt{1+i\cot\theta}.$$

The kernel $K_\theta(t,x)$, which is clearly symmetric in x and t, is parameterized by an angle $\theta \in \mathbb{R}$ and p is an integer. For simplicity, we may write a,b,c or $a_\theta, b_\theta, c_\theta$ instead of

$a(\theta), b(\theta)$ and $c(\theta)$. The special cases where $\theta = 0, \pi/2$, and π yield the following FrFT of f:

$$F_0[f](u) = f(u), \quad F_{\pi/2}[f](u) = \hat{f}(u), \quad F_\pi[f](u) = f(-u),$$

where $\hat{f}$ denotes the ordinary Fourier transform of f; see Eq. (4.1). From now on we shall focus our attention on the cases where $\alpha \neq n\pi$.

Existence: It is evident that the integral in (4.4) exists if $f \in L^1(\mathbb{R})$ and it can be easily extended to $L^p(\mathbb{R})$ for $1 \leq p \leq 2$, using arguments similar to those used for extending the conventional Fourier transform to $L^p(\mathbb{R})$ for $1 \leq p \leq 2$. For example, since $L^1 \bigcap L^2$ is dense in L^2, for $f \in L^2$, choose a sequence $\{f_n\}_{n=0}^\infty$ such that $f_n \to f$, and define $\mathcal{F}_\theta[f]$ as

$$\mathcal{F}_\theta[f] = \lim_{n\to\infty} \mathcal{F}_\theta[f_n].$$

In [160], Kerr defined the fractional Fourier transform of $f \in L^2(\mathbb{R})$ as

$$\mathcal{F}_\theta[f](x) = \lim_{R\to\infty} \int_{-R}^{\mathbb{R}} f(t) K_\theta(t,x)\, dt,$$

where the limit is taken in the mean. In fact, $\mathcal{F}_\theta$ is not only a homeomorphism on $L^2(\mathbb{R})$, but a unitary operator on it as well, [69].

Moreover, in view of the relation

$$\left| e^{iax^2} F_\theta(x) \right| = |F_\theta(x)| \leq |c(\theta)| \int_{\mathbb{R}} |f(t)|\, dt = |c(\theta)|\, \|f\|_1;$$

we have

$$\|F_\theta\|_\infty \leq |c(\theta)|\, \|f\|_1,$$

which shows that $\mathcal{F}_\theta$ is a bounded linear transformation from $L^1(\mathbb{R})$ into $L^\infty(\mathbb{R})$. In fact, $F_\theta(x)$ is continuous and goes to zero as $|x| \to \infty$, (see Property 3 below); i.e., $F_\theta \in C_0(\mathbb{R})$.

To show the continuity of F_θ, we have

$$\begin{aligned} \Delta F(x,t) &= F_\theta(x+\Delta x) - F_\theta(x) = \int_{\mathbb{R}} f(t) \left(K_\theta(x+\Delta x, t) - K_\theta(x,t)\right) dt \\ &= c(\theta) \int_{\mathbb{R}} f(t) \exp\left\{-ia(\theta)(x^2+t^2) + ib(\theta)tx\right\} I(x,t)\, dt, \end{aligned}$$

where

$$I(x,t) = e^{i(\Delta x)[-a(\theta)(\Delta x) - 2a(\theta)x + b(\theta)t]} - 1.$$

Therefore,

$$|\Delta F(x,t)| \leq |c(\theta)| \int_{\mathbb{R}} |f(t)|\, |I(x,t)|\, dt \leq 2|c(\theta)|\, \|f\|_1 < \infty,$$

and by the Lebesgue dominated convergence theorem, we may take the limit inside the integral to obtain $\lim_{\Delta x\to 0} |\Delta F(x,t)| = 0$ since $\lim_{\Delta x\to 0} |I(x,t)| = 0$.

The fact that $\lim_{|x|\to\infty} F_\theta(x) = 0$ will follow from the Riemann-Lebesgue Lemma; see item 3 below.

It should be emphasized that if $f \in L^1(\mathbb{R})$, there is no guarantee that $F_\theta \in L^1(\mathbb{R})$. For example, let

$$f(t) = e^{iat^2 - t} \quad \text{for } t \geq 0, \text{ and } 0, \text{ otherwise.}$$

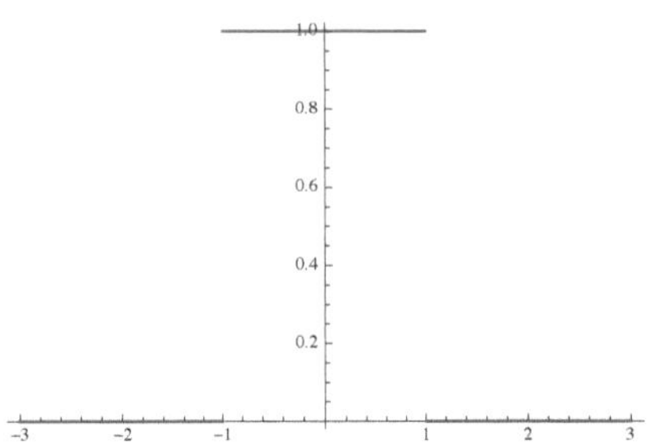

Figure 4.1 Zero FT

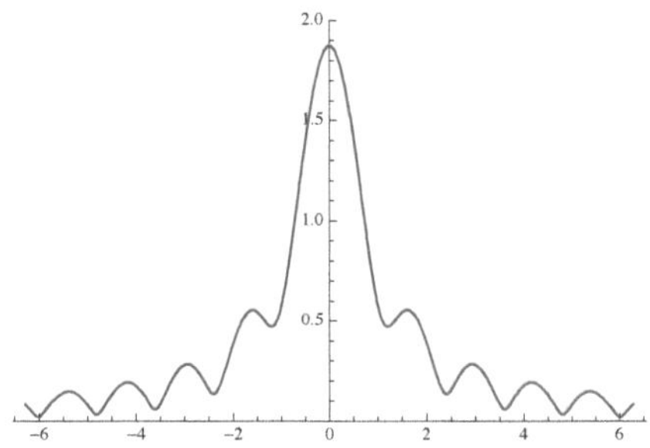

Figure 4.2 One quarter FT

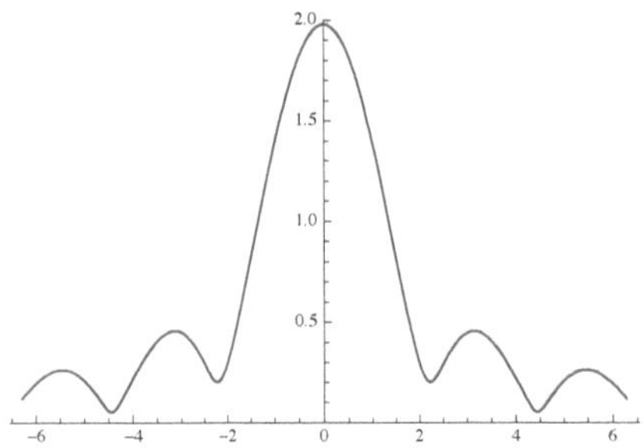

Figure 4.3 One half FT

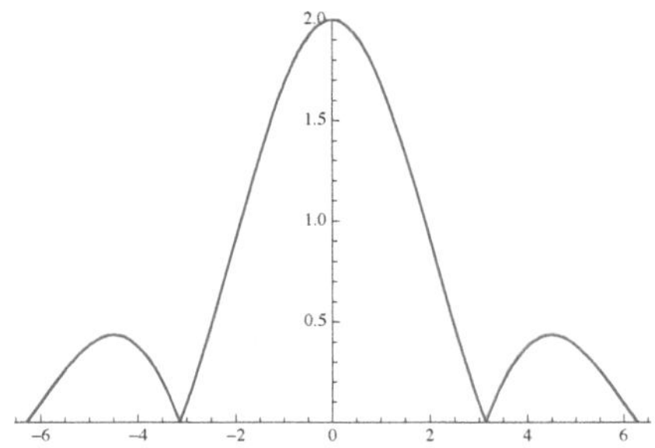

Figure 4.4 One FT

Hence

$$\begin{aligned} F_\theta(x) &= c(\theta)\int_0^\infty \exp\left\{-ia(x^2+t^2)+ibxt\right\} e^{iat^2-t}dt \\ &= c(\theta)e^{-iax^2}\int_0^\infty e^{-t(1-ibx)}dt \\ &= c(\alpha)\frac{e^{-iax^2}}{(1-ibx)} \notin L^1(\mathbb{R}). \end{aligned}$$

It is worth noting here that the inverse transform of (4.4) is given by

$$f(t) = \int_{-\infty}^{\infty} \widehat{F}_\theta(x)\, K_{-\theta}(t,x)\, dx, \tag{4.6}$$

provided that the integral exists.

To demonstrate the relationship between the fractional and the conventional Fourier transforms of a real-valued function, we will use a very simple function $f(t) = \chi_{[-1,1]}$, the characteristic function of the interval $[-1,1]$, whose Fourier transform, or in other words, its Fourier transform of order one ($\theta = \pi/2$) is $\sin x/x$, apart from a multiplicative constant. Since the fractional Fourier transform F_θ is in general a complex-valued function, we will plot only its absolute value, $|F_\theta|$ for $\theta = 0, \pi/8, \pi/4, \pi/2$, that is the fractional Fourier transform of f of order, $0, 1/4, 1/2, 1$. See Figs. 4.1–4.4.

In the following table we list the fractional Fourier transform of some functions. But recall that $c_\alpha = \sqrt{1+i\cot\alpha}/\sqrt{2\pi}$.

The Fractional Fourier Transform

$No.$	$f(t)$	$F_\alpha(x)$
1)	$\phi_0(t) = \exp(-t^2/2)$	$\exp(-x^2/2)$
2)	$\phi_n(t) = H_n(t)\exp(-t^2/2)$	$e^{in\alpha} H_n(x)\exp(-x^2/2)$
3)	$\delta(t)$	$c_\alpha \exp\left(-\frac{ix^2}{2}(\cot\alpha)\right)$
4)	$\delta(t-k)$	$c_\alpha \exp\left(-i\frac{(x^2+k^2)}{2}(\cot\alpha) + ikx\csc\alpha\right)$
5)	1	$\sqrt{1-i\tan\alpha}\exp\left(i\frac{x^2}{2}\tan\alpha\right)$
6)	e^{ikt}	$\sqrt{1-i\tan\alpha}\exp\left(i\frac{x^2+k^2}{2}\tan\alpha + ik\sec\alpha\right)$
7)	$\exp(-t^2/2+kt)$	$\exp\left(-\frac{x^2}{2} - \frac{ik^2}{2}e^{i\alpha}\sin\alpha + kxe^{i\alpha}\right)$

The proof of the first two entries follows from the fact that ϕ_0 and ϕ_n are eigenfunctions of the FrFT. The proof of entries (3) and (4) follows from the definition of the delta function, while that of the last three entries follows from Eq. (4.8) below.

4.3 OPERATIONAL CALCULUS

For the reader's convenience, we start this section by listing some of the main properties of the fractional Fourier transform. Some of these properties are easy to prove and will be left to the reader. But the ones that are more technical, we will restate as theorems and provide them with proof.

1. **Linearity:** The Fractional Fourier Transform is linear, i.e.,

$$\mathcal{F}_\theta[\alpha f+\beta g]=\alpha\mathcal{F}_\theta[f]+\beta\mathcal{F}_\theta[g]$$

where α and β are constants.

2. **Eigenvalues and Eigenfunctions:** The eigenfunctions of the fractional Fourier transform are the Hermite functions h_n (see Chapter 3), and the corresponding eigenvalues are $e^{in\theta}$, that is

$$\mathcal{F}_\theta[h_n(t)](x)=e^{in\theta}h_n(x),$$

3. **Riemann-Lebesgue Lemma:** If $f\in L^1(\mathbb{R})$, its fractional Fourier transform $F_\theta(x)$ is continuous and tends to zero as $|x|\to\infty$, i.e., $\lim_{|x|\to\infty}F_\theta(x)=0$

4. **Additivity:** $\mathcal{F}_\theta\mathcal{F}_\phi=\mathcal{F}_{\theta+\phi}$,

5. **Commutativity:** $\mathcal{F}_\theta\mathcal{F}_\phi=\mathcal{F}_\phi\mathcal{F}_\theta$,

6. **Associativity:** $\mathcal{F}_{\theta_1}(\mathcal{F}_{\theta_2}\mathcal{F}_{\theta_3})=(\mathcal{F}_{\theta_1}\mathcal{F}_{\theta_2})\mathcal{F}_{\theta_3}$.

7. **Inverse:** $(\mathcal{F}_\theta)^{-1}=\mathcal{F}_{-\theta}$.

8. **Parseval's Relation:**

$$\langle f,g\rangle=\langle F_\theta,G_\theta\rangle;$$

where F_θ, G_θ are the fractional Fourier transforms of f and g, respectively. Hence,

$$\|f\|_2=\|F_\theta\|_2.$$

Moreover,

$$\langle f,G_\theta\rangle=\langle F_\theta,g\rangle,$$

9. **Shift:**

$$\mathcal{F}_\theta(T_\tau f)=\exp\left[-i\tau^2\cos\theta\sin\theta/2+i\tau x\sin\theta\right]F_\theta(x-\tau\cos\theta),$$

where $T_\tau f(t)=f(t-\tau)$ is the shift operator.

10. **Modulation:**

$$\begin{aligned}\mathcal{F}_\theta\left(e^{i\beta t}f(t)\right)(x) &= \exp\left(i\frac{\beta^2}{2}\cos\theta\sin\theta+i\beta x\cos\theta\right)F_\theta(\beta\sin\theta+x)\\ &= \exp\left[i\beta\cos\theta\left(x+\frac{\beta}{2}\sin\theta\right)\right]F_\theta(\beta\sin\theta+x)\end{aligned}$$

11. **Multiplication by the input variable:**

$$\mathcal{F}_\theta(tf(t))(x) = -i\sin\theta\frac{d}{dx}F_\theta(x) + x\cos\theta F_\theta(x),$$

or more generally

$$\mathcal{F}_\theta(t^n f(t))(x) = \left(-i\sin\theta\frac{d}{dx} + x\cos\theta\right)^n F_\theta(x).$$

12. **Differentiation:**

$$\mathcal{F}_\theta(f'(t))(x) = \cos\theta\frac{d}{dx}F_\theta(x) - ix\sin\theta F_\theta(x).$$

More generally,

$$\mathcal{F}_\theta\left(f^{(n)}(t)\right)(x) = \left(\cos\theta\frac{d}{dx} - ix\sin\theta\right)^n F_\theta(x).$$

Hence,

$$\frac{d}{dx}F_\theta(x) = \cos\theta\mathcal{F}_\theta(f'(t))(x) + i\sin\theta F_\theta(tf(t))(x),$$

and

$$xF_\theta(x) = \cos\theta\mathcal{F}_\theta(tf(t))(x) + i\sin\theta F_\theta(f'(t))(x).$$

13. **Convolution:** Let

$$\overleftarrow{f}(t) = f(t)e^{-ia(\theta)t^2},$$

and set

$$h(t) = (f \star g)(t) = d(\theta)e^{ia(\theta)t^2}\left(\overleftarrow{f} * \overleftarrow{g}\right)(t),$$

then

$$H_\theta(x) = e^{ia(\theta)x^2}F_\theta(x)G_\theta(x),$$

where $*$ is the standard convolution operation for the Fourier transform.

14. **Product:** Let

$$\overrightarrow{f}(t) = f(t)e^{ia(\theta)t^2},$$

and define

$$h(t) = (f \otimes g)(t) = d(-\theta)e^{-ia(\theta)t^2}\left(\overrightarrow{f} * \overrightarrow{g}\right)(t),$$

then

$$\mathcal{F}_\theta\left[f(t)g(t)e^{-ia(\theta)t^2}\right](x) = (F_\theta \otimes G_\theta)(x); \quad \text{see [348].}$$

15. **Poisson Summation Formula for the Fractional Fourier Transform:** The Poisson summation formula for the fractional Fourier transform takes the form

$$\begin{aligned} & c(\theta)\sum_{k=-\infty}^{\infty} f(t+2\sigma k)e^{-ia(\theta)(t+2\sigma k)^2} \\ = & \frac{1}{2\sigma}\sum_{k=-\infty}^{\infty} F_\theta(k\pi\sin\theta/\sigma)e^{ia(\theta)(k\pi\sin\theta/\sigma)^2}e^{-ik\pi t/\sigma} \\ = & \frac{1}{2\sigma}\sum_{k=-\infty}^{\infty} F_\theta(u_k)e^{ia(\theta)u_k^2}e^{-it_k t} \end{aligned}$$

where $t_k = k\pi/\sigma$ and $u_k = t_k\sin\theta$.

16. **Sampling Theorem:** If $f(t) \in L^2(\mathbb{R})$ is bandlimited to $[-\sigma,\sigma]$, that is the support of $F_\theta(x)$ is $[-\sigma,\sigma]$; hence

$$f(t) = \int_{-\sigma}^{\sigma} F_\theta(x)K_{-\theta}(t,x)dx,$$

then f can be reconstructed from its samples at the points $\{u_n\}_{n\in\mathbb{Z}}$ via the formula

$$f(t) = e^{ia(\theta)t^2}\sum_{n=-\infty}^{\infty} f(u_n)e^{-ia(\theta)u_n^2}\frac{\sin\sigma b(\theta)(t-u_n)}{\sigma b(\theta)(t-u_n)},$$

where $u_n = (n\pi/\sigma)\sin\theta$; [325, 339, 345, 350]

17. **Wigner Distribution:** If we denote the Wigner distribution of f by $W_f(u,v)$, then the Wigner distribution of F_θ is related to the Wigner distribution of f by the formula

$$W_{F_\theta}(u,v) = W_f(u\cos\theta - v\sin\theta, u\sin\theta + v\cos\theta).$$

That is the Wigner distribution of F_θ is obtained from the Wigner distribution of f by a θ-rotation in the clockwise direction.

We now turn our attention to the proofs of the above mentioned properties. Properties (1), (2) and (5) are easy to prove and left to the reader. We begin by proving the additive property of the transform which is essential in establishing that the fractional Fourier transforms form a one-parameter periodic group.

Theorem 6 (Addition Formula). *Let F_α and F_β be the fractional Fourier transform of f with respect to the angles α and β, respectively. Then*

$$F_\beta F_\alpha = F_{\alpha+\beta}.$$

Proof *For simplicity let us denote $a(\alpha), b(\alpha), c(\alpha)$ by $a_\alpha, b_\alpha, c_\alpha$. Since*

$$F_\alpha(x) = \int_{\mathbb{R}} f(t)K_\alpha(x,t)dt,$$

it follows that

$$\begin{aligned} F_\beta[F_\alpha](u) &= \int_{\mathbb{R}} F_\alpha(x)K_\beta(u,x)dx \\ &= \int_{\mathbb{R}} K_\beta(u,x)dx \int_{\mathbb{R}} f(t)K_\alpha(x,t)dt \\ &= c_\alpha c_\beta e^{-ia_\beta u^2} \int_{\mathbb{R}} f(t)e^{-ia_\alpha t^2} I(x,t)dt, \end{aligned} \tag{4.7}$$

where

$$I(x,t) = \int_{\mathbb{R}} e^{-ix^2(a_\alpha+a_\beta)+ix(b_\alpha t+b_\beta u)}dx.$$

with the help of the relation [96, P. 121]

$$\int_{\mathbb{R}} e^{-p^2x^2+iwx}dx = \frac{\sqrt{\pi}}{p}e^{-w^2/4p^2}, \tag{4.8}$$

we have

$$\begin{aligned} I(x,t) &= \frac{\sqrt{\pi}}{\sqrt{i(a_\alpha+a_\beta)}} \exp\left\{\frac{i\left(b_\alpha^2t^2+b_\beta^2u^2+2b_\alpha b_\beta ut\right)}{4(a_\alpha+a_\beta)}\right\} \\ &= \frac{\sqrt{\pi}}{\sqrt{i(a_\alpha+a_\beta)}} \exp\left\{\frac{i\left[(\csc^2\alpha)t^2+(\csc^2\beta)u^2+2(\csc\alpha\csc\beta)ut\right]}{2(\cot\alpha+\cot\beta)}\right\} \\ &= \frac{\sqrt{\pi}}{\sqrt{i(a_\alpha+a_\beta)}} \exp\left\{\frac{i(\sin\alpha\sin\beta)\left[(\csc^2\alpha)t^2+(\csc^2\beta)u^2+2(\csc\alpha\csc\beta)ut\right]}{2\sin(\alpha+\beta)}\right\} \\ &= \frac{\sqrt{\pi}}{\sqrt{i(a_\alpha+a_\beta)}} \exp\left\{\frac{i\left[\frac{\sin\beta}{\sin\alpha}t^2+\frac{\sin\alpha}{\sin\beta}u^2+2ut\right]}{2\sin(\alpha+\beta)}\right\} \\ &= \frac{\sqrt{\pi}}{\sqrt{i(a_\alpha+a_\beta)}} \exp\left\{i(Au^2+Bt^2)+i\csc(\alpha+\beta)ut\right\}, \end{aligned}$$

where

$$A = \frac{\sin\alpha}{2\sin\beta\sin(\alpha+\beta)}, \quad B = \frac{\sin\beta}{2\sin\alpha\sin(\alpha+\beta)}.$$

Substituting $I(x,t)$ back into Eq. (4.7), we have

$$\begin{aligned} F_\beta[F_\alpha](u) &= \frac{\sqrt{\pi}c_\alpha c_\beta}{\sqrt{i(a_\alpha+a_\beta)}} \\ &\times \int_{\mathbb{R}} f(t)\exp\left\{i\left[(A-a_\beta)u^2+(B-a_\alpha)t^2+\csc(\alpha+\beta)ut\right]\right\}dt. \end{aligned} \tag{4.9}$$

But

$$\begin{aligned}(A-a_\beta) &= \frac{\sin\alpha}{2\sin\beta\sin(\alpha+\beta)} - \frac{\cot\beta}{2}\\ &= \frac{1}{2\sin\beta}\left[\frac{\sin\alpha}{\sin(\alpha+\beta)} - \cos\beta\right]\\ &= \frac{1}{2\sin\beta}\left[\frac{\sin\alpha(1-\cos^2\beta)-\cos\alpha\sin\beta\cos\beta}{\sin(\alpha+\beta)}\right]\\ &= -\frac{\cot(\alpha+\beta)}{2} = -a_{\alpha+\beta}.\end{aligned}$$

Similarly, $(B-a_\alpha) = -\frac{\cot(\alpha+\beta)}{2} = -a_{\alpha+\beta}$. *Therefore, the exponential factor in Eq. (4.9) is reduced to the desired one*

$$-i\left\{a_{\alpha+\beta}(u^2+t^2)-\csc(\alpha+\beta)ut\right\} = -ia_{\alpha+\beta}(u^2+t^2)+ib_{\alpha+\beta}ut.$$

As for the constant term outside the integral, we have

$$\begin{aligned}\frac{\sqrt{\pi}c_\alpha c_\beta}{\sqrt{i(\cot\alpha+\cot\beta)/2}} &= \frac{\sqrt{\pi}\sqrt{1+i\cot\alpha}\sqrt{1+i\cot\beta}}{2\pi\sqrt{i(\cot\alpha+\cot\beta)/2}}\\ &= \frac{1}{\sqrt{2\pi}}\frac{\sqrt{\sin\alpha\sin\beta}\sqrt{1+i(\cot\alpha+\cot\beta)-\cot\alpha\cot\beta}}{\sqrt{i\sin(\alpha+\beta)}}\\ &= \frac{1}{\sqrt{2\pi}}\frac{\sqrt{\sin\alpha\sin\beta+i(\cos\alpha\sin\beta+\sin\alpha\cos\beta)-\cos\alpha\cos\beta}}{\sqrt{i\sin(\alpha+\beta)}}\\ &= \frac{1}{\sqrt{2\pi}}\frac{\sqrt{i\sin(\alpha+\beta)-\cos(\alpha+\beta)}}{\sqrt{i\sin(\alpha+\beta)}}\\ &= \frac{1}{\sqrt{2\pi}}\sqrt{1+i\cot(\alpha+\beta)} = c_{\alpha+\beta}.\end{aligned}$$

Therefore,

$$F_\beta\left[F_\alpha\right](u) = F_{\alpha+\beta}(u).$$

As a corollary of Theorem 6, we obtain the commutative and associative relations for the fractional Fourier transform, $F_\beta[F_\alpha] = F_\alpha[F_\beta]$ and $F_\alpha\left(F_\beta F_\gamma\right) = \left(F_\alpha F_\beta\right)F_\gamma$.

Next we derive the inversion formula for the FrFT.

Theorem 7. *The inverse of* $\mathcal{F}_\alpha$ *is* $\mathcal{F}_{-\alpha}$ *that is* $\mathcal{F}_\alpha\mathcal{F}_{-\alpha} = \mathcal{I}$. *In other words* $F_{-\alpha}[F_\alpha](u) = f(u)$.

Proof

$$\begin{aligned}
F_{-\alpha}[F_\alpha](u) &= \int_{\mathbb{R}} F_\alpha(x)K_{-\alpha}(u,x)dx \\
&= \int_{\mathbb{R}} K_{-\alpha}(u,x)dx \int_{\mathbb{R}} f(t)K_\alpha(x,t)dt \\
&= \int_{\mathbb{R}} f(t)dt \int_{\mathbb{R}} K_\alpha(x,t)K_{-\alpha}(u,x)dx \\
&= c_\alpha c_{-\alpha} \int_{\mathbb{R}} f(t)\left(e^{ia_\alpha(u^2-t^2)}\right)dt \int_{\mathbb{R}} e^{ib_\alpha x(t-u)}dx \\
&= \frac{2\pi|c_\alpha|^2}{b}\int_{\mathbb{R}} f(t)\left(e^{ia_\alpha(u^2-t^2)}\right)\delta(t-u)dt = f(u).
\end{aligned}$$

The last equation follows from Eq. (1.27) and the fact that $2\pi|c_\alpha|^2 = b_\alpha$.

Theorem 8. *(***Parseval's Relation:***) Let* F_α, G_α *denote the fractional Fourier transform of* f, g*, respectively. Then*

$$\langle f, g\rangle = \langle F_\alpha, G_\alpha\rangle;$$

hence

$$\|f\|_2 = \|F_\alpha\|_2.$$

Moreover,

$$\langle f, G_\alpha\rangle = \langle F_\alpha, g\rangle,$$

Proof

$$\begin{aligned}
\langle F_\alpha, G_\alpha\rangle &= \int_{\mathbb{R}} F_\alpha(x)\overline{G}_\alpha(x)dx \\
&= \int_{\mathbb{R}} dx\left(\int_{\mathbb{R}} f(t)K_\alpha(x,t)dt\right)\left(\int_{\mathbb{R}} \bar{g}(y)\overline{K}_\alpha(x,y)dy\right) \\
&= \int_{\mathbb{R}} f(t)dt \int_{\mathbb{R}} \bar{g}(y)dy \int_{\mathbb{R}} K_\alpha(x,t)\overline{K}_\alpha(x,y)dx \\
&= |c(\alpha)|^2 \int_{\mathbb{R}} f(t)dt \int_{\mathbb{R}} \bar{g}(y)dy \int_{\mathbb{R}} e^{-ia(t^2-y^2)}e^{ibx(t-y)}dx \\
&= \frac{2\pi|c(\alpha)|^2}{b}\int_{\mathbb{R}} f(t)e^{-iat^2}dt \int_{\mathbb{R}} e^{iay^2}\bar{g}(y)\delta(t-y)dy \\
&= \int_{\mathbb{R}} f(t)\bar{g}(t)dt = \langle f, g\rangle.
\end{aligned}$$

By putting $f = g$*, we obtain* $\|f\|_2 = \|F_\alpha\|_2$*. The proof of the last relation is smiliar and left to the reader.*

Theorem 9 (Shift and Modulation Rules). *The following relations hold;*

$$\mathcal{F}_\alpha(T_\tau f) = \exp\left[-i\tau^2\cos\alpha\sin\alpha/2 + i\tau x\sin\alpha\right]F_\alpha(x-\tau\cos\alpha),$$

and

$$\mathcal{F}_\alpha\left(e^{i\beta t}f(t)\right)\exp\left[i\beta\cos\alpha\left(x+\frac{\beta}{2}\sin\alpha\right)\right]F_\alpha(\beta\sin\alpha + x)$$

Proof *With some computations, the proof of the shift and modulation formulas is straightforward. For example, for the shift rule, observe that*

$$\begin{aligned}\mathcal{F}_\alpha(T_\tau f)(x) &= \int_{\mathbb{R}} f(t-\tau)K_\alpha(x,t)dt \\ &= \int_{\mathbb{R}} f(t)K_\alpha(x,t+\tau)dt,\end{aligned}$$

where

$$K_\alpha(x,t+\tau) = c(\alpha)\exp\left\{-ia\left[x^2+t^2+\tau^2+2t\tau\right]+ibx(t+\tau)\right\}.$$

On the other hand

$$\mathcal{F}_\alpha(x-\tau\cos\alpha) = \int_{\mathbb{R}} f(t)K_\alpha(x-\tau\cos\alpha,t)dt,$$

where

$$\begin{aligned}&K_\alpha(x-\tau\cos\alpha,t) = c(\alpha) \\ \times\ &\exp\left\{-ia\left[x^2+t^2+\tau^2\cos^2\alpha-2x\tau\cos\alpha\right]+ibt(x-\tau\cos\alpha)\right\}.\end{aligned}$$

By multiplying $K_\alpha(x-\tau\cos\alpha,t)$ by

$$\exp\left\{-i\frac{\tau^2}{2}\cos\alpha\sin\alpha+i\tau x\sin\alpha\right\}$$

and simplifying, we obtain $K_\alpha(x,t+\tau)$. The proof of the modulation rule is similar.

The next theorem deals with the multiplication by the input variable and the differentiation of the input function.

Theorem 10 (Multiplication by the input variable and Differentiation:). *Let $f \in \mathcal{S}$, the Schwartz space of functions. We have*

$$\mathcal{F}_\alpha(tf(t))(x) = -i\sin\alpha\frac{d}{dx}F_\alpha(x)+x\cos\alpha F_\alpha(x) \tag{4.10}$$

or more generally

$$\mathcal{F}_\alpha(t^n f(t))(x) = \left(-i\sin\alpha\frac{d}{dx}+x\cos\alpha\right)^n F_\alpha(x). \tag{4.11}$$

Furthermore,

$$\mathcal{F}_\alpha(f'(t))(x) = \cos\alpha\frac{d}{dx}F_\alpha(x)-ix\sin\alpha F_\alpha(x). \tag{4.12}$$

More generally,

$$\mathcal{F}_\alpha\left(f^{(n)}(t)\right)(x) = \left(\cos\alpha\frac{d}{dx}-ix\sin\alpha\right)^n F_\alpha(x). \tag{4.13}$$

Hence,

$$\frac{d}{dx}F_\alpha(x) = \cos\alpha\mathcal{F}_\alpha(f'(t))(x)+i\sin\alpha F_\alpha(tf(t))(x), \tag{4.14}$$

and

$$xF_\alpha(x) = \cos\alpha\mathcal{F}_\alpha(tf(t))(x)+i\sin\alpha F_\alpha(f'(t))(x). \tag{4.15}$$

Proof *Equation (4.10) follows from the observation that*

$$\frac{\partial}{\partial x}K_\alpha(x,t) = (-2ia(\alpha)x + ib(\alpha)t)\,K_\alpha(x,t),$$

and that

$$tK_\alpha(x,t) = -i\sin\alpha\frac{\partial}{\partial x}K_\alpha(x,t) + x\cos\alpha K_\alpha(x,t).$$

Equation (4.11) follows by induction and writing $t^2 f(t)$ as $tg(t)$, where $g(t) = tf(t)$, and so on.

Equation (4.12) follows from the fact that

$$-\frac{\partial}{\partial t}K_\alpha(x,t) = \left(\cos\alpha\frac{\partial}{\partial x} - ix\sin\alpha\right)K_\alpha(x,t),$$

and (4.13) is obtained by induction. Equation (4.14) and (4.15) are obtained by solving (4.10) and (4.12).

4.3.1 Convolution Theorem

One of the earliest results on the convolution and product of the FrF is due to Almeida [14] whose main result reads as follows:

Let W be that subspace of the space of all integrable functions with the property that $f \in W$ if and only if the Fourier transform $\hat{f}$ of f is also in W. Let f and g be in W and denote their convolution h by $f * g$, i.e.,

$$h(x) = (f*g)(x) = \frac{1}{\sqrt{2\pi}}\int_{-\infty}^{\infty} f(t)g(x-t)\,dt. \tag{4.16}$$

Then the fractional Fourier transform of h, denoted by H_α, is given by

$$H_\alpha(u) = |\sec\alpha|e^{-j(u^2/2)\tan\alpha}\int_{-\infty}^{\infty} F_\alpha(v)g\left[(u-v)\sec\alpha\right]e^{j(v^2/2)\tan\alpha}dv; \tag{4.17}$$

see Eq. 5 in [14]. It should be noted that the space W is the same as the space $L^1(\mathbb{R})\bigcap\mathcal{W}$ in Almeida's notation, where $\mathcal{W}$ is the Wiener algebra consisting of functions that are Fourier transforms of functions in $L^1(\mathbb{R})$.

As for the FrFT of the product h of two functions f and g, i.e., $h(x) = f(x)g(x)$, we have (cf. Eq. 2 in [14])

$$H_\alpha(u) = \frac{|\csc\alpha|}{\sqrt{2\pi}}e^{j(u^2/2)\cot\alpha}\int_{-\infty}^{\infty} F_\alpha(v)G_{\pi/2}\left[(u-v)\csc\alpha\right]e^{-j(v^2/2)\cot\alpha}dv. \tag{4.18}$$

Unlike the convolution theorem for the Fourier transform, which states that the Fourier transform of the convolution of two functions is the product of their Fourier transforms, the one for the FrFT does not seem to be as nice or as practical as the one for the Fourier transform. The reason, in our opinion, is that the convolution operation defined by (4.16) is not the right sort of convolution for the FrFT.

In the general framework of convolution theory (see [352, Ch. 4]), it is known that to any integral transformation $\mathcal{L}$, one may, at least theoretically, associate with it a convolution operation, $\star$, such that

$$\mathcal{L}(f\star g)=\mathcal{L}(f)\mathcal{L}(g). \tag{4.19}$$

For example, the convolution operation associated with the Hankel transform is too complicated to be stated here, but the interested reader can find the details in [352, Section 21.6].

In [348] we proposed a new convolution structure for the FrFT that is different from those introduced in [14] and [225]. Unlike those introduced in [14] and [225], ours preserves property (4.19) and is easier to implement, in particular, in filter design.

Theorem 11 is particularly useful in filter design. For example, if we are interested only in the frequency spectrum of the fractional Fourier transform in the region $[\omega_1,\omega_2]$ of a signal f, we choose the filter impulse response, g, so that G_α is constant over $[\omega_1,\omega_2]$, and zero or of rapid decay outside that region. Passing the output of the filter through the chirp multiplier, $e^{ja(\alpha)u^2}$, yields that part of the spectrum of f over $[\omega_1,\omega_2]$. This is clearly easier to implement than the one suggested in [14]. Equation (4.17) does not seem to have an immediate application in signal processing, but products of similar nature have proved to be useful in optics, see [220].

Let us introduce the following definition.

Definition 12. *For any two integrable functions f and g, we define the convolution operation $\star$ as*

$$h(x)=(f\star g)(x)=d(\alpha)e^{ia(\alpha)x^2}\left(\overleftarrow{f}*\overleftarrow{g}\right)(x),$$

where $*$ is the standard convolution operation for the Fourier transform as defined by (4.16) and $\overleftarrow{f}(t)=f(t)e^{-ia(\theta)t^2}$. Likewise, we define the operation $\otimes$ by

$$(f\otimes g)(x)=d(-\alpha)e^{-ia(\alpha)x^2}\left(\overrightarrow{f}*\overrightarrow{g}\right)(x),$$

where $\overrightarrow{f}(t)=f(t)e^{ia(\theta)t^2}$,

See Fig. 4.5 for a realization of the convolution operation $\star$.

Proposition 1. *The operation $\star$ given in Definition 12 is both commutative, i.e., $f\star g=g\star f$ and associative,*

$$((f\star g)\star h)=(f\star(g\star h)) \tag{4.20}$$

Proof *That $\star$ is commutative is easy to prove and left to the reader. To prove that it is associative, let us first simplify the notation and have tilde signify multiplication by the chirp $e^{-ia(\alpha)t^2}$ so that $\tilde{f}(t)=e^{-ia(\alpha)t^2}f(t)$.*

Let $f\star g=u$ and $g\star h=v$. The left-hand side of (4.20) takes the form

$$((f\star g)\star h)(x)=(u\star h)(x)=d(\alpha)e^{ia(\alpha)x^2}(\tilde{u}*\tilde{h})(x),$$

but

$$u(x)=d(\alpha)e^{ia(\alpha)x^2}(\tilde{f}*\tilde{g})(x),$$

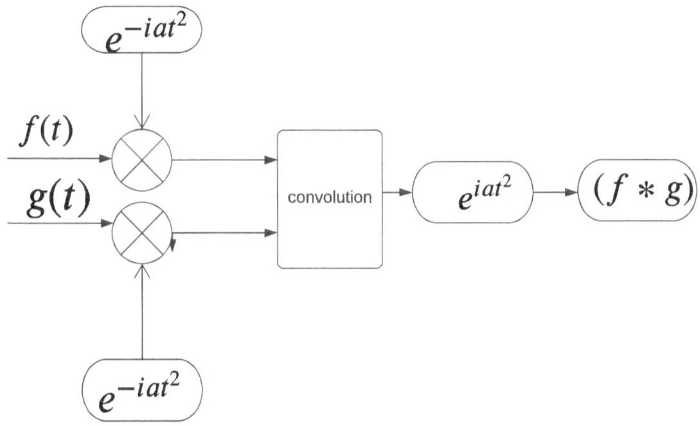

Figure 4.5 Convolution for the FRFT

hence

$$\tilde{u} = d(\alpha)(\tilde{f} * \tilde{g})(x).$$

It follows that

$$((f \star g) \star h)(x) = (u \star h)(x) = d^2(\alpha)e^{ia(\alpha)x^2}\left((\tilde{f} * \tilde{g}) * \tilde{h}\right)(x). \tag{4.21}$$

Similarly, for the right-hand side of (4.20) we have

$$v(x) = d(\alpha)e^{ia(\alpha)x^2}(\tilde{g} * \tilde{h})(x);$$

hence

$$\tilde{v}(x) = d(\alpha)(\tilde{g} * \tilde{h})(x),$$

and we have

$$(f \star (g \star h))(x) = (f \star v)(x) = d^2(\alpha)e^{ia(\alpha)x^2}\left(\tilde{f} * (\tilde{g} * \tilde{h})\right)(x). \tag{4.22}$$

Since the convolution operation $*$ is associative, it follows from (4.21) and (4.22) that so is the operation $\star$. ∎

Now we state and prove our convolution theorem.

Theorem 11. *Let $h(x) = (f \star g)(x)$ and $F_\alpha, G_\alpha, H_\alpha$ denote the FrFT of f, g and h respectively. Then*

$$H_\alpha(u) = e^{ia(\alpha)u^2}F_\alpha(u)G_\alpha(u). \tag{4.23}$$

Moreover,

$$\mathcal{F}_\alpha\left[f(x)g(x)e^{-ia(\alpha)x^2}\right](u) = (F_\alpha \otimes G_\alpha)(u). \tag{4.24}$$

Proof *From the definition of the FrFT and Definition 12, we have*

$$H_\alpha(u) = c(\alpha)\int_{-\infty}^{\infty} h(t)e^{-i[a(t^2+u^2)-but]}dt$$

$$= c^2(\alpha)\int_{-\infty}^{\infty} e^{-i[a(t^2+u^2)-but]}e^{iat^2}dt\int_{-\infty}^{\infty} f(x)e^{-iax^2}g(t-x)e^{-ia(t-x)^2}dx$$

$$= c^2(\alpha)\int_{-\infty}^{\infty}\int_{-\infty}^{\infty} f(x)g(t-x)\exp\left\{-i[a(t^2+u^2)-but+2ax^2-2atx]\right\}dx\,dt.$$

By making the change of variable, $t-x=v$, we obtain

$$H_\alpha(u) = c^2(\alpha)\int_{-\infty}^{\infty}\int_{-\infty}^{\infty} f(x)g(v)\exp\left\{-ia(x^2+u^2+v^2)+ibu(x+v)\right\}dx\,dv$$

$$= c^2(\alpha)e^{iau^2}\int_{-\infty}^{\infty} f(x)e^{\left(-ia(x^2+u^2)+ibux\right)}dx\int_{-\infty}^{\infty} g(v)e^{\left(-ia(v^2+u^2)+ibuv\right)}dv$$

$$= e^{iau^2}F_\alpha(u)G_\alpha(u),$$

which is (4.23).

As for (4.24), we have from Definition 12

$$(F_\alpha\otimes G_\alpha)(u) = e^{-ia(\alpha)u^2}d(-\alpha)\left(\vec{F}_\alpha * \vec{G}_\alpha\right)(u)$$

$$= e^{-ia(\alpha)u^2}c(-\alpha)\int_{-\infty}^{\infty} e^{ia(\alpha)x^2}F_\alpha(x)e^{ia(\alpha)(u-x)^2}G_\alpha(u-x)\,dx.$$

But from the definition of the FrFT, we obtain

$$\begin{aligned}(F_\alpha\otimes G_\alpha)(u) &= \\ & e^{-ia(\alpha)u^2}c(\alpha)c(-\alpha)\int_{-\infty}^{\infty} f(z)dz\int_{-\infty}^{\infty} G_\alpha(u-x) \\ \times\ & \exp\{i[a(u-x)^2-az^2+bxz]\}\,dx \\ &= e^{-ia(\alpha)u^2}c(\alpha)c(-\alpha)\int_{-\infty}^{\infty} e^{-i[2az^2-bzu]}f(z)dz \\ \times\ & \int_{-\infty}^{\infty} G_\alpha(v)\exp\{i[a(v^2+z^2)-bvz]\}\,dv \\ &= c(\alpha)\int_{-\infty}^{\infty} f(z)g(z)e^{-iaz^2}\exp\{-ia(z^2+u^2)+ibzu\}dz \\ &= \int_{-\infty}^{\infty} f(z)g(z)e^{-iaz^2}K_\alpha(u,z)dz = \mathcal{F}\left(f(z)g(z)e^{-iaz^2}\right)(u)\end{aligned}$$

which is Eq. (4.24).

In order to remove the exponential factor in (4.23), we may redefine the convolution of f and g as follows.

$$h(x) = (f\tilde{\star}g)(x) = \frac{d(\alpha)}{\sqrt{\pi}}e^{ia(\alpha)x^2}\left(\overleftarrow{f} * \overleftarrow{g}\right)(\sqrt{2}x),$$

where $*$ is the standard convolution operation for the Fourier transform.

Corollary 1. *Let* $h(x) = (f\tilde{\star}g)(x)$. *Then* $H_\alpha(u) = F_\alpha(u/\sqrt{2})G_\alpha(u\sqrt{2})$.

Proof

$$
\begin{aligned}
H_\alpha(u) &= c(\alpha)\int_{\mathbb{R}} h(x)\exp\left\{-ia(x^2+u^2)+ibux\right\}dx \\
&= \frac{c(\alpha)d(\alpha)}{\sqrt{\pi}}e^{-iau^2}\int_{\mathbb{R}} f(t)e^{-iat^2}dt\int_{\mathbb{R}} e^{ibxu}e^{-ia(\sqrt{2}x-t)^2}g(\sqrt{2}x-t)dx \\
&= c(\alpha)^2e^{-iau^2}\int_{\mathbb{R}} f(t)e^{-iat^2}dt\int_{\mathbb{R}} e^{ibu(y+t)/\sqrt{2}}e^{-iay^2}g(y)dy \\
&= c(\alpha)^2\int_{\mathbb{R}} f(t)\exp\left\{-i(at^2+a(\frac{u}{\sqrt{2}})^2)+ibt(\frac{u}{\sqrt{2}})\right\}dt \\
&\times \int_{\mathbb{R}} g(y)\exp\left\{-i(ay^2+a(\frac{u}{\sqrt{2}})^2+iby(\frac{u}{\sqrt{2}})\right\}dy \\
&= F_\alpha(u/\sqrt{2})G_\alpha(u/\sqrt{2}).
\end{aligned}
$$

4.3.2 Poisson Summation Formula for the Fractional Fourier Transform

The Poisson summation formula for the Fourier transform and the WSK sampling theorem are closely related. In fact, it has been shown [57] that the sampling theorem may be derived from the Poisson summation formula; however, we will not pursue this relationship here and we will treat them separately. For related work, see [180].

The Poisson summation formula for the fractional Fourier transform is a formula that relates a Fourier-type series obtained from a periodic extension of a function $f \in L^1(\mathbb{R})$ to a series involving samples of the fractional Fourier transform F_α of f.

Theorem 12 (The Poisson Summation Formula for the Fractional Fourier transform). *Let* $f \in L^1(\mathbb{R})$, *and* $\sigma > 0$. *The Poisson Summation Formula for the fractional Fourier transform takes the form*

$$
\begin{aligned}
c(\alpha)\sum_{k=-\infty}^{\infty} f(t+2\sigma k)e^{-ia(\alpha)(t+2\sigma k)^2} &= \frac{1}{2\sigma}\sum_{k=-\infty}^{\infty} F_\alpha(\frac{k\pi\sin\alpha}{\sigma})e^{ia(\alpha)(k\pi\sin\alpha/\sigma)^2}e^{-ik\pi t/\sigma} \\
&= \frac{1}{2\sigma}\sum_{k=-\infty}^{\infty} F_\alpha(u_k)e^{ia(\alpha)u_k^2}e^{-it_kt}
\end{aligned}
$$

where $t_k = k\pi/\sigma$ *and* $u_k = t_k\sin\alpha$, *provided that the series converge.*

Proof *First, let us recall from Chapter 1, the following Poisson summation formula for the Fourier transform*

$$
\sum_{k=-\infty}^{\infty} f(t+2\sigma k) = \frac{\sqrt{2\pi}}{2\sigma}\sum_{k=-\infty}^{\infty} \hat{f}(k\pi/\sigma)e^{-ik\pi t/\sigma}, \quad \sigma > 0. \tag{4.25}
$$

Let

$$
F_\alpha(x) = c(\alpha)\int_{\mathbb{R}} f(t)\exp\left\{-ia(x^2+t^2)+ibxt\right\}dt,
$$

which we may write as

$$\tilde{G}(x) = F_\alpha(x)e^{iax^2} = \frac{1}{\sqrt{2\pi}}\int_{\mathbb{R}} g(t)e^{ibxt}dt,$$

where $g(t) = d(\alpha)f(t)e^{-iat^2}$. *Since* f *is in* $L^1(\mathbb{R})$, *so is* g, *and we have*

$$\hat{G}(x) = \tilde{G}(x/b) = \frac{1}{\sqrt{2\pi}}\int_{\mathbb{R}} g(t)e^{ixt}dt,$$

that is $\hat{G}$ *is the Fourier transform of* g. *By substituting* g *and* $\hat{G}$ *into the Poisson summation formula for the Fourier transform Eq. (4.25), we obtain the result.*

4.3.3 Sampling Theorem for the Fractional Fourier Transform

The Whittaker-Shannon-Kotel'nikov (WSK) sampling theorem [357], commonly known as Shannon sampling theorem, enables us to reconstruct a bandlimited function from its samples taken at the Nyquist rate. Likewise, the sampling theorem for the fractional Fourier transform will enable us to reconstruct a function bandlimited in the fractional Fourier transform domain from its samples. The sampling theorems for band-limited and time-limited signals in the fractional Fourier transform domain may be deduced from the WSK sampling theorem; see [37, 106, 339, 345, 350].

Let $\Omega = [-\sigma, \sigma]$ and

$$B_\Omega = \left\{f : f \in L^2(\mathbb{R}) \text{ such that } \mathrm{Supp}F_\alpha \subseteq \Omega\right\}$$

be the space of all $L^2(\mathbb{R})$ functions bandlimited to Ω in the fractional Fourier transform domain. First, we will show that the space B_Ω is a reproducing-kernel Hilbert space and then derive its sampling theorem. To this end, we have

$$\begin{aligned} f(t) &= \int_\Omega F_\alpha(x)K_{-\alpha}(t,x)dx \\ &= \int_\Omega K_{-\alpha}(t,x)dx \int_{\mathbb{R}} f(y)K_\alpha(x,y)dy \\ &= \int_{\mathbb{R}} f(y)dy \int_\Omega K_{-\alpha}(t,x)K_\alpha(x,y)dx \\ &= c(\alpha)c(-\alpha)\int_{\mathbb{R}} f(y)e^{-ia(\alpha)(y^2-t^2)}dy \int_{-\sigma}^{\sigma} e^{ib(\alpha)x(y-t)}dx \\ &= \frac{1}{2\pi\sin\alpha}\int_{\mathbb{R}} f(y)e^{-ia(\alpha)(y^2-t^2)}\frac{2\sin b_\alpha\sigma(y-t)}{b_\alpha(y-t)}dy \\ &= \int_{\mathbb{R}} f(y)R(t,y)dy, \end{aligned}$$

where

$$R(t,y) = e^{-ia(\alpha)(y^2-t^2)}\frac{\sin b_\alpha\sigma(y-t)}{\pi(y-t)}, \qquad b_\alpha = b(\alpha),$$

is the reproducing kernel. As a special case, when $\alpha = \pi/2$, we obtain

$$R(t,y) = \frac{\sin\sigma(y-t)}{\pi(y-t)},$$

which is the reproducing kernel of the space of functions bandlimited to $[-\sigma,\sigma]$ in the Fourier transform domain.

Theorem 13. *Let $f(t)$ be bandlimited to $[-\sigma,\sigma]$, in the fractional Fourier transform domain, that is the support of $F_\alpha(x)$ is $[-\sigma,\sigma]$. Then*

$$\begin{aligned} f(t) &= e^{ia(\alpha)t^2}\sum_{n=-\infty}^{\infty} f(u_n)e^{-ia(\alpha)u_n^2}\frac{\sin\sigma(b_\alpha t-t_n)}{\sigma(b_\alpha t-t_n)} \\ &= e^{ia(\alpha)t^2}\sum_{n=-\infty}^{\infty} f(u_n)e^{-ia(\alpha)u_n^2}\frac{\sin\sigma b_\alpha(t-u_n)}{\sigma b_\alpha(t-u_n)}, \end{aligned}$$

where $u_n = n\pi/(\sigma b(\alpha)) = t_n\sin\alpha$.

Proof *First, note that if F is a square-integrable function with support in $[-\sigma,\sigma]$, then so are $\overleftarrow{F}(t)$ and $\overrightarrow{F}(t)$. Since $f(t)$ is bandlimited to $[-\sigma,\sigma]$, we have,*

$$f(t) = \int_{-\sigma}^{\sigma} F_\alpha(x)K_{-\alpha}(t,x)dx$$

or

$$g(t) = \frac{f(t)e^{-ia(\alpha)t^2}}{d(-\alpha)} = \frac{1}{\sqrt{2\pi}}\int_{-\sigma}^{\sigma} F_\alpha(x)e^{ia(\alpha)x^2-ib(\alpha)tx}dx.$$

By setting $\tilde{F}_\alpha(x) = F_\alpha(x)e^{ia(\alpha)x^2}$, we obtain

$$g(t) = \frac{1}{\sqrt{2\pi}}\int_{-\sigma}^{\sigma} \tilde{F}_\alpha(x)e^{-ib(\alpha)tx}dx. \tag{4.26}$$

On the account of the fact that $\left\{e^{in\pi x/\sigma}/\sqrt{2\sigma}\right\}_{n=-\infty}^{\infty}$ is an orthonormal system on $[-\sigma,\sigma]$, it is easy to see that

$$e^{-ibxt} = \sum_{n=-\infty}^{\infty}\frac{\sin\sigma\left[b(\alpha)t-t_n\right]}{\sigma\left[b(\alpha)t-t_n\right]}e^{-in\pi x/\sigma}, \quad \text{for } |x|\le\sigma,$$

where $t_n = \frac{n\pi}{\sigma}$. By substituting this last summation into (4.26) and interchanging the summation and integration signs which is valid because of the uniform convergence of the series, we obtain

$$\begin{aligned} g(t) &= \sum_{n=-\infty}^{\infty}\frac{\sin\sigma\left[b(\alpha)t-t_n\right]}{\sigma\left[b(\alpha)t-t_n\right]}\frac{1}{\sqrt{2\pi}}\int_{-\sigma}^{\sigma}\tilde{F}_\alpha(x)e^{-in\pi x/\sigma}dx \\ &= \sum_{n=-\infty}^{\infty}\frac{\sin\sigma\left[b(\alpha)t-t_n\right]}{\sigma\left[b(\alpha)t-t_n\right]}g\left(\frac{n\pi}{\sigma b(\alpha)}\right) \\ &= \sum_{n=-\infty}^{\infty}\frac{\sin\sigma\left[b(\alpha)t-t_n\right]}{\sigma\left[b(\alpha)t-t_n\right]}g(u_n). \end{aligned}$$

Therefore, we finally have

$$\begin{aligned} f(t) &= e^{ia(\alpha)t^2} \sum_{n=-\infty}^{\infty} f(u_n) e^{-ia(\alpha)u_n^2} \frac{\sin \sigma(b_\alpha t - t_n)}{\sigma(b_\alpha t - t_n)} \\ &= e^{ia(\alpha)t^2} \sum_{n=-\infty}^{\infty} f(u_n) e^{-ia(\alpha)u_n^2} \frac{\sin \sigma b_\alpha(t - u_n)}{\sigma b_\alpha(t - u_n)}. \end{aligned}$$

Shannon sampling theorem has been generalized in a number of different ways. For example, it has been shown that a bandlimited signal can be reconstructed from samples of the signal and its first p derivatives each taken at $1/(p+1)$ of the Nyquist rate. A more general extension of this result was introduced by A. Papoulis [229] in what is known as Papoulis sampling theorem or the generalized sampling theorem.

Papoulis posed the following question: If a bandlimited signal is fed simultaneously into a multi-channel system consisting of p linear systems, such as filters with system functions $H_1(\omega), \ldots, H_p(\omega)$, is it possible to reconstruct f from the samples of the p output signals, $g_1(t), \ldots, g_p(t)$?, where

$$g_k(t) = \frac{1}{\sqrt{2\pi}} \int_{-\sigma}^{\sigma} F(\omega) H_k(\omega) e^{it\omega} d\omega, \quad k = 1, \ldots, p,$$

and F is the Fourier transform of f.

He then answered that question in the affirmative under the condition that a system of linear equations involving the system functions has a solution. The samples of the output signals are each taken at $1/p$ of the Nyquist rate. As a special case of Papoulis theorem, one obtains the formula for reconstructing a bandlimited function from its samples and the samples of its first p derivatives each taken at $1/(p+1)$ of the Nyquist rate. For this reason, Papoulis theorem is sometimes referred to as the *Generalized Sampling Theorem*. An extension of Papoulis theorem to the one-dimensional fractional Fourier transform domain has been obtained by D. Wei [309], D. Wei and Q. Ran [311]. For more on sampling in the fractional Fourier transform domain, see [33, 37, 248, 260, 266, 270, 276, 296, 325].

4.3.4 The Wigner Distribution

The Wigner distribution of a signal $f(t)$ is a time-frequency representation of f. It has applications in quantum mechanics, signal analysis and optics [77, 115, 116, 202, 215, 265, 321]. It was introduced in quantum physics by E. P. Wigner in 1932 [317], but it was the Dutch mathematician N. G. de Bruijn [50] who introduced it in mathematics in 1973 and developed many of its properties. The Wigner distribution was introduced again in signal analysis in 1980 by T. Claasen and W. Mecklenbräuker [75, 76].

Recall that the Wigner distribution of a signal f is defined as

$$W_f(t,\omega) = \frac{1}{\sqrt{2\pi}} \int_{\mathbb{R}} f(t + x/2) \overline{f}(t - x/2) e^{i\omega x} dx.$$

It has been shown that the Wigner distribution of the Fourier transform $\hat{f}$ of f is related to the Wigner distribution of f by the relation

$$W_{\hat{f}}(t,\omega) = W_f(-\omega, t).$$

Hence, it may be said that $W_{\hat{f}}(t,\omega)$ is obtained from $W_f(t,\omega)$ by a rotation of $\pi/2$ in the clockwise direction. Thus, one may ask what does correspond to a rotation of the Wigner distribution by an angle $\pi/4$ in the clockwise direction? Whatever it is, one may call it the one half Fourier transform, which as we will soon see, is consistent with the definition of the FrFT.

More generally, what does correspond to a rotation of the Wigner distribution by an angle θ in the clockwise direction? In other words, find g such that

$$W_g(t,\omega) = W_f(t\cos\theta - \omega\sin\theta, t\sin\theta + \omega\cos\theta).$$

It turns out that g is the fractional Fourier transform of f with angle θ. In some engineering literature t is usually used to denote time and ω to denote frequency; however, we may not adopt this convention to allow for more general interpretation.

Definition 13. *The cross–Wigner distribution function $W_{f,g}(u,v)$ of two functions f and g is defined as*

$$\begin{aligned} W_{f,g}(t,\omega) &= \frac{1}{\sqrt{2\pi}}\int_{-\infty}^{\infty} f\left(t+\frac{x}{2}\right)\overline{g}\left(t-\frac{x}{2}\right)e^{i\omega x}\,dx \\ &= \frac{2}{\sqrt{2\pi}}e^{-2it\omega}\int_{-\infty}^{\infty} f(x)\overline{g}(2t-x)e^{2i\omega x}\,dx. \end{aligned}$$

The auto-Wigner distribution of a function f is defined as $W_f(t,\omega) = W_{f,f}(t,\omega)$.

We will show that the cross-Wigner distribution of the fractional Fourier transforms F_α and G_α of f and g is obtained from the cross-Wigner distribution of f and g by a rotation with an angle α in the clockwise direction.

Theorem 14. *Let $W_{F_\alpha,G_\alpha}(u,v)$ denote the cross-Wigner distribution of F_α, G_α. Then*

$$W_{F_\alpha,G_\alpha}(u,v) = W_{f,g}\left(u\cos\alpha - v\sin\alpha, u\sin\alpha + v\cos\alpha\right).$$

In particular, for $\alpha = \pi/2$, we have

$$W_{\hat{f},\hat{g}}(u,v) = W_{f,g}(-v,u) \quad \text{or equivalently } W_{f,g}(u,v) = W_{\hat{f},\hat{g}}(v,-u).$$

Proof

$$\begin{aligned} W_{F_\alpha,G_\alpha}(u,v) &= \frac{2}{\sqrt{2\pi}}e^{-2iuv}\int_{\mathbb{R}} F_\alpha(y)\overline{G}_\alpha(2u-y)e^{2ivy}dy \\ &= \frac{2}{\sqrt{2\pi}}|c_\alpha|^2 e^{-2iuv}\int_{\mathbb{R}} e^{2ivy}dy\int_{\mathbb{R}} f(x)\exp\left[-ia(x^2+y^2)+ibxy\right]dx \\ &\times \int_{\mathbb{R}} \overline{g}(t)\exp\left[ia(t^2+(2u-y)^2)-ibt(2u-y)\right]dt \\ &= \frac{2}{\sqrt{2\pi}}e^{-2iuv}|c_\alpha|^2\int_{\mathbb{R}} f(x)e^{-iax^2}dx\int_{\mathbb{R}} \overline{g}(t)\exp\left[iat^2-2ibtu\right]I(x,t;u,v)dt \end{aligned}$$

where

$$\begin{aligned} I(x,t;u,v) &= \int_{\mathbb{R}} \exp\left\{2ivy - iay^2 + ibxy + ia(2u-y)^2 + ibty\right\} dy \\ &= e^{4iau^2} \int_{\mathbb{R}} \exp\{-iy(-2v - bx + 4au - bt)\} dy \\ &= (2\pi)e^{4iau^2} \delta(4au - 2v - bx - bt) \\ &= (2\pi)e^{4iau^2} \delta(2bU - bx - bt) \end{aligned}$$

and

$$U = \frac{2au}{b} - \frac{v}{b} = u\cos\alpha - v\sin\alpha.$$

Therefore, we have

$$W_{F_\alpha,G_\alpha}(u,v) = \frac{2}{\sqrt{2\pi}} |c_\alpha|^2 e^{-2iuv} \int_{\mathbb{R}} f(x) e^{-iax^2} J(x,u,v) dx, \tag{4.27}$$

where

$$\begin{aligned} J(x,u,v) &= \int_{\mathbb{R}} \overline{g}(t) \exp\left[iat^2 - 2ibtu\right] I(x,t;u,v) dt \\ &= (2\pi)e^{4iau^2} \int_{\mathbb{R}} \overline{g}(t) \exp\left[iat^2 - 2ibtu\right] \delta(2bU - bx - bt) dt \\ &= \frac{2\pi}{b} e^{4iau^2} \overline{g}(2U - x) \exp\left\{ia(2U-x)^2 - 2ibu(2U-x)\right\}. \end{aligned}$$

Substituting $J(x;u,v)$ into Eq. (4.27), we obtain

$$\begin{aligned} W_{F_\alpha,G_\alpha}(u,v) &= \frac{2}{\sqrt{2\pi}} e^{-2iuv} |c_\alpha|^2 \left(\frac{(2\pi)}{b} e^{4iau^2}\right) \int_{\mathbb{R}} f(x) e^{-iax^2} \\ &\times \overline{g}(2U - x) \exp\left\{ia(2U-x)^2 - 2ibu(2U-x)\right\} dx. \end{aligned}$$

By replacing x by $U + x$, in the last integral and keeping in mind that

$$\frac{|c_\alpha|^2}{b} = 1/(2\pi),$$

we have

$$\begin{aligned} W_{F_\alpha,G_\alpha}(u,v) &= \frac{2}{\sqrt{2\pi}} e^{-2iuv} e^{4iau^2} \int_{\mathbb{R}} f(U+x) \overline{g}(U-x) \\ &\times \exp\left\{-ia(U+x)^2 + ia(U-x)^2 - 2ibu(U-x)\right\} dx \\ &= 2e^{-2iuv} e^{4iau^2} \int_{\mathbb{R}} f(U+x) \overline{g}(U-x) \\ &\times \exp\{2ix(bu - 2aU) - 2ibuU\} dx. \end{aligned}$$

In view of the fact that $-2ibuU = -2iu(2au - v) = -4iau^2 + 2iuv$, and

$$\begin{aligned} 2ix(bu - 2aU) &= 2ix(bu - \frac{4a^2}{b}u + \frac{2av}{b}) \\ &= 2ix\left[u\left(\frac{b^2 - 4a^2}{b}\right) + \frac{2a}{b}v\right] = 2ix(u\sin\alpha + v\cos\alpha), \end{aligned}$$

it follows that

$$W_{F_\alpha,G_\alpha}(u,v) = \frac{2}{\sqrt{2\pi}} \int_{\mathbb{R}} f(U+x)\overline{g}(U-x)\exp(2ixV)\,dx,$$

where

$$V = u\sin\alpha + v\cos\alpha.$$

Finally, by replacing x by $x/2$ in the last equation, we obtain

$$\begin{aligned} W_{F_\alpha,G_\alpha}(u,v) &= \frac{1}{\sqrt{2\pi}} \int_{\mathbb{R}} f(U+x/2)\overline{g}(U-x/2)e^{iVx}dx = W_{f,g}(U,V) \\ &= W_{f,g}(u\cos\alpha - v\sin\alpha, u\sin\alpha + v\cos\alpha), \end{aligned}$$

which completes the proof.

4.4 THE FRACTIONAL HILBERT TRANSFORM

The Hilbert transform has many interesting and important engineering applications ranging from airfoil [352, Ch. 14] and optical filter designs to signal processing [49, 229, 243]. Of particular interest to us here are its applications in communication theory.

The role that the Hilbert transform plays in single side-band **SSB** suppressed carrier modulation and compatible single side-band **CSSB** modulation is well known [229, 243]. The Hilbert transformation is also used in the construction of analytic signals, which in turn can be used to construct the complex envelop of a real signal. The complex envelop method is very useful in finding the output of bandpass filters. Moreover, the Hilbert transform of a bandpass signal is used in the reconstruction of the signal by sampling the signal and its Hilbert transform, each at half the Nyquist rate [357, p. 67].

The Hilbert transform of a signal $f(t)$ is defined as

$$H[f](t) = \tilde{f}(t) = \frac{1}{\pi} \int_{-\infty}^{\infty} \frac{f(x)}{x-t} dx\,,$$

and the analytic part of f is defined as

$$F(t) = f(t) + i\tilde{f}(t) = f(t) + iH[f](t).$$

One of the most important properties of analytic signals is that they contain no negative frequency components. In other words, the analytic part of a signal is obtained by suppressing the signal's negative frequencies. For this to hold, the Hilbert transform, $\tilde{f}$, of the signal f must be closely related to the Fourier transform, $\hat{f}$, of f, and indeed it is. In fact, if

$$\hat{f}(\omega) = R(\omega) + iX(\omega) = \frac{1}{\sqrt{2\pi}} \int_{-\infty}^{\infty} f(t)e^{i\omega t}dt\,, \tag{4.28}$$

then

$$f(t) = \sqrt{2/\pi} \int_0^\infty (R(\omega)\cos t\omega - X(\omega)\sin t\omega)\,d\omega\,, \tag{4.29}$$

$$\tilde{f}(t) = \sqrt{2/\pi} \int_0^\infty (X(\omega)\cos t\omega + R(\omega)\sin t\omega)d\omega\,; \tag{4.30}$$

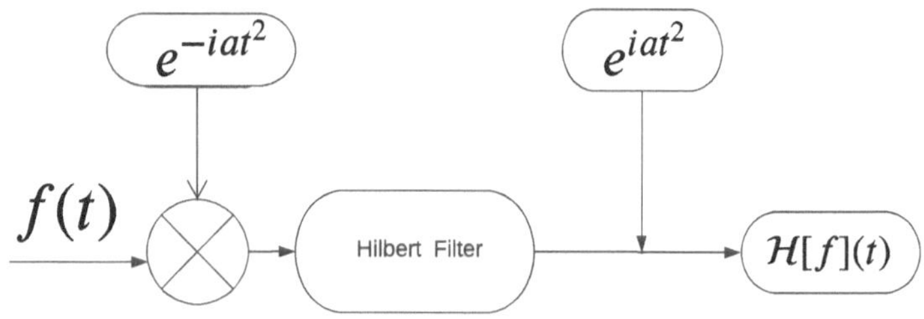

Figure 4.6 Fractional Hilbert filter

and

$$F(t) = \sqrt{2/\pi} \int_0^\infty \hat{f}(\omega) e^{-it\omega} d\omega \,. \tag{4.31}$$

An important relation between the Fourier and Hilbert transforms is the following [352, p.277]:

$$(\widehat{H[f]})(\omega) = -i\mathrm{sgn}(\omega)\hat{f}(\omega). \tag{4.32}$$

Examples of how analytic signals can be used in signal analysis applications and modulations can be found in [49, p. 269] and [229, pp. 251-255].

The goal of this section is to extend Equations (4.28–4.32) to the fractional Fourier transform. In other words, we shall define an analytic representation of a signal f that is obtained by suppressing the negative frequency content of the fractional Fourier transform, $F_\alpha[f](\omega)$, of f. To this end, we have to modify and generalize the definition of the Hilbert transform.

Other generalizations of the Hilbert transform that are related to the FrFT have been introduced in [82, 185]. However, ours is essentially different from theirs.

Definition 14. *The fractional Hilbert transform* **FrHT** *of a signal $f(t)$ is defined as*

$$H_\alpha[f](t) = \frac{e^{ia(\alpha)t^2}}{\pi} \int_{-\infty}^{\infty} \frac{f(x)e^{-ia(\alpha)x^2}}{x-t} dx \,, \qquad \textit{for } \alpha \neq n\pi.$$

Computing the FrHT of a signal f is equivalent to multiplying it by a chirp, $e^{-ia(\alpha)t^2}$, then passing the product $f_1(t)$ through a standard Hilbert filter and finally multiplying the output $\tilde{f}_1(t)$ by the chirp $e^{ia(\alpha)t^2}$; see Fig. 4.6.

The generalized Hilbert transform does not possess the semi-group property, $H_\beta H_\alpha = H_{\beta+\alpha}$. To see that, let $f(x) = e^{ia(\alpha)x^2}$, then $H_\alpha[f] = 0$ since the Hilbert transform of the function $g(t) = 1$ is zero, while it is easy to see that $H_{\beta+\alpha}[f] \neq 0$.

Definition 15. *We define the generalized analytic part of a signal $f(t)$ as*

$$\tilde{F}_\alpha(t) = f(t) + iH_\alpha[f](t) \,.$$

Notice that the generalized Hilbert transform and the generalized analytic part of f both reduce to the standard ones when $\alpha = \pi/2$. The two main results of this section will be stated as theorems.

Theorem 15. *[347] Let $F_\alpha(\omega)$ be the FrFT with angle α of a signal $f(t)$, and let $H_\alpha(t)$ and $\tilde{F}_\alpha(t)$ be its fractional Hilbert transform and generalized analytic part respectively. Then*

$$\tilde{F}_\alpha(t) = 2\int_0^\infty F_\alpha(\omega)K_{-\alpha}(\omega,t)d\omega, \quad \text{for } \alpha \neq n\pi.$$

This means $\tilde{F}_\alpha(t)$ is obtained from f by suppressing the negative frequency content of $F_\alpha(\omega)$. Similarly, to suppress the positive frequencies of the FrFT, we may take the function $G_\alpha(t) = f(t) - iH_\alpha[f](t)$, instead of $\tilde{F}_\alpha(t)$.

Proof *Let*

$$I(t) = \int_0^\infty F_\alpha(\omega)K_{-\alpha}(\omega,t)d\omega,$$

hence

$$\begin{aligned} I(t) &= \int_0^\infty K_{-\alpha}(\omega,t)d\omega\left[\int_{-\infty}^\infty f(u)K_\alpha(u,\omega)du\right] \\ &= \int_{-\infty}^\infty f(u)du\left[\int_0^\infty K_\alpha(u,\omega)K_{-\alpha}(\omega,t)d\omega\right]. \end{aligned}$$

But since $a(-\alpha) = -a(\alpha), b(-\alpha) = b(\alpha)$ and $d(\alpha)d(-\alpha) = \csc\alpha$, it is easy to see that

$$K_\alpha(u,\omega)K_{-\alpha}(\omega,t) = \frac{\csc\alpha}{2\pi}\exp\left\{ia(\alpha)\left[t^2 - u^2\right] + ib(\alpha)(u-t)\omega\right\}.$$

Therefore,

$$I(t) = \frac{\csc\alpha}{2\pi}\int_{-\infty}^\infty f(u)du\int_0^\infty \exp\left\{ia(\alpha)\left[t^2-u^2\right] + ib(\alpha)(u-t)\omega\right\}d\omega,$$

or

$$I(t) = \frac{\csc\alpha}{2\pi}e^{ia(\alpha)t^2}\int_{-\infty}^\infty f(u)e^{-ia(\alpha)u^2}J(u,t)\,du,$$

where

$$J(u,t) = \int_0^\infty \exp\{ib(\alpha)(u-t)\omega\}\,d\omega.$$

If we set $\lambda = b(u-t)$, we can reduce $J(u,t)$ to

$$\begin{aligned} J(u,t) = \int_0^\infty e^{i\lambda\omega}d\omega &= \frac{1}{2}\int_{-\infty}^\infty (1+sgn\,\omega)e^{i\lambda\omega}d\omega \\ &= \frac{1}{2}\left[2\pi\delta(\lambda) + \frac{2i}{\lambda}\right]. \end{aligned}$$

Therefore, $I = \frac{\csc\alpha}{2\pi}e^{ia(\alpha)t^2}\left[I_1 + I_2\right]$, where

$$I_1 = \pi\int_{-\infty}^\infty f(u)e^{-ia(\alpha)u^2}\delta(bu-bt)\,du \quad \text{and } I_2 = i\int_{-\infty}^\infty \frac{f(u)e^{-ia(\alpha)u^2}}{b(u-t)}du.$$

Upon replacing u by v/b, in I_1 we obtain

$$\begin{aligned} I_1 &= (\pi/b)\int_{-\infty}^\infty f(v/b)e^{-ia(\alpha)v^2/b^2}\delta(v-bt)\,dv \\ &= (\pi/b)f(t)e^{-ia(\alpha)t^2}. \end{aligned}$$

Now it follows that

$$\begin{aligned} I(t) &= \frac{\csc\alpha}{2\pi}e^{ia(\alpha)t^2}\left[(\pi/b)f(t)e^{-ia(\alpha)t^2}+\frac{i}{b}\int_{-\infty}^{\infty}\frac{f(u)e^{-ia(\alpha)u^2}}{(u-t)}du\right] \\ &= \frac{1}{2}\left[f(t)+\frac{i}{\pi}e^{ia(\alpha)t^2}\int_{-\infty}^{\infty}\frac{f(u)e^{-ia(\alpha)u^2}}{(u-t)}du\right] \\ &= \frac{1}{2}\{f(t)+iH_\alpha[f](t)\}; \end{aligned}$$

and therefore

$$\tilde{F_\alpha}(t)=2I(t)=2\int_0^\infty F_\alpha(\omega)K_{-\alpha}(\omega,t)d\omega.$$

The proof of the last statement is similar. ∎

The next theorem generalizes Equation (4.32).

Theorem 16. *Let F_α and H_α be defined as above. Then*

$$F_\alpha[H_\alpha[f]](\omega)=\begin{cases} -isgn(\omega)F_\alpha[f](\omega), & if\, 0<\alpha<\pi \\ isgn(\omega)F_\alpha[f](\omega), & if\, \pi<\alpha<2\pi \end{cases}$$

Proof *Let*

$$g(t)=\frac{1}{\pi}\int_{-\infty}^{\infty}\frac{f(x)e^{-ia(\alpha)x^2}}{x-t}dx$$

so that $H_\alpha[f](t)=e^{ia(\alpha)t^2}g(t)$. Thus,

$$\begin{aligned} F_\alpha[H_\alpha[f]](\omega) &= \int_{-\infty}^{\infty}K_\alpha(t,\omega)e^{ia(\alpha)t^2}g(t)dt \\ &= \frac{c(\alpha)}{\pi}\int_{-\infty}^{\infty}e^{-ia(t^2+\omega^2)+ibt\omega}dt\, e^{iat^2}\int_{-\infty}^{\infty}\frac{f(x)e^{-iax^2}}{x-t}dx \\ &= \frac{c(\alpha)}{\pi}\int_{-\infty}^{\infty}f(x)e^{-ia(x^2+\omega^2)}\left(\int_{-\infty}^{\infty}\frac{e^{ibt\omega}}{x-t}dt\right)dx \\ &= \frac{c(\alpha)}{\pi}\int_{-\infty}^{\infty}f(x)e^{-ia(x^2+\omega^2)}\left(\int_{-\infty}^{\infty}\frac{e^{ib\omega(x-y)}}{y}dy\right)dx \\ &= \int_{-\infty}^{\infty}f(x)K_\alpha(x,\omega)\left(\frac{1}{\pi}\int_{-\infty}^{\infty}\frac{e^{-ib\omega y}}{y}dy\right)dx \\ &= iF_\alpha(\omega)sign(-b\omega) \end{aligned}$$

4.5 FRACTIONAL TIME-FREQUENCY REPRESENTATIONS

The representation of a signal by means of its spectrum or Fourier transform is a powerful tool in solving many problems in signal processing and optics. However, in some instances this is not the most useful way of representing signals. For instance, we often think of music or speech as signals in which the spectrum evolves over time in a significant way. We imagine that at each instant we hear a certain combination of frequencies, and that they are constantly changing. This time evaluation of the frequencies is not reflected directly in the Fourier transform.

A signal can be reconstructed from its Fourier transform, but the transform contains information about the frequencies of the signal over all times instead of showing how the frequencies vary with time. Many of the signals encountered in practical applications such as speech, music and biological signals, have time-varying frequencies that are not adequately represented by the Fourier transform.

Time-frequency analysis provides a characterization of signals in terms of joint time and frequency content. One of the earliest methods of achieving time-dependent frequency analysis is the Gabor transform which is named after D.Gabor who introduced it in his fundamental work [104]. His work led to the development of the Windowed (short-time) Fourier transform which works by first dividing a signal into short consecutive segments using a window function and then computing the Fourier transform of each segment.

On the other hand, this method also has its limitation which is attributed to the fact that shortening the time window comes at the expense of the frequency resolution because the uncertainty principle of the Fourier transform asserts a reciprocal relation between the spread of a function and the spread of its Fourier transform; see Section 1.8.

Nowadays there are several important time-frequency representations. such as the Wigner distribution, radar and cross ambiguity functions, short-time (sliding widow) Fourier transform and wavelets.

In this section, we will examine the relationship between the fractional Fourier transform and a number of time-frequency representations, such as the Wigner distribution, the cross-ambiguity function and the windowed Fourier transform [15, 16, 28, 41, 64, 265, 340].

We will present three fractional time-frequency representations: the fractional Wigner distribution (FWD), the fractional ambiguity function and the fractional Windowed Fourier transform, all depending on an angular parameter α, and which, for $\alpha = \frac{\pi}{2}$, reduce to the corresponding conventional representations.

Because all these fractional time-frequency representations are closely related to the uncertainty principle for the fractional Fourier transform, we will close our study of these different representations by a discussion of the uncertainty principle for the fractional Fourier transform.

4.5.1 Fractional Wigner Distributions

The Wigner distribution appeared in the work E. P. Wigner [317] on quantum mechanics and was later introduced into signal analysis by Ville [301]. The Wigner-Ville distribution (WVD), which considers an analytic version of the signal, presents a unique signature of the signal and possesses all the desirable properties of a time-frequency representation [147]. It is closely related to the *Radar Ambiguity Function* [318]

We begin by defining the fractional Wigner distribution which may also be called (FrWD) of two functions f and g.

Definition 16. *The fractional (or weighted) Wigner distribution (FrWD) of two functions f and g is defined as*

$$W_{f,g}^{\alpha}(t,\omega)=\int_{\mathbb{R}} f\left(t+\frac{1}{2}x\right)\bar{g}\left(t-\frac{1}{2}x\right)K_{\alpha}(x,\omega)\,dx. \tag{4.33}$$

When $f=g$, we shall write $W_{f,f}^{\alpha}=W_{f}^{\alpha}$ for short.

For $\alpha=\frac{\pi}{2}$, the FrWD, $W_{\alpha}(t,\omega)$, coincides with the Wigner distribution (WD) $W_{f,g}(t,\omega)$, of f and g [186, 215] defined by

$$W_{f,g}(t,\omega)=\frac{1}{\sqrt{2\pi}}\int_{\mathbb{R}} f\left(t+\frac{1}{2}x\right)\bar{g}\left(t-\frac{1}{2}x\right)e^{i\omega x}\,dx.$$

and for all $\alpha\neq\frac{(2k+1)\pi}{2}$, we obtain new classes of time-frequency representations (TFR).

If $f,g\in L^{2}(\mathbb{R})$, then $W_{f,g}^{\alpha}$ exists. For, by the Cauchy-Schwarz inequality we have

$$\begin{aligned}\left|W_{f,g}^{\alpha}(t,\omega)\right|^{2} &= |c(\alpha)|^{2}\left(\int_{\mathbb{R}}\left|f\left(t+\frac{1}{2}x\right)\right|\left|\bar{g}\left(t-\frac{1}{2}x\right)\right|dx\right)^{2}\\ &\leq |c(\alpha)|^{2}\left(\int_{\mathbb{R}}\left|f\left(t+\frac{1}{2}x\right)\right|^{2}dx\right)\left(\int_{\mathbb{R}}\left|g\left(t-\frac{1}{2}x\right)\right|^{2}dx\right)\\ &\leq 4|c(\alpha)|^{2}\left(\int_{\mathbb{R}}|f(x)|^{2}dx\right)\left(\int_{\mathbb{R}}|g(y)|^{2}dy\right)=4|c(\alpha)|^{2}\|f\|_{2}^{2}\|g\|_{2}^{2}<\infty.\end{aligned}$$

In fact, we shall soon show that $W_{f,g}^{\alpha}\in L^{2}(\mathbb{R}^{2})$.

Theorem 17. *Let $f,g\in L^{2}(\mathbb{R})$ and F_{α}, G_{α} be their fractional Fourier transforms, respectively. Then*

1.
$$\langle f,g\rangle=\langle F_{\alpha},G_{\alpha}\rangle=c(-\alpha)\int_{\mathbb{R}}\int_{\mathbb{R}} W_{f,g}^{\alpha}(t,\omega)\exp\{ia(\alpha)\omega^{2}\}d\omega\,dt.$$

   In particular,

$$\|f\|_{2}^{2}=\|F_{\alpha}\|_{2}^{2}=c(\alpha)\int_{\mathbb{R}}\int_{\mathbb{R}} W_{f}^{\alpha}(t,\omega)\exp\{ia(\alpha)\omega^{2}\}d\omega\,dt.$$

2. *The following generalization of Moyal's formula holds*

$$\langle W_{f_1,g_1}^{\alpha},W_{f_2,g_2}^{\alpha}\rangle_{L^{2}(\mathbb{R}^{2})}=\langle f_1,f_2\rangle_{L^{2}(\mathbb{R})}\overline{\langle g_1,g_2\rangle}_{L^{2}(\mathbb{R})}. \tag{4.34}$$

   In particular,

$$\left\|W_{f,g}^{\alpha}\right\|=\|f\|\,\|g\|. \tag{4.35}$$

   Hence, $W_{f,g}^{\alpha}\in L^{2}(\mathbb{R}^{2})$ if f and g are in $L^{2}(\mathbb{R})$.

Proof *(1) For fixed t, it follows from the inversion formula of the fractional Fourier transform that*

$$f(t+\frac{x}{2})\bar{g}(t-\frac{x}{2}) = \int_{\mathbb{R}} W_{f,g}^{\alpha}(t,\omega)\, K_{-\alpha}(x,\omega)\, d\omega.$$

Setting $t_1 = t + x/2$ and $t_2 = t - x/2$, we obtain

$$f(t_1)\bar{g}(t_2) = \int_{\mathbb{R}} W_{f,g}^{\alpha}\left(\frac{t_1+t_2}{2},\omega\right) K_{-\alpha}(t_1-t_2,\omega)\, d\omega.$$

In particular, if $x = 0$, i.e., $t_1 = t_2 = t$, then

$$\begin{aligned} f(t)\bar{g}(t) &= \int_{\mathbb{R}} W_{f,g}^{\alpha}(t,\omega)\, K_{-\alpha}(0,\omega)\, d\omega \\ &= c(-\alpha)\int_{\mathbb{R}} W_{f,g}^{\alpha}(t,\omega)\exp\{i(\cot\alpha/2)\omega^2\}d\omega. \end{aligned}$$

For $f = g$, the FrWD satisfies the time-frequency marginals

$$|f(t)|^2 = \int_{\mathbb{R}} W_f^{\alpha}(t,\omega)\, K_{-\alpha}(0,\omega)\, d\omega.$$

Moreover, we have

$$\begin{aligned} \int_{\mathbb{R}} f(t)\bar{g}(t)\, dt &= \int_{\mathbb{R}} F_{\alpha}(\omega)\bar{G}_{\alpha}(\omega)\, d\omega \\ &= c(-\alpha)\int_{\mathbb{R}}\int_{\mathbb{R}} W_{f,g}^{\alpha}(t,\omega)\exp\{ia(\alpha)\omega^2\}d\omega\, dt. \end{aligned}$$

In particular,

$$\int_{\mathbb{R}} |f(t)|^2 dt = \int_{\mathbb{R}} |F_{\alpha}(\omega)|^2\, d\omega = c(-\alpha)\int_{\mathbb{R}}\int_{\mathbb{R}} W_f^{\alpha}(t,\omega)\exp\{ia(\alpha)\omega^2\}d\omega\, dt.$$

(2) With some computations, it is easy to see that

$$\int_{\mathbb{R}} W_{f_1,g_1}^{\alpha}(t,\omega)\bar{W}_{f_2,g_2}^{\alpha}(u,\omega)\, d\omega = \int_{\mathbb{R}} f_1(t+x/2)\bar{g}_1(t-x/2)\bar{f}_2(u+x/2)g_2(u-x/2)\, dx, \tag{4.36}$$

from which, when $u = t$, we obtain

$$\int_{\mathbb{R}} W_{f_1,g_1}^{\alpha}(t,\omega)\bar{W}_{f_2,g_2}^{\alpha}(t,\omega)\, d\omega = \int_{\mathbb{R}} f_1(t+x/2)\bar{g}_1(t-x/2)\bar{f}_2(t+x/2)g_2(t-x/2)\, dx, \tag{4.37}$$

and hence

$$\begin{aligned} &\int_{\mathbb{R}}\int_{\mathbb{R}} W_{f_1,g_1}^{\alpha}(t,\omega)\bar{W}_{f_2,g_2}^{\alpha}(t,\omega)\, d\omega\, dt = \\ &\int_{\mathbb{R}}\int_{\mathbb{R}} f_1(t+x/2)\bar{g}_1(t-x/2)\bar{f}_2(t+x/2)g_2(t-x/2)\, dx\, dt. \end{aligned} \tag{4.38}$$

Therefore, upon substituting $x = t + x/2$ and $y = t - x/2$, in the right-hand side, we have

$$\int_{\mathbb{R}}\int_{\mathbb{R}} f_1(x)\bar{g}_1(y)\bar{f}_2(x)g_2(y)\, dx\, dy = \langle f_1,\, f_2\rangle_{L^2(\mathbb{R})}\overline{\langle g_1,\, g_2\rangle}_{L^2(\mathbb{R})}, \tag{4.39}$$

and by combining this last equation and (4.38) we obtain

$$\langle W^{\alpha}_{f_1,g_1}, W^{\alpha}_{f_2,g_2}\rangle_{L^2(\mathbb{R}^2)} = \langle f_1, f_2\rangle_{L^2(\mathbb{R})}\overline{\langle g_1, g_2\rangle}_{L^2(\mathbb{R})}. \tag{4.40}$$

In particular, if $f_1 = f_2 = f$ *and* $g_1 = g_2 = g$, *we have*

$$\left\| W^{\alpha}_{f,g} \right\| = \|f\| \, \|g\| .$$

Furthermore, if $f_1 = f_2 = g_1 = g_2 = f$, *we have*

$$\left\| W^{\alpha}_{f} \right\| = \|f\|_2^2 .$$

■

4.5.2 Fractional Time and Frequency Shifts

In this section, we discuss fractional time and frequency shifts . A signal $f(t)$ that is shifted by a time t_0 and frequency ω_0 in the fractional Fourier transform domain is given by

$$g(t) = \tilde{f}(t) = \exp\left\{-ib(\alpha)\omega_0 t + ia(\alpha)\left[\omega_0^2 + t^2\right]\right\} f(t-t_0).$$

We now show how the fractional Wigner distribution of $g = \tilde{f}$ is related to that of f.

Theorem 18. *Let* f, g *be defined as above. Then*

$$W^{\alpha}_{g}(t,\omega) = H_\alpha(t,\omega)\, W^{\alpha}_{f}(t-t_0, \omega-\omega_0 + t\cos\alpha),$$

where

$$H_\alpha(t,\omega) = \exp\left\{ia(\alpha)\left[\omega_0^2 + t^2\cos^2\alpha - 2\omega\omega_0 - -2t\cos\alpha\,(\omega-\omega_0)\right]\right\}.$$

Proof
By substituting g *in (4.33), we have*

$$\begin{aligned}
W^{\alpha}_{g}(t,\omega) &= \int_{\mathbb{R}} g(t+x/2)\overline{g}(t-x/2)K_\alpha(x,\omega)dx \\
&= \int_{\mathbb{R}} f\left(t-t_0+\frac{1}{2}x\right)\bar{f}\left(t-t_0-\frac{1}{2}x\right) \\
\times & \exp\left\{-ib(\alpha)\omega_0\left(t+\frac{1}{2}x\right) + ia(\alpha)\left[\omega_0^2 + \left(t+\frac{1}{2}x\right)^2\right]\right\} \\
\times & \exp\left\{ib(\alpha)\omega_0\left(t-\frac{1}{2}x\right) - ia(\alpha)\left[\omega_0^2 + \left(t-\frac{1}{2}x\right)^2\right]\right\} K_\alpha(x,\omega)\,dx.
\end{aligned}$$

After some simplifications, we obtain

$$\begin{aligned}
W^{\alpha}_{g}(t,\omega) &= c(\alpha)\int_{\mathbb{R}} f\left(t-t_0+\frac{1}{2}x\right)\bar{f}\left(t-t_0-\frac{1}{2}x\right) \\
\times & \exp\left[-ib(\alpha)x\omega_0 + 2ia(\alpha)xt\right]K_\alpha(x,\omega)dx \\
= & \; c(\alpha)\int_{\mathbb{R}} f\left(t-t_0+\frac{1}{2}x\right)\bar{f}\left(t-t_0-\frac{1}{2}x\right)E(t,\omega,x)dx,
\end{aligned} \tag{4.41}$$

where

$$E(t,\omega,x) = \exp\left\{-ia(\alpha)(x^2+\omega^2)+ib(\alpha)x(\omega-\omega_0)+2ia(\alpha)tx\right\}.$$

But it is easy to see that

$$\begin{aligned} & K_\alpha(x,\omega-\omega_0+t\cos\alpha) = \\ & \exp\left\{-ia(\alpha)\left[x^2+\omega^2+\omega_0^2+t^2\cos^2\alpha-2\omega\omega_0+2t\cos\alpha(\omega-\omega_0)\right]\right\} \\ \times\ & \exp\left\{ib(\alpha)x(\omega-\omega_0+t\cos\alpha)\right\}. \end{aligned}$$

By choosing

$$H_\alpha(t,\omega) = \exp\left\{ia(\alpha)\left[\omega_0^2+t^2\cos^2\alpha-2\omega\omega_0+2t\cos\alpha(\omega-\omega_0)\right]\right\},$$

it is evident that

$$E(t,\omega,x) = H_\alpha(t,\omega)\,K_\alpha(x,\omega-\omega_0+t\cos\alpha),$$

which is the exponential factor in Eq. (4.41), and this completes the proof.

In other words, we have shown that, apart from the modulation factor, H_α, the FrWD of $\tilde{f}$ is equal to the FrWD of f shifted by t_0 on the time axis and by $\omega_0 - t\cos\alpha$ on the frequency axis. In particular, when $\alpha = \pi/2$, we obtain the known result for the standard Wigner distribution.

4.5.3 The Fractional Cross-Ambiguity Function

Another time-frequency representation that plays an important role in radar, sonar signal processing and radio astronomy is the ambiguity function. Its idea was originally introduced in signal processing by E. P. Wigner in 1948 [317] but it was P. Woodward who formulated it in the context of radar theory [322]. Woodward proposed treating the question of radar signal ambiguity as part of the question of target resolution. To do that, he introduced a function that described the correlation between a radar signal and its Doppler-shifted and time-translated version. This function exhibits the interplay between measurement ambiguity and target resolution, and for this reason it is called the radar ambiguity function. The ambiguity function of a signal $f(t)$ is defined as

$$A(\tau,\omega) = \frac{1}{\sqrt{2\pi}}\int_{\mathbb{R}} f(t)f(t-\tau)e^{i\omega t}dt.$$

Its absolute value is called the uncertainty function as it is related to the uncertainty principle of radar signals.

The study of the radar ambiguity function in the context of pure mathematics and, in particular, in group theory and harmonic analysis, was pioneered by C. Wilcox [318], L. Auslander and R. Tolimieri [22]. For applications of the ambiguity functions in optics, see [230]. Most of the material presented in this section is based on the author's work in [265].

The cross-ambiguity function $A_{f,g}(u,v)$ of two functions f,g is defined as

$$A_{f,g}(u,v) = \frac{1}{\sqrt{2\pi}}\int_{\mathbb{R}} f(t+u/2)\,\bar{g}(t-u/2)\,e^{-ivt}dt.$$

The function $R_{f,g}(u) = A_{f,g}(u,0)$ is called the cross-correlation function of f and g and $R_f(u) = A_f(u,0)$ is called the auto-correlation function of f.

In this section, we generalize the cross-ambiguity function by introducing a fractional ambiguity function .

Definition 17. *The fractional (or weighted) ambiguity function of two integrable functions f and g is defined as*

$$A^{\alpha}_{f,g}(u,v) = \int_{\mathbb{R}} f(t+u/2)\,\bar{g}(t-u/2)\,K_{-\alpha}(t,v)\,dt. \tag{4.42}$$

When $f = g$, we denote $A^{\alpha}_{f,f}$ by A^{α}_{f}.

The fractional ambiguity function $A^{\alpha}_{f,g}$ exists, for example, if $f,g \in L^2(\mathbb{R})$. For, by the Cauchy-Schwarz inequality we have

$$\begin{aligned}
\left|A^{\alpha}_{f,g}(u,v)\right|^2 &= |c(\alpha)|^2 \left(\int_{\mathbb{R}} \left|f\left(t+\frac{1}{2}u\right)\right|\left|\bar{g}\left(t-\frac{1}{2}u\right)\right| dt\right)^2 \\
&\leq |c(\alpha)|^2 \left(\int_{\mathbb{R}} \left|f\left(t+\frac{1}{2}u\right)\right|^2 dt\right)\left(\int_{\mathbb{R}} \left|g\left(t-\frac{1}{2}u\right)\right|^2 dt\right) \\
&\leq |c(\alpha)|^2 \left(\int_{\mathbb{R}} |f(x)|^2\,dx\right)\left(\int_{\mathbb{R}} |g(y)|^2\,dy\right) \\
&= |c(\alpha)|^2 \|f\|_2^2 \|g\|_2^2 < \infty.
\end{aligned}$$

In fact, we shall soon show that $A^{\alpha}_{f,g} \in L^2(\mathbb{R}^2)$; see Eq. (4.51) below.

Theorem 19. *Let $f,g \in L^2(\mathbb{R})$ and F_α and G_α be their fractional Fourier transforms, respectively. Then*

1. $$\langle f,g\rangle = \langle F_\alpha, G_\alpha\rangle = \int\int A^{\alpha}_{f,g}(0,v)\,K_\alpha(t,v)\,dv\,dt.$$

   In particular,

   $$\|f\|^2 = \|F_\alpha\|^2 = \int\int A^{\alpha}_{f}(0,v)\,K_\alpha(t,v)\,dv\,dt.$$

2. *The following generalization of Moyal's formula holds*

   $$\langle A^{\alpha}_{f_1,g_1}, A^{\alpha}_{f_2,g_2}\rangle_{L^2(\mathbb{R}^2)} = \langle f_1, f_2\rangle_{L^2(\mathbb{R})}\overline{\langle g_1, g_2\rangle}_{L^2(\mathbb{R})}. \tag{4.43}$$

   In particular,

   $$\left\|A^{\alpha}_{f,g}\right\| = \|f\|\,\|g\|, \tag{4.44}$$

   which shows that $A^{\alpha}_{f,g} \in L^2(\mathbb{R}^2)$ if f and g are in $L^2(\mathbb{R})$.

Proof *1) Fix u. Then it follows from the inversion formula for the FrFT that*

$$f(t+u/2)\,\bar{g}(t-u/2) = \int_{\mathbb{R}} A^{\alpha}_{f,g}(u,v)\,K_\alpha(t,v)\,dv,$$

which, upon substituting $t_1 = t + u/2$ and $t_2 = t - u/2$, leads to

$$f(t_1)\bar{g}(t_2) = \int_{\mathbb{R}} A^{\alpha}_{f,g}(t_1 - t_2, v)\, K_\alpha\left(\frac{t_1+t_2}{2}, v\right) dv. \tag{4.45}$$

For $t = t_1 = t_2$, we have

$$f(t)\bar{g}(t) = \int_{\mathbb{R}} A^{\alpha}_{f,g}(0, v)\, K_\alpha(t, v)\, dv;$$

thus,

$$\int_{\mathbb{R}} f(t)\bar{g}(t)dt = \int_{\mathbb{R}} F_\alpha(\omega)G_\alpha(\omega)d\omega = \int_{\mathbb{R}}\int_{\mathbb{R}} A^{\alpha}_{f,g}(0, v)\, K_\alpha(t, v)\, dv\, dt,$$

which leads to

$$\|f\|_2^2 = \|F_\alpha\|_2^2 = \int_{\mathbb{R}}\int_{\mathbb{R}} A^{\alpha}_{f}(0, v)\, K_\alpha(t, v)\, dv\, dt.$$

2) Similarly, and in view of the fact that

$$\int_{\mathbb{R}} K_\alpha(t, v)K_{-\alpha}(x, v)dv = e^{ia(x^2-t^2)}\delta(x - t), \tag{4.46}$$

and because $2\pi|c(\alpha)|^2/b(\alpha) = 1$, we have

$$\int_{\mathbb{R}} A^{\alpha}_{f_1,g_1}(u, v)\bar{A}^{\alpha}_{f_2,g_2}(w, v)\, dv = \int_{\mathbb{R}} f_1(t+u/2)\bar{g}_1(t-u/2)\bar{f}_2(t+w/2)g_2(t-w/2)\, dt, \tag{4.47}$$

from which, when $w = u$, we obtain

$$\int_{\mathbb{R}} A^{\alpha}_{f_1,g_1}(u, v)\bar{A}^{\alpha}_{f_2,g_2}(u, v)\, dv = \int_{\mathbb{R}} f_1(t+u/2)\bar{g}_1(t-u/2)\bar{f}_2(t+u/2)g_2(t-u/2)\, dt, \tag{4.48}$$

and hence

$$\begin{aligned}&\int_{\mathbb{R}}\int_{\mathbb{R}} A^{\alpha}_{f_1,g_1}(u, v)\bar{A}^{\alpha}_{f_2,g_2}(u, v)\, du\, dv = \\ &\int_{\mathbb{R}}\int_{\mathbb{R}} f_1(t+u/2)\bar{g}_1(t-u/2)\bar{f}_2(t+u/2)g_2(t-u/2)\, du\, dt.\end{aligned} \tag{4.49}$$

Therefore, upon substituting $x = t + u/2$ and $y = t - u/2$, in the right-hand side of (4.49), we have

$$\langle A^{\alpha}_{f_1,g_1}, A^{\alpha}_{f_2,g_2}\rangle_{L^2(\mathbb{R}^2)} = \langle f_1, f_2\rangle_{L^2(\mathbb{R})}\overline{\langle g_1, g_2\rangle}_{L^2(\mathbb{R})}. \tag{4.50}$$

In particular, if $f_1 = f_2 = f$ and $g_1 = g_2 = g$, we have

$$\left\|A^{\alpha}_{f,g}\right\| = \|f\|\,\|g\|. \tag{4.51}$$

Moreover, if $f_1 = f_2 = g_1 = g_2 = f$, we have $\left\|A^{\alpha}_{f}\right\| = \|f\|^2$. ∎

Next we show how the fractional ambiguity function is related to the fractional Wigner distribution function.

$$\begin{aligned}
&\int_{\mathbb{R}}\int_{\mathbb{R}} W_{f,g}^{\alpha}(t,\omega)\, K_{-\alpha}(t,\eta)K_{-\alpha}(\tau,\omega)dt\,d\omega \\
=\;& \int_{\mathbb{R}}\int_{\mathbb{R}} K_{-\alpha}(t,\eta)K_{-\alpha}(\tau,\omega)dt\,d\omega \int f(t+u/2)\bar{g}(t-u/2)K_{\alpha}(u,\omega)\,du \\
=\;& \int_{\mathbb{R}} f(t+\tau/2)\bar{g}(t-\tau/2)K_{-\alpha}(\eta,t)\,dt = A_{f,g}^{\alpha}(\tau,\eta),
\end{aligned}$$

where the last step follows by interchanging the integral signs and using (4.46).

In the rest of this section we derive more properties of the fractional cross–ambiguity function $A_{f,g}^{\alpha}(u,v)$.

Let $g=f$ in (4.45), then

$$f(t)\overline{f}(\tau) = \int_{\mathbb{R}} A_f^{\alpha}(t-\tau,v)K_{\alpha}(\frac{t+\tau}{2},v)\,dv. \tag{4.52}$$

Set

$$H_f(t,\tau) = f(t)\overline{f}(\tau).$$

It is easy to see that H_f satisfies the following properties:

1. $\overline{H}(t,\tau) = H(\tau,t)$
2. $H(t,t) \geq 0$
3. $H(x,x)H(t,\tau) = H(t,x)H(x,\tau)$.

Let $\tilde{L}^2(\mathbb{R}^2)$ denote that subset of $L^2(\mathbb{R}^2)$ consisting of all functions $H(t,\tau)$ satisfying conditions (1)–(3). Consider the mapping $U[A_f^{\alpha}] = H_f$ defined by (4.52). It is easy to see that

$$\|A_f^{\alpha}\| = \|H_f\| = \|f\|^2,$$

where $\|H_f\|$ and $\|A_f^{\alpha}\|$ are defined in the usual way, e.g.,

$$\|H_f\|^2 = \int_{\mathbb{R}}\int_{\mathbb{R}} |H_f(t,\tau)|^2\, dt d\tau.$$

For any $F(t,\tau) \in \tilde{L}^2(\mathbb{R}^2)$, there exists $f \in L^2(\mathbb{R})$ such that $U[A_f^{\alpha}] = F$. In fact, if F is given, we can choose $f(t) = F(t,\tau_0)/\sqrt{F(\tau_0,\tau_0)}$, where τ_0 is chosen so that $F(\tau_0,\tau_0) > 0$. For,

$$\begin{aligned}
A_f^{\alpha}(u,v) &= \int_{\mathbb{R}} f(t+\frac{u}{2})\overline{f}(t-\frac{u}{2})K_{-\alpha}(t,v)\,dt \\
&= \frac{1}{F(\tau_0,\tau_0)}\int_{\mathbb{R}} F(t+\frac{u}{2},\tau_0)\overline{F}(t-\frac{u}{2},\tau_0)K_{-\alpha}(t,v)\,dt.
\end{aligned}$$

The integral exists because $f \in L^2(\mathbb{R})$ if $F \in L^2(\mathbb{R}^2)$. The last equation leads to

$$\frac{1}{F(\tau_0,\tau_0)}F(t+\frac{u}{2},\tau_0)\overline{F}(t-\frac{u}{2},\tau_0) = \int_{\mathbb{R}} A_f^{\alpha}(u,v)K_{\alpha}(t,v)\,dt$$

or

$$\frac{1}{F(\tau_0,\tau_0)}F(t,\tau_0)\overline{F}(\tau,\tau_0) = \int_{\mathbb{R}} A_f^\alpha(t-\tau,v)K_\alpha(\frac{t+\tau}{2},v)\,dv$$
$$= f(t)\overline{f}(\tau) = H_f(t,\tau).$$

Furthermore, since F satisfies properties (1)–(3)

$$\begin{aligned} H_f(t,\tau) &= f(t)\overline{f}(\tau) \\ &= \frac{F(t,\tau_0)\overline{F}(\tau,\tau_0)}{F(\tau_0,\tau_0)} = \frac{F(t,\tau_0)F(\tau_0,\tau)}{F(\tau_0,\tau_0)} \\ &= \frac{F(\tau_0,\tau_0)F(t,\tau)}{F(\tau_0,\tau_0)} = F(t,\tau). \end{aligned}$$

Recall from the Definition of the cross–ambiguity transformation that $A_{f,g}^\alpha$ is defined as a mapping from $L^2(\mathbb{R})\times L^2(\mathbb{R})\to L^2(\mathbb{R}^2)$, given by $\mathcal{A}^\alpha(f,g) = A_{f,g}^\alpha(u,v)$. Clearly, this mapping is bilinear. If $\{f_n\}_{n=1}^\infty$ and $\{g_n\}_{n=1}^\infty$ are two sequences in $L^2(\mathbb{R})$ such that $f_n\to f$ and $g_n\to g$ in $L^2(\mathbb{R})$, then by writing $f_ng_n - fg$ as $f_ng_n - f_ng + f_ng - fg$ and using the Cauchy-Schwarz inequality, we conclude the following theorem

Theorem 20. *The cross–ambiguity transformation A^α is a continuous transformation from $L^2(\mathbb{R})\times L^2(\mathbb{R})$ into $L^2(\mathbb{R}^2)$, in the sense that if $f_n\to f$ and $g_n\to g$ in $L^2(\mathbb{R})$, then $A_{f_n,g_n}^\alpha(u,v)\to A_{f,g}^\alpha(u,v)$.*

Let $\{\psi_n(t)\}_{n=1}^\infty$ be an orthonormal basis of $L^2(\mathbb{R})$ and set $\psi_{m,n}(u,v) = A_{\psi_m,\psi_n}^\alpha(u,v)$. From (4.50), we have

$$\langle\psi_{k,l},\psi_{m,n}\rangle = \langle\psi_k,\psi_m\rangle\overline{\langle\psi_l,\psi_n\rangle} = \delta_{k,m}\delta_{l,n}$$

which shows that $\{\psi_{m,n}(u,v)\}_{m,n=1}^\infty$ is an orthonormal set in $L^2(\mathbb{R}^2)$ and hence it is a basis for the closure of its span, which we shall denote by $\mathcal{AL}$. Now let $f\in L^2(\mathbb{R})$. Then

$$f(t) = \sum_{n=1}^\infty \hat{f}_n\psi_n(t) \quad \text{with} \quad \|f\|^2 = \sum_{n=1}^\infty \left|\hat{f}_n\right|^2.$$

Similarly, if $F\in\mathcal{AL}$, we have

$$F(u,v) = \sum_{m,n=1}^\infty \hat{F}_{m,n}\psi_{m,n}(u,v), \quad \text{with} \quad \|F\|^2 = \sum_{m,n=1}^\infty \left|\hat{F}_{m,n}\right|^2 < \infty.$$

Therefore, if $A_f^\alpha\in\mathcal{AL}$, then

$$A_f^\alpha(u,v) = \sum_{m,n=1}^\infty c_{m,n}\psi_{m,n}(u,v),$$

with

$$c_{m,n} = \langle A_f^\alpha,\psi_{m,n}\rangle = \langle A_{f,f}^\alpha, A_{\psi_m,\psi_n}^\alpha\rangle = \langle f,\psi_m\rangle\overline{\langle f,\psi_n\rangle} = \hat{f}_m\overline{\hat{f}_n}.$$

Conversely, if $c_{m,n} = \hat{f}_m\overline{\hat{f}_n}$, then $f = \sum\hat{f}_m\psi_m\in L^2(\mathbb{R})$. Thus, a function $F(u,v)\in\tilde{L}^2(\mathbb{R}^2)$ is an ambiguity function if and only if $\hat{F}_{m,n} = \hat{f}_m\overline{\hat{f}_n}$.

4.5.4 Fractional Windowed (Sliding-Window)-Fourier Transform

First, let us recall the definition of the windowed Fourier transform $F_h(t,w)$ of f with respect to a window function h, and recall some of its properties.

$$F_h(t,w) = \frac{1}{\sqrt{2\pi}} \int_{\mathbb{R}} f(x)\overline{h}(x-t)e^{iwx}dx,$$

where $f, h \in L^2(\mathbb{R})$.

Let $f, g \in L^2(\mathbb{R})$, and denote their windowed-Fourier transforms with respect to a window function h by F and G, respectively. Then Parseval's relation for the transform takes the form

$$\langle F, G\rangle_{L^2(\mathbb{R}^2)} = \|h\|^2 \langle f, g\rangle \,, \tag{4.53}$$

where

$$\langle F, G\rangle_{L^2(\mathbb{R}^2)} = \int_{\mathbb{R}} \int_{\mathbb{R}} F(t,w)\overline{G}(t,w)dtdw \,.$$

In particular, if $\|h\| = 1$, then the transformation is an isometry from $L^2(\mathbb{R})$ into $L^2(\mathbb{R}^2)$.

The inversion formula for the Windowed-Fourier transform is

$$f(x) = \frac{1}{\sqrt{2\pi}\|h\|^2} \int_{\mathbb{R}} \int_{\mathbb{R}} F(w,t)h(x-t)e^{-ixw}dtdw \,.$$

We now introduce the fractional windowed Fourier transform and then derive some of its properties.

Definition 18. *Let $h \in L^2(\mathbb{R})$ be the window function and assume that $f \in L^2(\mathbb{R})$. We define the fractional windowed Fourier transform of f with angle α and window function h as*

$$\mathcal{F}_{\alpha;h}[f](t,w) = F_{\alpha;h}(t,w) = \int_{\mathbb{R}} f(x)\overline{h}(x-t)K_\alpha(x,w)dx.$$

Theorem 21. *Let $F_{\alpha;h}$ and $G_{\alpha;h}$ be the fractional windowed Fourier transforms of f and g, respectively. Then we have*

$$\langle F_{\alpha;h}, G_{\alpha;h}\rangle_{L^2(\mathcal{R}^2)} = \|h\|^2 \langle f, g\rangle,$$

where

$$\langle F_{\alpha;h}, G_{\alpha;h}\rangle_{L^2(\mathbb{R}^2)} = \int_{\mathbb{R}} \int_{\mathbb{R}} F_{\alpha;h}(t,w)G_{\alpha;h}(t,w)dtdw.$$

In particular, $\|F_{\alpha;h}\|^2_{L^2(\mathbb{R}^2)} = \|h\|^2\,\|f\|^2$. Moreover, the inversion formula for the fractional windowed Fourier transform is given by

$$f(x) = \frac{1}{\|h\|^2} \int_{\mathbb{R}} \int_{\mathbb{R}} h(x-t)F_{\alpha;h}(t,w)K_{-\alpha}(x,w)dwdt$$

Proof *For fixed t, $F_{\alpha;h}$ is the fractional Fourier transform of $f(x)\overline{h}_t(x)$, where $h_t(x) = h(x-t)$. Therefore, by Parseval's relation for the FrFT, we have*

$$\begin{aligned}
\langle F_{\alpha;h}, G_{\alpha;h}\rangle_{L^2(\mathbb{R})} &= \langle \mathcal{F}_\alpha[fh_t], \mathcal{F}_\alpha[gh_t]\rangle_{L^2(\mathbb{R})} \\
&= \int_{\mathbb{R}} F_{\alpha;h}(w)\overline{G}_{\alpha;h}(w)dw \\
&= \langle fh_t, gh_t\rangle = \int_{\mathbb{R}} f(x)\overline{g}(x)h(x-t)\overline{h}(x-t)dx.
\end{aligned}$$

By integrating over t, we obtain the result. The inversion formula follows from the inversion formula of the FrFT

$$f(x)\overline{h}(x-t) = \int_{\mathbb{R}} F_{\alpha;h}(w)K_{-\alpha}(x,w)dw,$$

and multiplying by $h(x-t)$ and then integrating with respect to t over $\mathbb{R}$.

4.6 UNCERTAINTY PRINCIPLE FOR THE FRACTIONAL FOURIER TRANSFORM

The uncertainty principle for the Fourier transform asserts a reciprocal relation between the spread of a function and the spread of its Fourier transform. In practice this implies that shortening the time window comes at the expense of the frequency resolution [102, 277]. In this section, we present an uncertainty principle for the fractional Fourier transform .

Theorem 22. *Let $f(t)$ be a real-valued function such that*

$$\lim_{|t|\to\infty} tf^2(t) = 0, \quad \textit{and} \lim_{|\omega|\to\infty} \omega F_\alpha^2(\omega) = 0,$$

and let

$$T = \int t^2|f(t)|^2dt, \quad \sigma = \int \omega^2|F_\alpha(\omega)|^2d\omega,$$
$$U = \int |f'(t)|^2dt, \quad V = \int |F'_\alpha(\omega)|^2d\omega,$$

where the integrals are taken over the whole real line. Then

$$\frac{1}{2}E^2 \leq (TV+\sigma U)\cos^2\alpha + 2\sigma T\sin^2\alpha, \tag{4.54}$$

where E is the energy of the signal defined as

$$E = \int |f(t)|^2dt = \int |F_\alpha(\omega)|^2d\omega.$$

In particular, if $\alpha = \pi/2$, we obtain the uncertainty principle for the standard Fourier transform

$$\frac{1}{4}E^2 \leq \sigma T.$$

Proof *The assumption on f implies that*

$$\begin{aligned}\frac{1}{2}\int tdf^2(t) &= \frac{1}{2}\left\{tf^2(t)\Big|_{-\infty}^{\infty} - \int f^2(t)dt\right\}\\ &= -\frac{1}{2}E = \int tf(t)f'(t)dt.\end{aligned}$$

By the Cauchy-Schwartz inequality, we have

$$\begin{aligned}\frac{1}{4}E^2 &= \left|\int tf(t)f'(t)dt\right|^2\\ &\leq \left(\int t^2|f(t)|^2dt\right)\left(\int |f'(t)|^2dt\right).\end{aligned} \tag{4.55}$$

But by property (4.12) , we have

$$\int f'(t)K_\alpha(t,\omega)\,dt = F'_\alpha(\omega)\cos\alpha - i\omega F_\alpha(\omega)\sin\alpha,$$

and it follows from Parseval's relation that

$$\int |f'(t)|^2\,dt = \int \left(|F'_\alpha(\omega)|^2\cos^2\alpha + \omega^2|F_\alpha(\omega)|^2\sin^2\alpha\right)d\omega. \tag{4.56}$$

By substituting (4.56) into (4.55), we obtain

$$\frac{1}{4}E^2 \le T\left(V\cos^2\alpha + \sigma\sin^2\alpha\right) \tag{4.57}$$

Similarly,

$$\frac{1}{2}\int \omega dF_\alpha^2(\omega) = -\frac{1}{2}E = \int \omega F_\alpha(\omega)F'_\alpha(\omega)\,d\omega,$$

which yields

$$\frac{1}{4}E^2 \le \left(\int \omega^2|F_\alpha(\omega)|^2\,d\omega\right)\left(\int |F'_\alpha(\omega)|^2\,d\omega\right). \tag{4.58}$$

With some computations one can show that

$$\int |F'_\alpha(\omega)|^2\,d\omega = U\cos^2\alpha + T\sin^2\alpha,$$

which upon its substitution in (4.58), reduces (4.58) to

$$\frac{1}{4}E^2 \le \sigma\left(U\cos^2\alpha + T\sin^2\alpha\right). \tag{4.59}$$

Adding (4.57) and (4.59), we obtain (4.54).

4.7 FRACTIONAL FOURIER TRANSFORM OF GENERALIZED FUNCTIONS

Among a number of different extensions of the FrFT, is its extension to spaces of generalized functions (distributions). The aim of this section is to obtain fractional Fourier transform of generalized functions. We will develop this extension using two different approaches, one analytic and one algebraic.

Extending an integral transform to a class of generalized functions can be achieved in a number of different ways. Chief among them is the embedding method, in which the kernel of the integral transform is embedded in a suitable space of test-functions. The dual space of a testing-function space is a space of generalized functions. The integral transform of an element of the dual space is defined as the function given by taking the action of that element of the dual space on the kernel of the transform. That is if $\mathcal{A}$ is a space of test functions, $\mathcal{A}^*$ is its dual, and $k(t,x)$ is the kernel of the integral transform under consideration, then under the assumption that $k(t,x) \in \mathcal{A}$, we can define the integral transform of any element $f \in \mathcal{A}^*$ as

$$F(x) = \langle f(t), k(t,x)\rangle.$$

Other methods, such as the adjoint and sequence methods, can also be used; see [52, 108, 232, 346, 352] for details.

The algebraic method we shall introduce in this section is different from all of the above since it uses the algebraic structure of the integral transform under consideration rather than its analytic properties. This is particularly useful if such a structure is available. The method is elegant and the extension is natural once an appropriate space of generalized functions is constructed. In this approach, the way generalized functions are constructed resembles the way in which the rational numbers are constructed from the integers.

In the next subsection, we shall extend the FrFT to a space of generalized functions using the embedding method. This will be done very briefly since most of the derivation is routine. In Subsection 2, we construct a space of generalized functions using the algebraic approach, and then in Subsection 3, we define the fractional Fourier transforms of elements of that space.

4.7.1 The Embedding Method

In this section, we extend the FrFT to a space of generalized functions using the embedding method, which depends on analytic properties of the FrFT. Since this method is well known, we shall outline only the main steps of how to use it to extend the FrFT to generalized functions. The procedure is very similar to the one that Zemanian used to extend the Laplace transform to generalized functions [361, Ch. 3].

Since the kernel of the FrFT, $K_\alpha(t,x)$, is an infinitely differentiable function in both t and x, it belong to the space $\mathcal{E}$ consisting of all infinitely differentiable functions. The dual space $\mathcal{E}^*$ of $\mathcal{E}$ is the space of generalized functions with compact supports. We have the following definition.

Definition 19. *The FrFT of a generalized function f with compact support is defined as*

$$F_\alpha(x) = \langle f(t), K_\alpha(t,x)\rangle.$$

The FrFT, $F_\alpha(x)$, of a generalized function with compact support f is an infinitely differentiable function that grows no faster than a polynomial on the real axis as $|x| \to \infty$. In fact,

$$F^{(1)}(x) = -2ia(\alpha)x\langle f(t), K_\alpha(t,x)\rangle + ib(\alpha)\langle f(t), tK_\alpha(t,x)\rangle,$$

Hence

$$|F^{(1)}(x)| \leq C|x| \text{ for all } x \text{ and sufficiently large } C.$$

By induction, one can show that

$$|F^{(n)}(x)| \leq C|x|^n \text{ for all } x \text{ and sufficiently large } C.$$

The following theorem gives an inversion formula for the FrFT of a generalized function with compact support.

Theorem 23. *Let $\mathcal{F}_\alpha[f](x) = F_\alpha(x)$ be the FrFT of a generalized function with compact support f. Then*

$$f(t) = \lim_{r\to\infty} \int_{-r}^{r} F_\alpha(x) K_{-\alpha}(t,x)\, dx,$$

where the limit is taken in the space $\mathcal{D}^*$ of Schwartz distributions.

Proof *We need to show that*

$$\langle f(t), \phi \rangle = \lim_{r \to \infty} \langle \int_{-r}^{r} F_\alpha(x) K_{-\alpha}(t,x)\, dx, \phi \rangle ,$$

for all $\phi \in \mathcal{D}$, *where* $\mathcal{D}$ *is the space of all infinitely differentiable functions that vanish outside some compact set. We have*

$$\begin{aligned}
\langle \int_{-r}^{r} F_\alpha(x) K_{-\alpha}(t,x)\, dx, \phi(t) \rangle &= \int_{-\infty}^{\infty} \phi(t)\, dt \int_{-r}^{r} F_\alpha(x) K_{-\alpha}(t,x)\, dx \\
&= \int_{-\infty}^{\infty} \phi(t)\, dt \int_{-r}^{r} \langle f(u), K_\alpha(u,x) \rangle K_{-\alpha}(t,x)\, dx \\
&= \langle f(u), \int_{-\infty}^{\infty} \phi(t)\, dt \int_{-r}^{r} K_\alpha(u,x) K_{-\alpha}(t,x)\, dx \rangle
\end{aligned}$$

Changing the order of integration is possible because ϕ *is bounded and has compact support and the integrand is a continuous function of* t *and* x. *Therefore,*

$$\begin{aligned}
&\lim_{r \to \infty} \langle \int_{-r}^{r} F_\alpha(x) K_{-\alpha}(t,x)\, dx, \phi(t) \rangle = \\
&\qquad \lim_{r \to \infty} \langle f(u), \int_{-\infty}^{\infty} \phi(t)\, dt \int_{-r}^{r} K_\alpha(u,x) K_{-\alpha}(t,x)\, dx \rangle \\
&= \langle f(u), \lim_{r \to \infty} \int_{-\infty}^{\infty} \phi(t)\, dt \int_{-r}^{r} K_\alpha(u,x) K_{-\alpha}(t,x)\, dx \rangle \\
&= \langle f(u), \int_{-\infty}^{\infty} \phi(t)\, dt \int_{-\infty}^{\infty} K_\alpha(u,x) K_{-\alpha}(t,x)\, dx \rangle .
\end{aligned}$$

In view of Theorem (7), we have

$$\begin{aligned}
&\lim_{r \to \infty} \langle \int_{-r}^{r} F_\alpha(x) K_{-\alpha}(t,x)\, dx, \phi(t) \rangle = \\
&= \langle f(u), \int_{-\infty}^{\infty} \phi(t) K_0(u,t)\, dt \rangle = \langle f(u), \mathcal{F}_0[\phi](u) \rangle = \langle f(u), \phi(u) \rangle . \blacksquare
\end{aligned}$$

4.7.2 The Space of Boehmians

In this section, we introduce a space of generalized functions that will be needed in the extension of the FrFT. The elements of that space, which are called Boehmians, are more general than Schwartz distributions and regular operators [207, 208]. They were first introduced by J. Mikusiński in [209] as an extension of the notion of regular operators that was developed earlier by T. Boehme in [43].

Unlike the theory of generalized functions developed by Schwartz [258], Gel'fand and Shilov [108] and Zemanian [361], the theory of Boehmians treats generalized functions more as algebraic objects than as elements of the dual space of some testing-function space.

Simply put, Boehmians are equivalence classes of "quotients of sequences" constructed from an integral domain of continuous functions as rational numbers are constructed from the integers.

For Boehmians, the integral domain operations are addition and a binary operation called *convolution*. This convolution operation is not necessarily the same as the standard convolution operation denoted earlier by $*$. This flexibility in choosing the convolution operation is what makes the extension of the FrFT to Boehmians possible.

To make this section self-contained, we shall give a brief introduction to Boehmians since they are not very well known.

Boehmians. Let $\mathcal{G}$ be a complex linear space and $\mathcal{H}$ be a subspace of $\mathcal{G}$. Let $\otimes$ be a binary operation from $\mathcal{G} \times \mathcal{H}$ into $\mathcal{G}$ such that the following conditions are satisfied:

$$\text{If } \phi, \psi \in \mathcal{H}, \text{ then } \phi \otimes \psi \in \mathcal{H} \text{ and } \phi \otimes \psi = \psi \otimes \phi. \tag{4.60}$$

$$\text{If } f \in \mathcal{G} \text{ and } \phi, \psi \in \mathcal{H}, \text{ then } (f \otimes \phi) \otimes \psi = f \otimes (\phi \otimes \psi). \tag{4.61}$$

If $f, g \in \mathcal{G}, \phi \in \mathcal{H}$ and $\lambda \in C$, then

$$(f+g) \otimes \phi = f \otimes \phi + g \otimes \phi \text{ and } \lambda(f \otimes \phi) = (\lambda f \otimes \phi).$$

Let Δ be a family of sequences of elements of $\mathcal{H}$ such that:

$$\text{if } f \in \mathcal{G}, \{\phi_n\} \in \Delta, \text{ and } f \otimes \phi_n = 0 \text{ for all } n \in N, \text{ then } f = 0 \tag{4.62}$$

where N is the set of natural numbers, and

$$\text{if } \{\phi_n\}, \{\psi_n\} \in \Delta, \text{ then } \{\phi_n \otimes \psi_n\} \in \Delta. \tag{4.63}$$

A pair of sequences (f_n, ϕ_n) is called a *quotient of sequences*, and will be denoted by f_n/ϕ_n, if $f_n \in \mathcal{G}$ for $n = 1, 2, ..., \{\phi_n\} \in \Delta$, and $f_n \otimes \phi_m = f_m \otimes \phi_n$ for all $m, n \in N$. Two quotients of sequences f_n/ϕ_n and g_n/ψ_n are *equivalent* if $f_n \otimes \psi_n = g_n \otimes \phi_n$ for every $n \in N$. The equivalence class of f_n/ϕ_n will be denoted by $[f_n/\phi_n]$. The space of equivalence classes of quotients of sequences will be denoted by $\mathcal{B}(\mathcal{G}, \mathcal{H}, \otimes, \Delta)$ and its elements will be called *Boehmians*.

Addition and multiplication by scalars are defined in $\mathcal{B}(\mathcal{G}, \mathcal{H}, \otimes, \Delta)$ as follows:

$$\lambda[f_n/\phi_n] = [\lambda f_n/\phi_n];$$

and

$$[f_n/\phi_n] + [g_n/\psi_n] = [(f_n \otimes \psi_n + g_n \otimes \phi_n)/(\phi_n \otimes \psi_n)].$$

Clearly, $\mathcal{B}(\mathcal{G}, \mathcal{H}, \otimes, \Delta)$ is a vector space. The mapping $f \to [(f \otimes \phi_n)/\phi_n]$ is an isomorphism of $\mathcal{G}$ with a subspace of $\mathcal{B}(\mathcal{G}, \mathcal{H}, \otimes, \Delta)$. It is convenient to treat $\mathcal{G}$ as a subspace of $\mathcal{B}(\mathcal{G}, \mathcal{H}, \otimes, \Delta)$.

If $F = [f_n/\phi_n] \in \mathcal{B}(\mathcal{G}, \mathcal{H}, \otimes, \Delta)$ and $G = [g_n/\psi_n] \in \mathcal{B}(\mathcal{H}, \mathcal{H}, \otimes, \Delta)$, then we can define $F \otimes G$ as follows:

$$F \otimes G = [(f_n \otimes g_n)/(\psi_n \otimes \phi_n)].$$

It is easy to see that $F \otimes G \in \mathcal{B}(\mathcal{G}, \mathcal{H}, \otimes, \Delta)$ and that it is an extension of $\otimes$ from $\mathcal{G} \times \mathcal{H}$

onto $\mathcal{B}(\mathcal{G},\mathcal{H},\otimes,\Delta)\times\mathcal{B}(\mathcal{H},\mathcal{H},\otimes,\Delta)$. Using the identification of $\mathcal{H}$ with a subspace of $\mathcal{B}(\mathcal{G},\mathcal{H},\otimes,\Delta)$, we can write

$$[f_n/\phi_n]\otimes\psi = [(f_n\otimes\psi)/\phi_n],$$

and, in particular, $[f_n/\phi_n]\otimes\phi_k = f_k$.

If $\mathcal{G}$ is equipped with a notion of convergence, then we can define a convergence in $\mathcal{B}$. We say that a sequence of Boehmians $\{F_n\}$ is Δ-*convergent* to a Boehmian F, and we write $F_n \xrightarrow{\Delta} F$, if there exists $\{\delta_n\}\in\Delta$ such that $(F_n - F)\otimes\delta_n\in\mathcal{G}$ for every $n\in N$ and the sequence $\{(F_n - F)\otimes\delta_n\}$ converges to zero in $\mathcal{G}$. It can be proved that, under certain conditions, $\mathcal{B}$ equipped with Δ-convergence is a complete quasi-normed space.

It is often convenient to use a second type of convergence: δ-convergence. We say that a sequence of Boehmians $\{F_n\}$ is δ-*convergent* to a Boehmian F, and we write $F_n\xrightarrow{\delta} F$, if there exists $\{\delta_n\}\in\Delta$ such that $F_n\otimes\delta_k, F\otimes\delta_k\in\mathcal{G}$, for every $k,n\in N$, and the sequence $\{F_n\otimes\delta_k\}$ converges to $F\otimes\delta_k$ in $\mathcal{G}$, for every $k\in N$. It can be proved that the sequence $\{F_n\}$ is Δ-convergent to F if and only if every subsequence of $\{F_n\}$ contains a subsequence that is δ-convergent to F [206, 208].

4.7.3 The Algebraic Method

Now we are in a position to define the fractional Fourier transform of members of a class of generalized functions that was constructed in the preceding subsection. For related work, see [153].

Let $\mathcal{D}$ be the space of all infinitely differentiable functions with compact supports and $\mathcal{S}$ be the Schwartz space of infinitely differentiable, rapidly decreasing functions. In the notation of the preceding subsection, let us take $\mathcal{G}$ to be $L^1(\mathbb{R}), \mathcal{H}=\mathcal{D}$, and $\otimes$ to be the convolution operation $\star$. Furthermore, let Δ be the family of all sequences of real-valued functions, $\{\phi_n(x)\}_{n=1}^{\infty}\subset\mathcal{D}$ such that

1. $\int_{-\infty}^{\infty}\phi_n(x)e^{-ia(\alpha)x^2}\,dx = \frac{1}{c(\alpha)}$, for all n,
2. $\int_{-\infty}^{\infty}|\phi_n(x)|\,dx < M$, for some M *and* all n,
3. $\sigma_n(\delta)=\max_{|x|\geq\delta}|\phi_n(x)|\to 0$ as $n\to\infty$, for each $\delta>0$.

Lemma 1. *Let $\{\phi_n\}$ and $\{\psi_n\}$ be two sequences in Δ. Then, $\{\phi_n\star\psi_n\}$ is also in Δ.*

Proof *1) Let $k_n(x)=(\phi_n\star\psi_n)(x)$. From the definition of the FrFT, it follows that*

$$\mathcal{F}_\alpha[f](0)=\int_{-\infty}^{\infty}f(t)K_\alpha(t,0)\,dt = c(\alpha)\int_{-\infty}^{\infty}f(t)e^{-ia(\alpha)t^2}dt.$$

Thus, by Theorem 11, we have

$$\mathcal{F}_\alpha[k_n](u)=e^{ia(\alpha)u^2}\mathcal{F}_\alpha[\phi_n](u)\mathcal{F}_\alpha[\psi_n](u),$$

and therefore

$$\mathcal{F}_\alpha[k_n](0)=\mathcal{F}_\alpha[\phi_n](0)\mathcal{F}_\alpha[\psi_n](0)=c(\alpha)\int_{-\infty}^{\infty}k_n(t)e^{-ia(\alpha)t^2}dt=1,$$

which is equivalent to (1).

2) From the definition of k_n it follows

$$\int_{-\infty}^{\infty} |k_n(u)|\,du \leq \int_{-\infty}^{\infty} du \int_{-\infty}^{\infty} |\phi_n(t)||\psi_n(u-t)|\,dt \leq M^2, \textit{ for all } n.$$

3) The proof of (3) is similar. ■

In view of Lemma 1, one can verify that the space $\mathcal{B}(L^1(\mathbb{R}),\mathcal{D},\star,\Delta)$ is a space of *Boehmians* and will be called the space of integrable *Boehmians*.

Now we can define the fractional Fourier transform of an integrable *Boehmian*.

Definition 20. *The fractional Fourier transform with angle α of an integrable Boehmian, $[f_n/\phi_n]$ is defined as*

$$\mathcal{F}_\alpha[f_n/\phi_n] = [\mathcal{F}_\alpha[f_n]/\mathcal{F}_\alpha[\phi_n]].$$

Before we show that this definition makes sense, let us observe that if f is in $L^1(\mathbb{R})$, then $e^{\pm ia(\alpha)x^2} f$ is also in $L^1(\mathbb{R})$, and similarly if ϕ is in $\mathcal{D}$, then $e^{\pm ia(\alpha)x^2}\phi$ is also in $\mathcal{D}$. Since the FrFT of an integrable function f is a continuous function that vanishes at $\pm\infty$, it is easy to see that the FrFT of a function ϕ in $\mathcal{D}$ is in the Schwartz space $\mathcal{S}$.

Again in the notation of the preceding subsection, let us take $\mathcal{G}$ to be the space of all continuous functions, $C(\mathbb{R})$ and $\mathcal{H} = \mathcal{S}$. The operation $\otimes$ will be defined as follows:

$$F(u) \otimes G(u) = e^{ia(\alpha)u^2} F(u)G(u).$$

The set Δ, which will be denoted by Δ_p in this case, is defined as the set of all sequences $\{\Phi_n\}$, where $\Phi_n \in \mathcal{S}$ and Φ_n is the FrFT of a function $\phi_n \in \mathcal{D}$ with $\{\phi_n\} \in \Delta$. Now one can verify that $(\mathcal{S},\otimes)$ is a commutative ring with the mapping $\otimes : C(\mathbb{R}) \times \mathcal{S} \to C(\mathbb{R})$ satisfying properties (4.60)– (4.63) listed in preceding subsection.

Furthermore, we need to show that if $\{\Phi_n\}$ and $\{\Psi_n\}$, are in Δ_p, then $\{\Phi_n \otimes \Psi_n\}$, is also in Δ_p. But this follows immediately from the fact that $e^{\pm ia(\alpha)u^2}\Phi_n(u)\Psi_n(u)$ is in $\mathcal{S}$, whenever $\Phi_n(u)$ and $\Psi_n(u)$ are in $\mathcal{S}$ and that

$$(\Phi_n \otimes \Psi_n)(u) = e^{ia(\alpha)u^2} \mathcal{F}_\alpha[\Phi_n](u)\mathcal{F}_\alpha[\Psi_n](u) = \mathcal{F}_\alpha[\phi_n \star \psi_n].$$

Finally, we must show that Definition 20 is independent of the representation of a *Boehmian*. Let $[f_n/\phi_n]$ and $[g_n/\psi_n]$ be two different representations of the same integrable *Boehmians*, i.e., $f_n \star \psi_n = g_n \star \phi_n$ for all n. Then

$$\mathcal{F}_\alpha[f_n \star \psi_n](u) = e^{ia(\alpha)u^2} \mathcal{F}_\alpha[f_n]\mathcal{F}_\alpha[\psi_n],$$

and

$$\mathcal{F}_\alpha[g_n \star \phi_n](u) = e^{ia(\alpha)u^2} \mathcal{F}_\alpha[g_n]\mathcal{F}_\alpha[\phi_n],$$

which implies that

$$\mathcal{F}_\alpha[f_n \otimes \psi_n](u) = \mathcal{F}_\alpha[g_n \otimes \phi_n](u),$$

or equivalently

$$\mathcal{F}_\alpha[f_n/\phi_n] = \mathcal{F}_\alpha[g_n/\psi_n].$$

Now putting all these facts together, we obtain the following theorem.

Theorem 24. *The class $\mathcal{A} = (C(\mathbb{R}), \mathcal{S}, \otimes, \Delta_p)$ is a class of Boehmians and that the fractional Fourier transform is a linear transformation from the class of integrable Boehmians, $\mathcal{B}$, into the class $\mathcal{A}$.*

4.8 APPLICATIONS OF THE FRACTIONAL FOURIER TRANSFORM

We close this chapter by highlighting some applications of the fractional Fourier transform and refer the interested reader to other sources for details. As we pointed out before, the fractional Fourier transform was originally introduced by Wiener and Namias to solve problems in quantum mechanics, but in the early 1990's some physical implementations of the transform were realized in the field of optics. After its introduction in optics, the concept of fractional transforms started to receive considerable attention in signal analysis and other fields.

The Fourier transform plays an important role in both signal processing and optics, and as such, it is no surprise that the fractional Fourier transform, which is a generalization of the Fourier transform, has found many applications in those two fields.

In signal processing, the Fourier transform $\hat{f}(w)$ of a signal $f(t)$ represents the frequency content of the signal which may be displayed as the spectrum analyzer of the signal on hi-fi systems. It is also used in a number of time-frequency representations of signals [3, 4, 15, 16, 249, 268, 366].

In optics, historically two of the most important integral transforms used were the Fourier and the Fresnel integral transforms, but more recently the fractional Fourier transform has proved to be another important transform in the field. It has been shown that the Fresnel transform is equivalent to a scaled version of the FrFT. The Fourier transform models diffraction patterns such as the Fraunhofer diffraction, while the Fresnel transform models the Fresnel-Kirchoff diffraction. Both transforms can be realized by using lenses and free space propagation.

In an optical system with several lenses and using a point source for illumination, one first observes the Fourier transform of the object at the image of the point source, then an inverted image of the object, followed by an inverted Fourier transform, and then the image of the object and so on. This is a physical implementation of Eq. (4.3).

In the simplest case, when the Fourier transform is observed at the focal plane, one may argue that whatever is being observed halfway between the lens and the focal plane may be called the one half Fourier transform. This was one of the key ideas behind the physical representation of the fractional Fourier transform. This simple idea led to considerable progress in the field that has been demonstrated by several optical implementations of the fractional Fourier transform using lenses [203]; see also [9, 10, 11, 44, 148, 201, 320]

Although the fractional Fourier transform has been experimentally realized by optical systems consisting of lenses and filters, its main utility has been accentuated in modeling ray propagation in graded-index media (GRIN) [202]. In a quadratic graded index media (GRIN) the refractive index distribution of such a medium is given by

$$n^2(r) = n_1^2\left[1-(n_2/n_1)r^2\right],$$

where $r^2 = x^2 + y^2$ is the radial distance from the optical axis and n_1 and n_2 are the GRIN medium parameters. It is known that a parallel bundle of rays will be focused at a distance $L = (\pi/2)\sqrt{n_1/n_2}$ away from the input plane. Therefore, if an image represented by a function $f(x,y)$ is presented at the input plane $z = 0$, then the Fourier transform $\hat{f}(u,v)$ is observed at the plane $z = L$. That is the plane $z = L$ is the focal plane . Since the system is uniform in the axis direction, it is reasonable to call the light distribution at distance $z = \alpha L$, where $0 < \alpha < 1$, the fractional Fourier transform of order α. In other words, the fractional Fourier transform $\mathcal{F}_\alpha[f]$ can be physically observed as the functional form of the scalar light distribution at $z = \alpha L$; see [202].

In addition to its applications in optics, the FrFT has many applications in digital and image processing, in particular in filtering and image restoration. Filtering in the fractional Fourier transform domain is especially useful when the noise or distortion is of a chirp nature and the Wigner distribution of the signal and the noise overlap in some configuration but can be separated by rotation. In other words, the FrFT is used to separate a signal from its noisy component if the Wigner distributions of the signal and the noise are distinct and can be separated by appropriate rotations in the time-frequency domain. Furthermore, the FrFT is also used in optical encryption [280].

CHAPTER 5

Shift-Invariant and Sampling Spaces of the Fractional Fourier Transform

5.1 INTRODUCTION

Shift-invariant spaces play an important role in sampling theory, multiresolution analysis, and many other areas of signal and image processing. A special class of the shift-invariant spaces is the class of sampling spaces in which functions are determined by their values on a discrete set of points. One of the tools used in the study of sampling spaces is the Zak transform which is also related to the Poisson summation formula. A common thread among all these notions is the Fourier transform.

In this chapter, we extend some of these concepts to the fractional Fourier transform (FrFT) domain. First, we introduce two definitions of the discrete fractional Fourier transform and two semi-discrete fractional convolutions associated with them. We employ these definitions to derive necessary and sufficient conditions pertaining to FrFT domain, under which integer shifts of a function form an orthogonal basis or a Riesz basis for a shift-invariant space. The study of shift-invariant spaces and Riesz basis in the fractional Fourier transform domain leads us naturally to introduce the fractional Zak transform.

5.2 BASIC DEFINITIONS

Shift-invariant spaces (SIS) have been the focus of many research papers in recent years because of their close connection to wavelets and multiresolution analysis [84, 144, 302, 303]. They are subspaces of $L^2(\mathbb{R})$ that are invariant under integer translates. Typically, they are taken as the closure of sets of the form

$$\mathcal{V}(\phi_1,\ldots,\phi_r) = \left\{ f : f(t) = \sum_{p=1}^{r} \sum_{n=-\infty}^{+\infty} c_p[n]\,\phi_p(t-n)\,,\ \phi_p \in L^2(\mathbb{R}), \sum_{p=1}^{r} \sum_{n=-\infty}^{+\infty} |c_p[n]|^2 < \infty \right\}.$$

DOI: 10.1201/9781003089353-5

The functions ϕ_p are called the generators of the space. In many signal processing applications, it is of interest to represent a signal as a linear combination of shifted versions of some basic function φ that generates a stable basis for a space. Therefore, for simplicity, we consider spaces generated by one generator. That is

$$\mathcal{V}(\varphi)=\left\{f(t)=\sum_{n=-\infty}^{+\infty} c[n]\varphi(t-n), \varphi\in L^2(\mathbb{R}), \{c[n]\}\in\ell_2\right\}. \tag{5.1}$$

The closure of $\mathcal{V}(\varphi)$ is a closed subspace of $L^2(\mathbb{R})$. Furthermore, it is shift-invariant in the sense that for any $f\in\mathcal{V}(\varphi)$, its shifted version, $f(\cdot-k)\in\mathcal{V}(\varphi)$, $k\in\mathbb{Z}$, where $\mathbb{Z}$ denotes the set of integers.

For the basis functions to be stable, it is imperative that the family of functions $\{\varphi(t-n)\}_{n=-\infty}^{\infty}$ forms a Riesz basis or equivalently, there exists two positive constants $0<\eta_1,\eta_2<\infty$ such that

$$\forall c\in\ell_2,\quad \eta_1\|c\|_{\ell_2}^2\leqslant\left\|\sum_{n=-\infty}^{\infty} c[k]\varphi(t-k)\right\|_{L^2}^2\leqslant\eta_2\|c\|_{\ell_2}^2 \tag{5.2}$$

where $\|c\|_{\ell_2}^2$ is the squared ℓ_2-norm of the sequence. We define the Fourier transform of $h(t)$ by $\widehat{h}(\omega)=\frac{1}{\sqrt{2\pi}}\int_{-\infty}^{+\infty}h(t)e^{i\omega t}dt$. Recall, the Fourier domain equivalent of (5.2) is

$$\eta_1\leqslant\sum_{n=-\infty}^{+\infty}|\widehat{\varphi}(\omega+2\pi n)|^2\leqslant\eta_2. \tag{5.3}$$

The ratio $\rho=\frac{\eta_2}{\eta_1}$ is called the *condition number* of the Riesz basis. The basis is shift-orthonormal or a tight frame if $\rho=1$, [94, 114], where shift-orthonormality means that $\langle\varphi,\varphi(\cdot-k)\rangle=\delta_k$, where $\langle x,y\rangle=\int_{-\infty}^{+\infty}x(t)y^*(t)dt$ is the L^2-inner product and

$$\delta_k=\begin{cases}1, & \text{if } k=0\\ 0, & \text{if } k\neq 0\end{cases}.$$

If $\{\varphi(t-n)\}_{n=-\infty}^{\infty}$ is an orthonormal basis and the space $\mathcal{V}(\varphi)$ is a reproducing-kernel Hilbert space with reproducing kernel $K(x,y)$, then

$$K(x,y)=\sum_{n=-\infty}^{+\infty}\varphi(x-n)\varphi(y-n), \tag{5.4}$$

so that $f(x)=\int_{\mathbb{R}}f(y)K(x,y)dy$.

The following presentation is based on the authors' work in [36].

Definition 21. (Semi-discrete Convolution) *Let $\phi\in L^2(\mathbb{R})$. The semi-discrete convolution operator $*$ is defined as a linear map from ℓ_2 into $L^2(\mathbb{R})$ such that,*

$$\phi*: c\longmapsto(\phi*c)(t)=\sum_{n=-\infty}^{+\infty}c[n]\phi(t-n).$$

The range of this map is the shift-invariant space with the generator ϕ. In the Fourier domain,

$$F\{\phi * c\} = \widehat{\phi}(\omega)\,\widehat{C}(\omega), \text{where } \widehat{C}(\omega) = \sum_{k=-\infty}^{+\infty} c[k]\, e^{i\omega k},$$

where F denotes the Fourier transformation.

As a result, any function in the space $\mathcal{V}(\phi)$ can be viewed as a semi-discrete convolution of a sequence $\{c[k]\} \in \ell_2$ and $\phi \in L^2(\mathbb{R})$,

$$f(t) = (\phi * c)(t) = \sum_{n=-\infty}^{+\infty} c[n]\,\phi(t-n). \tag{5.5}$$

In light of the shift-invariant space model in (5.1), any signal of interest $f(t) \in \mathcal{V}(\varphi)$ can be represented in form of a linear combination of the integer translates of $\varphi(t)$ that are weighted by coefficients $c[k], k \in \mathbb{Z}$, and in case of the classical Shannon's sampling theorem, we have, $c[n] = f(n), n \in \mathbb{Z}$, where f is bandlimited to $[-\pi, \pi]$ and $\phi(t) = \operatorname{sinc}(t)$.

The generating function is often constrained by the application at hand. Closely related to SIS are sampling spaces which we are about to introduce.

Definition 22. *A sampling space is a space $\mathcal{V}(\phi) \subset L^2(\mathbb{R})$ satisfying the following conditions:*

1. *There exists a function $\psi \in \mathcal{V}(\phi)$ such that $\{\psi(.-k)\}_{k\in\mathbb{Z}}$ is a Riesz basis of $\mathcal{V}(\phi)$.*
2. *For any sequence $\{c_k\}_{k\in\mathbb{Z}} \in \ell_2$, the series $\sum_{k\in\mathbb{Z}} c_k \psi(.-k)$ converges pointwise.*
3. *For every $f \in V(\phi)$, $f(x) = \sum_{k=-\infty}^{+\infty} f(k)\,\psi(x-k)$ converges in the mean and uniformly.*

The last condition may be written as $f(x) = (f(k) * \psi)(x)$.

An example of a SIS that is also a sampling space is the space of bandlimited functions to $[-\pi, \pi]$, where $\phi(x) = \psi(x) = \operatorname{sinc}(x)$,

$$\operatorname{sinc}(z) = \frac{\sin(\pi z)}{\pi z}$$

and the expansion in item (3) above is just the Shannon sampling theorem.

$$f(x) = \sum_{k=-\infty}^{+\infty} f(k)\, \frac{\sin \pi(x-k)}{\pi(x-k)}, \qquad x \in \mathbb{R}, \tag{5.6}$$

where f is bandlimited to $[-\pi, \pi]$. The reproducing kernel is seen from (5.4) to be

$$K(x,y) = \frac{\sin[\pi(x-y)]}{\pi(x-y)}. \tag{5.7}$$

One of the important tools used in the study of sampling spaces is the Zak transform [84, 85, 144, 145, 353]. The Zak transform , which was introduced in quantum mechanics by J. Zak [331] to solve Schrödinger's equation for an electron subject to a periodic potential in a constant magnetic field, may be defined as

$$Z_f(t,\omega) = \sum_{k=-\infty}^{+\infty} f(t+k)e^{-2i\pi k\omega}, \quad f \in L^1(\mathbb{R})\bigcap L^2(\mathbb{R}).$$

It is evident that

$$Z_f(t,\omega+1) = Z_f(t,\omega),\ Z_f(t+1,\omega) = e^{2i\pi\omega} Z_f(t,\omega)$$

and that the Zak transform is a unitary transformation from $L^2(\mathbb{R})$ onto $L^2(Q)$, with

$$\|Z_f\|_{L^2(Q)} = \|f\|_{L^2(\mathbb{R})},$$

where Q is the unit square $Q = [0,1] \times [0,1]$.

To see the connection between the Zak transform and sampling spaces, let us for the sake of convenience define the Fourier transform of f as

$$\hat{f}(w) = \int_{-\infty}^{\infty} f(t)e^{-2\pi iwt}dt$$

so that the inverse transform, whenever it exists, is given by

$$f(t) = \int_{-\infty}^{\infty} \hat{f}(w)e^{2\pi iwt}dw.$$

If f belongs to a sampling space with sampling function $\psi \in L^2 \cap L^1$, it follows that since $f(t) = \sum_k f(k)\psi(t-k)$, we have

$$\hat{f}(w) = \hat{F}(w)\hat{\psi}(w), \text{where } \hat{F}(w) = \sum_k f(k)e^{-2\pi ikw}.$$

Clearly $\hat{F}(w)$ is periodic with period one; hence,

$$\left|\hat{f}(w+k)\right| = |P(w)| \left|\hat{\psi}(w+k)\right|,$$

which, in view of the fact that $P(w) = Z_f(0,w)$, implies that

$$G_f(w) = |Z_f(0,w)|^2 G_\psi(w),$$

where $G_g(w) = \sum_{k\in\mathbb{Z}} |\hat{g}(w+k)|^2$, which is called the Grammian of $g \in L^2(\mathbb{R})$. Therefore, for such a function, we have

$$A|Z_f(0,w)|^2 \leq G_f(w) \leq B|Z_f(0,w)|^2, \tag{5.8}$$

for some $A, B > 0$. From this relation, we obtain

$$A\sum_{k=-\infty}^{+\infty} |f(k)|^2 \leq \int_0^1 G_f(w)dw \leq B\sum_{k=-\infty}^{+\infty} |f(k)|^2, \tag{5.9}$$

and

$$\int_0^1 \frac{\sum_{k\in\mathbb{Z}} \left|\hat{f}(w+k)\right|}{|Z_f(0,w)|} dw = \int_{\mathbb{R}} \left|\hat{\psi}(w)\right| dw < \infty, \tag{5.10}$$

whenever $\hat{\psi} \in L^1(\mathbb{R})$. The Shannon sampling theorem is also known to follow from the Poisson summation formula

$$\sum_{k=-\infty}^{+\infty} f(t+k) = \sum_{k=-\infty}^{+\infty} \hat{f}(k)e^{2\pi ikt}, \tag{5.11}$$

or equivalently

$$\sum_k \hat{f}(w+k) = \sum_k f(k)e^{-2\pi ikw}, \tag{5.12}$$

which, when f is band-limited to $(-1/2, 1/2)$, leads to

$$\sum_k f(k) = \int_{\mathbb{R}} f(t)dt. \tag{5.13}$$

Moreover, due to (5.12), $Z_f(0,w) = \sum_{k=-\infty}^{+\infty} \hat{f}(w+k)$.

Definition 23. *A function f is said to be an interpolating function if $f(0) = 1$ and $f(k) = 0$ for all $k = \pm 1, \pm 2, \ldots$.*

It follows from (5.12) that if f is an interpolating function, then

$$\sum_k \hat{f}(w+k) = 1. \tag{5.14}$$

In our discussion so far, we have presented some multi-faceted applications of systems formed by integer shifts of functions and in which the Fourier transform plays an important role.

In the rest of this chapter, we generalize these results to chirp-modulated shift-invariant spaces. These are spaces in which the generators are modulated by chirps and then translated by integral multiples of a fixed quantity Δ that depends on an angle θ. We study these systems in the sense of the fractional Fourier transform. As it will be apparent, most of these problems are related to finding the equivalent representation of shift-orthogonality and Riesz basis condition in the fractional Fourier transform (FrFT) domain.

In other words, we will derive the equivalent of the Riesz basis condition (5.3) but in the fractional Fourier domain. This will help us to look at different problems which use the same model as in (5.1), in a unified fashion. Having developed the fractional Fourier transform (FrFT) generalization of (5.3), our next step will be to extend some of the aforementioned notions, such as the Zak transform, to the fractional Fourier transform domain.

The rest of the chapter is organized as follows. In Section 5.3, we introduce new definitions that will be used later on, such as the generalized discrete fractional Fourier transform, fractional convolution and semi-discrete fractional convolution. In Section 5.4, we derive our first main result which gives a necessary and sufficient condition pertaining to the fractional Fourier transform domain, under which integer shifts of a function form an orthogonal basis or a Riesz basis. In Section 5.5, we introduce the fractional Zak transform, which is a generalization of the Zak transform, that is more suitable for the fractional Fourier transform domain. In the last section we briefly discuss fractional delay filtering.

5.3 DISCRETE FRACTIONAL FOURIER TRANSFORM AND CONVOLUTION

5.3.1 Discrete Fractional Fourier Transform

For the reader's convenience , let us recall the definition of the fractional Fourier transform or FrFT of a signal or a function, say $x(t) \in L^2(\mathbb{R})$,

$$F^{\theta}[x](\omega) = \widehat{x}_{\theta}(\omega) = \int_{-\infty}^{\infty} x(t)\mathcal{K}_{\theta}(t,\omega)\,dt \tag{5.15}$$

where

$$\mathcal{K}_{\theta}(t,\omega) = \begin{cases} c(\theta)\cdot e^{-ia(\theta)(t^2+\omega^2)+ib(\theta)\omega t}, & \theta \neq p\pi \\ \delta(t-\omega), & \theta = 2p\pi \\ \delta(t+\omega), & \theta = (2p-1)\pi \end{cases} \tag{5.16}$$

is the transformation kernel with $c(\theta) = \sqrt{\frac{1+i\cot\theta}{2\pi}}$, $a(\theta) = \cot\theta/2$ and $b(\theta) = \csc\theta$. The kernel $\mathcal{K}_{\theta}(t,\omega)$ is parameterized by an angle $\theta \in \mathbb{R}$ and p is some integer. For simplicity, we may write a, b, c instead of $a(\theta), b(\theta)$ and $c(\theta)$ The inverse-FrFT with respect to angle θ is the FrFT with angle $-\theta$, given by

$$x(t) = \int_{-\infty}^{\infty} \widehat{x}_{\theta}(\omega)\mathcal{K}_{-\theta}(t,\omega)\,d\omega. \tag{5.17}$$

Definition 24. ([239]) *The discrete time fractional Fourier transform or the DTFrFT of a sequence, $f[n] \in \ell_2$, $n \in \mathbb{Z}$ with sampling time $t = n, n \in \mathbb{Z}$, is given by,*

$$\widehat{F}_{\theta}(\omega) = c(\theta) \sum_{n=-\infty}^{+\infty} f[n] \underbrace{e^{-ia(\theta)(\omega^2+n^2)+ib(\theta)n\omega}}_{\mathcal{K}_{\theta}(n,\omega)}. \tag{5.18}$$

The inverse discrete time fractional Fourier transform (IDTFrFT) of $\widehat{F}_{\theta}(\omega)$ is,

$$f(n) = \int_{\langle\Omega\rangle} \widehat{F}_{\theta}(\omega)\mathcal{K}_{-\theta}(n,\omega)d\omega \tag{5.19}$$

and $\langle\Omega\rangle = 2\pi|b(\theta)|$ denotes the integral width.

Next, we define up-chirping and down-chirping operations.

Definition 25. (Chirp-modulation) *Let $\lambda_{\theta}(\cdot) = e^{ia(\cdot)^2}$ be a domain independent modulation function. The chirp modulated and demodulated versions of a signal $f(t)$ are respectively, denoted by,*

$$\underbrace{\overrightarrow{f}(t) = f(t)\lambda_{\theta}(t)}_{\text{Modulation/Up-chirping}} \text{ and } \underbrace{\overleftarrow{f}(t) = f(t)\lambda_{\theta}^{*}(t)}_{\text{Demodulation/Down-chirping}}$$

where $*$ in the super-script denotes the complex conjugation.

Note that, chirp-modulation is a unitary operation, that is,

$$\left\langle \overrightarrow{f}(t), \overrightarrow{g}(t) \right\rangle = \langle f(t), g(t) \rangle. \tag{5.20}$$

5.3.2 Fractional Convolution

In this section, we develop the semi-discrete fractional convolution operator. The fractional convolution operator will be denoted by "$*_\theta$." At $\theta = \frac{\pi}{2}$ the fractional convolution operator assumes its conventional definition. For simplification purposes, we will denote the (usual) convolution operator by "$*$" instead of "$*_{\pi/2}$." The following definition will be needed in the sequel.

Definition 26. *The fractional convolution of two input signals, $x(t)$ and $y(t)$ is defined as [348]:*

$$(x *_\theta y)(t) = d(\theta)\lambda_\theta(t) \cdot \underbrace{([x(t)\lambda_\theta^*(t)] * [y(t)\lambda_\theta^*(t)])}_{\text{convolution of modulated inputs}} \tag{5.21}$$

$$= d(\theta)\,\lambda_\theta(t)\left\{\overleftarrow{x}(t) * \overleftarrow{y}(t)\right\},$$

where $d(\theta) = \sqrt{1+i\cot\theta}$.

The definition of fractional convolution is consistent with the FrFT duality principal —fractional convolution in time domain results in multiplication of fractional spectrum followed by down-chirping operation in FrFT domain, that is

$$F^\theta\left[x(t) *_\theta y(t)\right] = \lambda_\theta(\omega) \cdot \widehat{x}_\theta(\omega)\widehat{y}_\theta(\omega); \quad \text{see [348]}\,.$$

By combining the fractional convolution and Eq. (5.5), we obtain the semi-discrete fractional convolution .

Definition 27. *Let $\phi(t) \in L^2(\mathbb{R})$ be a compactly supported function whose infinite linear shifts span a space $S(\phi) = \overline{span}\left\{\sum_{k=-\infty}^{+\infty} c(k)\phi(t-k)\right\}$. We define the semi-discrete fractional convolution $*_\theta$ of a sequence $p(n) \in \ell_2, n \in \mathbb{Z}$ and $\phi(t)$ by,*

$$h(t) = (p *_\theta \phi)(t) = c(\theta)\,\lambda_\theta(t) \sum_{n=-\infty}^{\infty} \overleftarrow{p}(n)\overleftarrow{\phi}(t-n). \tag{5.22}$$

This is a generalization of Eq. (5.5). We will show that the fractional Fourier transform spectrum of $h(t)$ is

$$\widehat{h}_\theta(\omega) = F^\theta\left[p(k) *_\theta \phi(t)\right] = \lambda_\theta(\omega)\,\widehat{P}_\theta(\omega)\,\widehat{\phi}_\theta(\omega), \tag{5.23}$$

with $\widehat{P}_\theta(\omega)$ being given by Eq. (5.18) and $\widehat{\phi}_\theta$ is the fractional Fourier transform of ϕ. In fact, we will prove a more general result that deals with non-uniform shifts by introducing an analogue of $\hat{P}_\theta(w)$, denoted by $\hat{\hat{P}}_\theta(w)$, for non-uniform shifts.

Definition 28. *Let $\{t_k\}_{k=-\infty}^{\infty}$ be a sequence of real numbers such that $t_k < t_{k+1}$, and*

$$\lim_{k\to\infty} t_k = \infty, \qquad \lim_{k\to-\infty} t_k = -\infty.$$

We define the Generalized Discrete Fractional Fourier Transform of a sequence $\{p(k)\}$ as

$$\hat{\hat{P}}_\theta(w) = \sum_{k=-\infty}^{\infty} p(k)\mathcal{K}_\theta(t_k, w) = c(\theta) \sum_{k=-\infty}^{\infty} p(k)\,e^{-ia(t_k^2+w^2)+ibt_k w} \tag{5.24}$$

and define the convolution of $\{p(k)\}$ *with* $\phi \in L^2(\mathbb{R})$ *as*

$$h(t) = (p(k) \star_\theta \phi)(t) = c(\theta)\, \lambda_\theta(t) \sum_{k=-\infty}^{\infty} \overleftarrow{p}(k) \overleftarrow{\phi}(t - t_k),$$

or

$$\overleftarrow{h}(t) = \lambda_\theta^*(t) h(t) = c(\theta) \sum_{k=-\infty}^{\infty} \overleftarrow{p}(k) \overrightarrow{\phi}(t - t_k),$$

where

$$\overleftarrow{p}(k) = p(k)^{-iat_k^2}, \overleftarrow{\phi}(t) = \phi(t) e^{-iat^2},$$

and $a = a(\theta)$.

Lemma 2. *Let* p, ϕ *and* h *have the same meaning as above. Then,*

$$\hat{h}_\theta(w) = \lambda_\theta(w) \hat{\hat{P}}_\theta(w) \hat{\phi}_\theta(w), \tag{5.25}$$

where $\hat{\hat{P}}_\theta(w)$ *is given by Eq. (5.24).*

Proof *Consider the fractional Fourier transform of* h

$$\begin{aligned}
\hat{h}_\theta(w) &= \int_{-\infty}^{\infty} h(t) \mathcal{K}_\theta(t, w) dt \\
&= c^2(\theta) \sum_{k=-\infty}^{\infty} \overleftarrow{p}(k) \int_{-\infty}^{\infty} \phi(t - t_k) e^{-ia(t-t_k)^2} e^{-iaw^2 + ibtw} dt \\
&= c^2(\theta) \sum_{k=-\infty}^{\infty} p(k) e^{-iat_k^2} \int_{-\infty}^{\infty} \phi(x) e^{-ia(x^2+w^2)+ibw(x+t_k)} dx \\
&= c(\theta) \sum_{k=-\infty}^{\infty} p(k) e^{-ia(t_k^2+w^2)+ibwt_k} e^{iaw^2} \hat{\phi}_\theta(w) \\
&= \lambda_\theta(w) \hat{\hat{P}}_\theta(w) \hat{\phi}_\theta(w),
\end{aligned}$$

where $\hat{\hat{P}}_\theta$ *is given by (5.24). When* $t_k = k$*, Eq. (5.25) reduces to (5.23) which is the usual DTFrFT.*

5.4 SHIFT-INVARIANCE IN THE FRFT DOMAIN

In this section, we derive conditions on the existence of orthonormal and Riesz bases in shift-invariant spaces in the fractional Fourier transform domain. The next theorem gives such a condition on the scaling function to generate a Riesz basis. Throughout the rest of this chapter, we will denote by t_k the sequence $t_k = 2k\pi \sin\theta$.

Theorem 25. *Let* $\{p(n)\} \in \ell_2, \phi \in L^2(\mathbb{R})$ *and consider the chirp-modulated shift-invariant subspaces of* $L^2(\mathbb{R})$

$$\mathcal{V}_\theta(\phi) = closure \left\{ c(\theta)\, \lambda_\theta(t) \sum_{k=-\infty}^{+\infty} \overleftarrow{p}[k] \overleftarrow{\phi}(t - k) \right\}$$

where $t_k = k\Delta$*, and* $\Delta = 2\pi \sin\theta = 2\pi / b(\theta)$.

Then $\{\overleftarrow{\phi}(t-k)\}$ *is a Riesz basis for* $\mathcal{V}_\theta$ *if and only if there exist two positive constants* $\eta_1, \eta_2 > 0$ *such that*

$$\eta_1 \leq \sum \left|\hat{\phi}_\theta(w+t_k)\right|^2 \leq \eta_2 \tag{5.26}$$

$\forall w \in [0, \Delta]$, *and* $\hat{\phi}_\theta$ *is the fractional Fourier transform of* ϕ.

Proof *Let* $f \in \mathcal{V}_\theta(\phi)$. *Hence* $f(t) = (p *_\theta \phi)(t)$, *and by Eq. (5.23) it follows that*

$$\hat{f}_\theta(w) = \lambda_\theta(w)\hat{P}_\theta(w)\hat{\phi}_\theta(w),$$

where $\hat{P}_\theta(w)$ *is given by (5.18), i.e.,*

$$\hat{P}_\theta(w) = c(\theta) \sum_{k=-\infty}^{+\infty} p(k)e^{-ia(\theta)(w^2+k^2)+ib(\theta)kw}.$$

Thus,

$$\begin{aligned}
\left\|\widehat{f}_\theta(\omega)\right\|^2_{L^2(R)} &= \int_{-\infty}^{+\infty} \left|\lambda_\theta(\omega)\widehat{P}_\theta(\omega)\widehat{\phi}_\theta(\omega)\right|^2 d\omega \\
&= \sum_{k=-\infty}^{+\infty} \int_{t_k}^{t_{k+1}} \left|\widehat{P}_\theta(\omega)\widehat{\phi}_\theta(\omega)\right|^2 d\omega \\
&= \sum_{k=-\infty}^{+\infty} \int_0^{\Delta} \left|\widehat{P}_\theta(\omega+t_k)\right|^2 \left|\widehat{\phi}_\theta(\omega+t_k)\right|^2 d\omega.
\end{aligned}$$

But since $e^{ib(\theta)mt_k} = e^{2i\pi mk} = 1$, *we have*

$$\begin{aligned}
\hat{P}_\theta(w+t_k) &= c(\theta) \sum_{=-\infty}^{\infty} p(m)e^{-ia\left[(w+t_k)^2+m^2\right]+ibm(w+t_k)} \\
&= c(\theta) \sum_{m=-\infty}^{\infty} p(m)e^{-ia\left[w^2+m^2\right]+ibmw}e^{-ia\left[2wt_k+t_k^2\right]} \\
&= \hat{P}_\theta(w)e^{-ia(w+t_k)^2+iaw^2}.
\end{aligned}$$

Hence $\left|\hat{P}_\theta(w+t_k)\right| = |\hat{P}_\theta(w)|$, *and it follows that*

$$\begin{aligned}
\left\|\hat{f}_\theta\right\|^2 &= \sum_{k=-\infty}^{+\infty} \int_0^{\Delta} \left|\hat{P}_\theta(w)\right|^2 \left|\hat{\phi}_\theta(w+t_k)\right|^2 dw \\
&= \int_0^{\Delta} \left|\hat{P}_\theta(w)\right|^2 \tilde{G}_{\phi;\theta}(w)dw,
\end{aligned}$$

where $\tilde{G}_{\phi;\theta}(w) = \sum_k |\hat{\phi}_\theta(w+t_k)|^2$ *is a shifted Grammian of* ϕ *with angle* θ*. Moreover*

$$\begin{aligned} ||\hat{P}_\theta||^2{}_{L^2_{(0,\Delta)}} &= \int_0^\Delta \sum_{k,m} p(k)p^*(m)\mathcal{K}_\theta(k,w)\mathcal{K}_{-\theta}(m,w)dw \\ &= |c(\theta)|^2 \sum_{k,m} p(k)p^*(m)e^{-ia(k^2-m^2)} \int_0^\Delta e^{-ibw(m-k)}dw \\ &= |c(\theta)|^2 \Delta \sum_k |p(k)|^2 = \sum_k \left|\overleftarrow{p}(k)\right|^2 = \|p(k)\|^2. \end{aligned}$$

Because $|c(\theta)|^2 = c(\theta)c^*(\theta) = c(\theta)c(-\theta) = \frac{\csc\theta}{2\pi}$*, we have* $|c(\theta)|^2 \Delta = 1$*. Thus*

$$\left\|\hat{f}_\theta\right\|^2 = \|f\|^2 = \int_0^\Delta \left|\hat{P}_\theta(w)\right|^2 \tilde{G}_{\phi;\theta}(w)dw,$$

and if $\eta_1 \leq \tilde{G}_{\phi;\theta}(w) \leq \eta_2$*, we have*

$$\eta_1 \|p(n)\|^2 = \eta_1 \left\|\overleftarrow{p}(n)\right\|^2 \leq \|f\|^2 \leq \eta_2 \|p(n)\|^2 = \eta_2 \left\|\overleftarrow{p}(n)\right\|^2,$$

which, in view of (5.2), shows that $\{\overleftarrow{\phi}(t-k)\}$ *is a Riesz basis.*

Corollary 2. *Let* $\{p(n)\} \in \ell_2$*,* $\phi \in L^2(\mathbb{R})$ *and consider the chirp-modulated shift-invariant subspaces of* $L^2(\mathbb{R})$

$$\begin{aligned} \mathcal{V}(\phi) = closure\Big\{ & f \in L^2 : f(t) = (p \star_\theta \phi)(t); \\ & = c(\theta)\lambda_\theta(t) \sum_{k=-\infty}^{+\infty} \overleftarrow{p}(k)\overleftarrow{\phi}(t-t_k) \Big\} \end{aligned}$$

Then $\{\overleftarrow{\phi}(t-t_k)\}$ *is a Riesz basis for* $\mathcal{V}(\phi)$ *if and only if there exist two positive constants* $\eta_1, \eta_2 > 0$ *such that*

$$\eta_1 \leq \sum_{k=-\infty}^{+\infty} \left|\hat{\phi}_\theta(w+k)\right|^2 \leq \eta_2 \tag{5.27}$$

$\forall w \in [0,\Delta]$*, and* $\hat{\phi}_\theta$ *is the fractional Fourier transform of* ϕ*.*

Proof *Since* $f(t) = (p \star_\theta \phi)(t)$*, we have from Eq. (5.25),*

$$\hat{f}_\theta(w) = \lambda_\theta(w)\hat{\hat{P}}_\theta(w)\hat{\phi}_\theta(w);$$

and hence $|\hat{f}_\theta(w)|^2 = |\hat{\hat{P}}_\theta(w)|^2|\hat{\phi}_\theta(w)|^2$*, where* $\hat{\hat{P}}_\theta(w)$ *is given by Eq. (5.24). The proof is very similar to that of Theorem 25, except one uses the fact that* $|\hat{\hat{P}}_\theta(w)|$ *is periodic with period one and that*

$$\int_0^1 \left|\hat{\hat{P}}_\theta(w)\right|^2 dw = \sum_{k=-\infty}^{+\infty} |p[k]|^2 = \|p\|^2,$$

and replaces the Grammian $\tilde{G}_{\phi,\theta}$ with $G_{\phi,\theta}$.

To generate an orthonormal basis for the space $\mathcal{V}(\phi)$ generated by $\{\overleftharpoon{\phi}(t-k)\}_{k=-\infty}^{\infty}$, we borrow the following idea from wavelet theory. Consider the function $g(t)$ given by

$$\hat{g}_\theta(w) = \frac{\hat{\phi}_\theta(\omega)}{\sqrt{\tilde{G}_{\phi;\theta}(w)}}.$$

It is easy to see that Δ–translates of g generate an orthonormal basis. For, since $\tilde{G}_{\phi;\theta}(w+t_k) = \tilde{G}_{\phi;\theta}(w)$, we have

$$\sum_{k=-\infty}^{\infty} |g_\theta(w+t_k)|^2 = \frac{1}{\tilde{G}_{\phi;\theta}(w)} \sum_{k=-\infty}^{\infty} |\hat{\phi}_\theta(w+t_k)|^2 = 1,$$

which in view of (5.3), completes the proof.

We close this section by an example of $\mathcal{V}(\phi)$. Let $\phi_n(t) = e^{ia(\theta)t^2}\,\text{sinc}(t-n)$ and call $\phi(t) = \phi_0(t)$. The family of functions $\{\phi_n(t)\}_{n\in\mathbb{Z}}$ has the following important properties:
♠ *Shift-Orthonormality:* $\{\phi_n(t)\}_{n\in\mathbb{Z}}$ forms an orthonormal basis. The orthonormality property is easily verified as

$$\langle\phi_n(t), \phi_k(t)\rangle = \delta_{n-k}.$$

♠ *Fractional Bandlimitedness:* The family of functions $\{\phi_n(t)\}_{n\in\mathbb{Z}}$ corresponds to a space of functions which are bandlimited in the FrFT domain. The fundamental member of this family, $\phi(t)$ has a fractional bandlimited spectrum since

$$F^\theta\left\{e^{ia(\theta)t^2}\,\text{sinc}(t)\right\} = c(\theta)\,e^{-ia(\theta)\omega^2}\chi_{[-\pi\sin(\theta),\pi\sin(\theta)]}(\omega),$$

where $\chi_I(\omega)$ is the characteristic function of the set I.

The above example enables us to present Shannon's sampling theorem for the fractional Fourier transform [325, 345, 350] from an approximation theoretic perspective. Let $\mathcal{V}_\theta(\phi)$ be the space spanned by *modulated, weighted* and shifted versions of $\phi(t)$, that is,

$$\mathcal{V}_\theta(\phi) = \left\{ e^{ia(\theta)t^2} \sum_{k=-\infty}^{+\infty} \overleftharpoon{p}[k]\,\text{sinc}(t-k) : \{p[k]\}_k \in \ell_2 \right\}.$$

Indeed, let f be bandlimited to $[-\pi,\pi]$ and $\mathcal{P}$ be the idempotent, orthogonal projection operator $\mathcal{P}_{\mathcal{V}_\theta(\phi)} : L^2 \to \mathcal{V}_\theta(\phi)$ defined by

$$\mathcal{P}_{\mathcal{V}_\theta(\phi)} f = \sum_{n=-\infty}^{+\infty} \langle f,\phi_n\rangle\phi_n, \quad \phi_n(t) = e^{ia(\theta)t^2}\,\text{sinc}(t-n).$$

The modulated coefficients of expansion are computed by,

$$\begin{aligned}\underbrace{\langle f,\phi_n\rangle}_{p[k]e^{-ia(\theta)k^2}} &= \int_{-\infty}^{\infty} f(t)\,\phi_n^*(t)\,dt = \int_{-\infty}^{\infty} \overleftharpoon{f}(t)\,\text{sinc}(t-n)\,dt \\ &= f(t)\,e^{-ia(\theta)t^2} * \text{sinc}(-t)\Big|_{t=n} = f(n)\,e^{-ia(\theta)n^2}\end{aligned}$$

which turns out to be $p[n] = f(n)$, i.e., samples of $f(t)$. Consequently, we have,

$$\begin{aligned}\mathcal{P}_{\mathcal{V}_\theta(\phi)} f &= \sum_{n=-\infty}^{+\infty} \langle f, \phi_n(\cdot)\rangle \phi_n(t) \\ &= e^{ia(\theta)t^2} \sum_{n=-\infty}^{+\infty} f(n) e^{-ia(\theta)n^2} \operatorname{sinc}(t-n),\end{aligned}$$

which is the Shannon's sampling formula for the fractional Fourier transform. Furthermore, due to the projection step, this approximation is optimal in the least square sense. Once the samples are obtained, the reconstruction is simply the semi-discrete fractional convolution between the samples (or coefficients) and the generating function.

5.5 THE FRACTIONAL ZAK TRANSFORM

In this section, we extend the Zak transform and some of its properties to the fractional Fourier transform domain.

Definition 29. *We define the fractional Zak transform with angle θ of a function $f \in L^2(\mathbb{R})$ as*

$$Z_{f,\theta}(t,w) = \frac{1}{\sqrt{\Delta}} \sum_{k\in\mathbb{Z}} f(t+k) e^{-ia(\theta)\left(w^2+k^2\right)+ib(\theta)kw}.$$

Since, $Z_{f,\theta}(t, w+\Delta) = e^{-ia(2w+\Delta)\Delta} Z_{f,\theta}(t,w)$, we have,

$$Z_{f,\theta}(t, w+m\Delta) = e^{-ia(2w+m\Delta)m\Delta} Z_{f,\theta}(t,w), \tag{5.28}$$

Moreover,

$$\begin{aligned}&\int_0^\Delta \left|Z_{f,\theta}(t,w)\right|^2 dw \\ &= \frac{1}{\Delta} \sum_{m,k=-\infty}^{\infty} f(t+k) f^*(t+m) e^{-ia\left(k^2-m^2\right)} \int_0^\Delta e^{-ibw(m-k)} dw \\ &= \sum_{k=-\infty}^{\infty} |f(t+k)|^2.\end{aligned} \tag{5.29}$$

Hence

$$\begin{aligned}\int_0^1 \int_0^\Delta \left|Z_{f,\theta}(t,w)\right|^2 dw dt &= \int_0^1 \sum_{k=-\infty}^{\infty} |f(t+k)|^2 dt \\ &= \sum_{k=-\infty}^{\infty} \int_k^{k+1} |f(x)|^2 dx = ||f||_{L_2}^2.\end{aligned}$$

Thus, $||Z_{f,\theta}||^2_{L^2(Q)} = ||f||^2_{L^2(\mathbb{R})}$, where $Q = [0,1] \times [0,\Delta]$. That is the fractional Zak transform is an isometry between $L^2(\mathbb{R})$ and $L^2(Q)$.

Theorem 26. *Let $\overleftarrow{f}$ belong to the sampling space generated by $\overleftarrow{\psi}$. Then there exists $A, B > 0$ such that*

$$A \leq \frac{\tilde{G}_{f,\theta}(w)}{|Z_{f,\theta}(0,w)|^2} \leq B, \tag{5.30}$$

which is the analogue of Eq. (5.8), and

$$A \sum_{k=-\infty}^{+\infty} |f(k)|^2 \leq \int_0^{\Delta} G_{f;\theta}(w)dw \leq B \sum_{k=-\infty}^{+\infty} |f(k)|^2, \tag{5.31}$$

which is the analogue of Eq. (5.9). Moreover, if $\hat{\psi}_\theta \in L^1(\mathbb{R})$, then

$$\int_0^1 \frac{\sum_{k\in\mathbb{Z}} |\hat{f}_\theta(w+t_k)|}{|Z_{f,\theta}(0,w)|} dw = \int_{\mathbb{R}} |\hat{\psi}_\theta(w)|dw < \infty, \tag{5.32}$$

holds and it is the analogue of Eq. (5.10).

Proof *Since $\overleftarrow{f}$ belongs to the sampling space generated by $\overleftarrow{\psi}$, we have*

$$\overleftarrow{f}(t) = e^{-iat^2} f(t) = c(\theta) \sum_{k=-\infty}^{\infty} \overleftarrow{f}(k)\overleftarrow{\psi}(t-k)$$

or

$$f(t) = c(\theta)\lambda_\theta(t) \sum_{k=-\infty}^{\infty} f(k)e^{-iak^2}\psi(t-k)e^{-ia(t-k)^2}.$$

From the last equation and (5.23), we obtain

$$\hat{f}_\theta(w) = \left(c(\theta) \sum_{k=-\infty}^{\infty} f(k)e^{-ia(k^2+w^2)+ibwk} \right) e^{iaw^2}\hat{\psi}_\theta(w),$$

or $e^{-iaw^2}\hat{f}_\theta(w) = \sqrt{\Delta}c(\theta)Z_{f,\theta}(0,w)\hat{\psi}_\theta(w)$.

Because $|\sqrt{\Delta}c(\theta)|^2 = 1$, and $|Z_f|$ is periodic in w with period Δ; see Eq. (5.28), we obtain

$$|\hat{f}_\theta(w+k\Delta)|^2 = |Z_{f,\theta}(0,w)|^2 \left|\hat{\psi}_\theta(w+k\Delta)\right|^2,$$

Setting $t_k = k\Delta$, we have $\tilde{G}_{f;\theta}(w) = |Z_{f,\theta}(0,w)|^2 \tilde{G}_{\psi;\theta}(w)$, where $\tilde{G}_{h;\theta}(w) = \sum_{k\in\mathbb{Z}} |\widehat{h}_\theta(w+t_k)|^2$. Since

$$\int_0^{\Delta} \tilde{G}_{\psi;\theta}(w)dw = \sum_{k=-\infty}^{+\infty} \int_0^{\Delta} |\hat{\psi}_\theta(w+t_k)|^2 dw = ||\hat{\psi}_\theta||^2_{L_2} < \infty,$$

it follows that $\tilde{G}_{\psi;\theta}(w) < \infty$, almost everywhere. But by (5.26), $\eta_1 \leq \tilde{G}_{\psi,\theta}(w) \leq \eta_2$, and hence (5.30) follows.

By integrating (5.30) and using (5.29), we obtain (5.31). Finally, it follows that,

$$|\hat{f}_\theta(w+t_k)| = |Z_{f,\theta}(0,w)||\hat{\psi}_\theta(w+t_k)|.$$

Hence,

$$\begin{aligned}\int_0^{\Delta} \frac{\sum_{k\in\mathbb{Z}} |\hat{f}_\theta(w+t_k)|}{|Z_{f,\theta}(0,w)|} dw &= \int_0^{\Delta} \sum_{k=-\infty}^{+\infty} \left|\hat{\psi}_\theta(w+t_k)\right| dw \\ &= \int_{\mathbb{R}} |\hat{\psi}_\theta(w)| dw < \infty,\end{aligned}$$

which completes the proof.

5.6 APPLICATIONS: FRACTIONAL DELAY FILTERING

In many problems of practical interest one is interested in the samples of τ–shifted version of a fractional bandlimited signal $f(t)$, that is $f(t-\tau)|_{t=nT}, \tau \in [0,T]$. This is known as fractional delay filtering (FDF) of $f(t)$. In view of the FrFT sampling theory, it is known that the Nyquist rate for sampling frequency modulated signals is lower than the one used in conventional Fourier theory. The τ–delayed samples of $f(t)$ can be obtained as follows:

$$\begin{aligned}f_\tau(k) &= f(t-\tau)|_{t=k} \\ &= e^{ia(\theta)(k-\tau)^2} \sum_{n=-\infty}^{\infty} f(n)\, e^{-ia(\theta)n^2} \operatorname{sinc}(k-\tau-n).\end{aligned}$$

A drawback of this methodology is that in practice one only has finite number of samples of the function of interest and in addition the series has a low convergence rate. The function $f(t)$ evaluated at $t=t_0$ leads to,

$$\begin{aligned}f(t_0)\, e^{-ia(\theta)t_0^2} = f(0) \operatorname{sinc}(t_0) + \\ \frac{\sin(\pi t_0)}{\pi} \sum_{n\neq 0, n\in\mathbb{Z}} (-1)^n \frac{f(n)\, e^{-ia(\theta)n^2}}{t_0-n}.\end{aligned}$$

For more details on this topic, the reader may consult [38].

CHAPTER 6

Two-Dimensional Coupled Fractional Fourier Transform (CFrFT)

6.1 INTRODUCTION

In Chapters 3 and 4, we discussed the historical development and derivation of the one-dimensional fractional Fourier transform, as well as, some of its main properties. The transform has been generalized in a number of different ways. For example, in one direction, the kernel of the transform was replaced by more general kernels leading to more general integral transforms, such as the linear canonical transform and the special affine Fourier transform. In another direction, the transform has been extended to higher dimensions [240, 253, 254].

In this chapter, we will introduce an extension of the fractional Fourier transform to two dimensions and then discuss several of its properties, such as its additivity, convolution theorem, Poisson summation formula, sampling theorem and an extension to a space of generalized functions. We conclude the chapter by introducing another integral transform that is related to the two-dimensional fractional Fourier transform and which is known as the Gyrator transform.

In later chapters, we will present more general transforms, such as the linear canonical and the special affine Fourier transforms which include the fractional Fourier transform as a special case. But first, let us recall that the N-dimensional Fourier transform, denoted by $\hat{f}$ or F of a function f, is defined as

$$F(\mathbf{w}) = \frac{1}{(\sqrt{2\pi})^N} \int_{\mathbb{R}^N} f(\mathbf{x}) e^{i(\mathbf{w}\cdot\mathbf{x})} d\mathbf{x},$$

so that its inverse is given by

$$f(\mathbf{x}) = \frac{1}{(\sqrt{2\pi})^N} \int_{\mathbb{R}^N} F(\mathbf{w}) e^{-i(\mathbf{w}\cdot\mathbf{x})} d\mathbf{w},$$

whenever the integrals exist, where

$$\mathbf{x}, \mathbf{w} \in \mathbb{R}^N, \quad \mathbf{w}\cdot\mathbf{x} = w_1 x_1 + \cdots + w_N x_N, \quad f(\mathbf{x}) = f(x_1, x_2, \ldots, x_N),$$

DOI: 10.1201/9781003089353-6

and

$$\int f(\mathbf{x})d\mathbf{x} = \int \cdots \int f(x_1, x_2, \ldots, x_N)dx_1 \cdots dx_N.$$

One of the useful properties of the Fourier transform is given by its convolution theorem which states that: if $f \in L^s(\mathbb{R}^N), g \in L^t(\mathbb{R}^N)$, then

$$h(\mathbf{x}) = (f * g)(\mathbf{x}) = \frac{1}{(\sqrt{2\pi})^N} \int_{\mathbb{R}^N} f(\mathbf{x} - \mathbf{y})g(\mathbf{y})d\mathbf{y},$$

exists almost everywhere and $h \in L^r(\mathbb{R}^N)$, where

$$\frac{1}{r} = \frac{1}{s} + \frac{1}{t} - 1, \quad r, s, t \geq 1,$$

and

$$H(\mathbf{w}) = F(\mathbf{w})G(\mathbf{w}), \tag{6.1}$$

where F, G, H are the Fourier transform of f, g, h, respectively.

Another important property associated with the Fourier transform is the Poisson summation formula, which in N dimensions, takes the form

$$\sum_{\mathbf{k} \in \mathbb{Z}^N} f(\mathbf{x} + 2\pi\mathbf{k}) = \frac{1}{(2\pi)^{N/2}} \sum_{\mathbf{k} \in \mathbb{Z}^N} e^{-i(\mathbf{k} \cdot \mathbf{x})} \hat{f}(\mathbf{k}), \tag{6.2}$$

where $\mathbf{k} = (k_1, k_2, \ldots, k_N) \in \mathbb{Z}^N$.

A sufficient condition for Eq. (6.2) to hold is

$$|f(\mathbf{x})| \leq C_1 (1 + |\mathbf{x}|)^{-N-\epsilon_1}, \quad \text{and } \left|\hat{f}(\mathbf{w})\right| \leq C_2 (1 + |\mathbf{w}|)^{-N-\epsilon_2}, \tag{6.3}$$

for some $C_1, C_2, \epsilon_1, \epsilon_2 > 0$, [114].

As a prelude to one of the main results of this chapter, let us recall Namias's original idea which we will generalize to higher dimensions. Namias' idea started with the observation that the Hermite functions

$$h_n(x) = e^{-x^2/2} H_n(x)$$

are the eigenfunctions of the Fourier transform with eigenvalues $e^{in\pi/2}$, that is

$$\mathcal{F}_{\pi/2}[h_n(x)] = \mathcal{F}[h_n(x)](\omega) = e^{in\pi/2} h_n(\omega), \tag{6.4}$$

where $H_n(x)$ is the Hermite polynomial of degree n. Namias looked for a family of integral transforms $\{\mathcal{F}_\theta\}$ indexed by a parameter θ such that when $\theta = \pi/2$, $\mathcal{F}_{\pi/2}$ coincides with the ordinary Fourier transform and, in addition, the eigenfunctions of $\mathcal{F}_\theta$ are the Hermite functions but with eigenvalues $e^{in\theta}$, that is

$$\mathcal{F}_\theta [h_n(x)](\omega) = e^{in\theta} h_n(\omega). \tag{6.5}$$

When $\theta = \pi/2$, Eq. (6.5) reduces to Eq. (6.4).

Although Namias' work was confined to the Fourier transform in one variable, his work has been extended to N dimensions in a straightforward fashion by taking the tensor product of N copies of the one-dimensional fractional Fourier transform. It was also shown that in two dimensions, the relationship between the fractional Fourier transform and the Wigner distribution is maintained to some extent but in a simple way. More on this topics can be found in the next chapter.

Therefore, the goals of this chapter are as follows:

1. To introduce a new definition of the two-dimensional fractional Fourier transform $F_{\alpha,\beta}$ that is not a tensor product of two one-dimensional copies of the fractional Fourier transform. We will refer to this transform as the *coupled fractional Fourier transform* (CFrFT). It is a by-product of a generalization of Eq. (6.5) that uses a relatively new family of Hermite functions, known as complex Hermite functions of two variables.

2. To derive properties of the new transform, such as its inversion formula, additivity, convolution theorem, Poisson summation formula, sampling theorem and an extension to a space of generalized functions.

3. To introduce the Gyrator transform and present some of its properties.

The chapter is organized as follows. In the next section we briefly discuss the standard n-dimensional fractional Fourier transform and some of its properties. In Section 6.3, we introduce the coupled two-dimensional fractional Fourier transform and its inversion formula, followed by a presentation of its additive property and convolution formula in Sections 6.4 and 6.5. The Poisson summation formula and the sampling theorem associated with the transform are introduced in Sections 6.6 and 6.7. In Section 6.8, we extend the coupled fractional Fourier transform to a space of generalized functions. The last section of the chapter, Section 6.9 is dedicated to the Gyrator transform which is closely related to the two-dimensional fractional Fourier transform.

6.2 FRACTIONAL FOURIER TRANSFORM IN HIGHER DIMENSIONS

The fractional Fourier transform can be extended to higher dimensions using different methods. In this section, we give a brief presentation of two these methods.

6.2.1 The Direct Product Representation

The fractional Fourier transform in n-variables is defined by taking the tensor product of n copies of the one-dimensional fractional Fourier transforms [220]. That is

$$F_{\theta_1,\ldots,\theta_n}(\omega_1,\ldots,\omega_n)=\int_{\mathbb{R}}\cdots\int_{\mathbb{R}} K_{\theta_1}(t_1,\omega_1)\cdots K_{\theta_n}(t_n,\omega_n)\, f(t_1,\ldots,t_n)dt_1\cdots dt_n, \tag{6.6}$$

where $K_{\theta_i}(t_i,\omega_i)$, $i=1,2,\ldots,n$, is the kernel of the one-dimensional fractional Fourier transform given by (5.16)

In particular, in two dimensions we have

$$F_{\theta_1,\theta_2}(\omega_1,\omega_2)=\int_{\mathbb{R}}\int_{\mathbb{R}} K_{\theta_1}(t_1,\omega_1)\, K_{\theta_2}(t_2,\omega_2)\, f(t_1,t_2)dt_1 dt_2, \tag{6.7}$$

The eigenvalue equation takes the form

$$\mathcal{F}_{\theta_1,\dots,\theta_n}\left[h_{k_1,\dots,k_n}(t_1,\dots,t_n)\right] = e^{i(\theta_1 k_1+\cdots+\theta_n k_n)} h_{k_1,\dots,k_n}(t_1,\dots,t_n), \tag{6.8}$$

where

$$h_{k_1,\dots,k_n}(t_1,\dots,t_n) = h_{k_1}(t_1)\cdots h_{k_n}(t_n),$$

and $h_{k_i}(t_i)$ is the Hermite function of order k_i. In particular, in two dimensions, we have the important relation

$$\mathcal{F}_{\theta_1,\theta_2}\left[h_{k_1,k_2}(t_1,t_2)\right] = e^{i(\theta_1 k_1+\theta_2 k_2)} h_{k_1,k_2}(t_1,t_2) = \left[e^{i(\theta_1 k_1)} h_{k_1}(t_1)\right]\left[e^{i(\theta_2 k_2)} h_{k_2}(t_2)\right]. \tag{6.9}$$

6.2.2 Metaplectic Representation

The fractional Fourier transform can also be viewed from a different perspective suggested by abstract harmonic analysis [101].

Let $p = (p_1,\dots,p_n)$, $q = (q_1,\dots,q_n)$ denote the momentum and position of a particle and (p,q) be its *Phase-Space Representation*. . The Symplectic form on $\mathbb{R}^{2n}$ is defined as

$$[(p,q),(p',q')] = pq' - qp' = \sum_{k=1}^{n} p_k q_k' - \sum_{k=1}^{n} p_k' q_k = (p,q)\cdot\mathcal{J}(p',q'),$$

where

$$\mathcal{J} = \begin{pmatrix} 0 & I \\ -I & 0 \end{pmatrix},$$

and I is the $n \times n$ identity matrix.

The set of all $2n \times 2n$ matrices T that satisfy

$$[T(p,q),T(p',q')] = [(p,q),(p',q')]$$

form the symplectic group $Sp(n,\mathbb{R})$.

Let

$$\mathcal{A} = \begin{pmatrix} A & B \\ C & D \end{pmatrix} \in Sp(n,\mathbb{R}),$$

where A,B,C,D are $n \times n$ matrices. Note that both $\mathcal{J}$, $\mathcal{A}$ are $2n \times 2n$ matrices. It is easy to show that $\mathcal{J}^{-1} = -\mathcal{J}$,

$$\mathcal{A}^* \in Sp(n,\mathbb{R}) \quad \text{and } \mathcal{A}^{-1} = \mathcal{J}\mathcal{A}^*\mathcal{J}^{-1},$$

where $\mathcal{A}^*$ is the Hermitian adjoint of $\mathcal{A}$ which is equal to the transpose of $\mathcal{A}$ when $\mathcal{A}$ is real. Moreover, we have

$$\mathcal{A}^{-1} = \begin{pmatrix} D^* & -B^* \\ -C^* & A^* \end{pmatrix},$$

which yields $A^*D - BC^* = I$.

It is known [101, P. 193] that there exists a unitary operator $\mu(\mathcal{A})$ called the *metaplectic representation* of $Sp(n,\mathbb{R})$ on $L^2(\mathbb{R}^n)$ whose action is given by

$$\mu(\mathcal{A})f(x) = (i)^{n/2}(\det{}^{-1/2}B)\int e^{2\pi i S(x,y)}f(y)dy,$$

where $S(x,y) = -(1/2)xDB^{-1}x + yB^{-1}x - (1/2)yB^{-1}Ay, \quad \det B \neq 0.$

As a special case if

$$\mathcal{A} = \mathcal{A}_\theta = e^{\theta\mathcal{J}} = \begin{pmatrix} \cos\theta I & \sin\theta I \\ -\sin\theta I & \cos\theta I \end{pmatrix}.$$

and $\sin\theta \neq 0$, then

$$\mu(\mathcal{A}_\theta)f(x) = (i\csc\theta)^{n/2}\int e^{-\pi i(\cot\theta)(x^2+y^2)+2\pi i(\csc\theta)xy}f(y)dy,$$

which is the *fractional Fourier transform,* for more details, see [101, 115].

6.3 THE TWO-DIMENSIONAL FRACTIONAL FOURIER TRANSFORM

In this section, we introduce a new two-dimensional fractional Fourier transform that is not a tensor product of two one-dimensional transforms. The construction of the new transform follows Namias's approach as given by Eq. (6.5), but uses a relatively new system of complex Hermite functions of two variables that was introduced by M. Ismail in [138]. To make the chapter self-contained, we will briefly present this system in the next subsection.

6.3.1 Complex Hermite Polynomials

The complex Hermite polynomials $H_{m,n}(z,\bar{z})$ may be defined as [138]

$$H_{m,n}(z_1,z_2) = \sum_{k=0}^{m\wedge n}(-1)^k k!\begin{pmatrix} m \\ k \end{pmatrix}\begin{pmatrix} n \\ k \end{pmatrix} z_1^{m-k}z_2^{n-k}. \tag{6.10}$$

Their generating function is given by

$$\sum_{m,n=0}^{\infty} H_{m,n}(z_1,z_2)\frac{t^m}{m!}\frac{s^n}{n!} = e^{tz_1+sz_2-ts},$$

and they satisfy the following orthogonality relation

$$\frac{1}{\pi}\int_{\mathbb{R}^2} H_{m,n}(x+iy,x-iy)\overline{H}_{p,q}(x+iy,x-iy)e^{-x^2-y^2}dxdy = m!n!\delta_{m,p}\delta_{n,q}. \tag{6.11}$$

They satisfy the three-term recurrence relations

$$\begin{aligned} zH_{p,q}(z,\bar{z}) &= qH_{p,q-1}(z,\bar{z}) + H_{p+1,q}(z,\bar{z}), \\ \bar{z}H_{p,q}(z,\bar{z}) &= pH_{p-1,q}(z,\bar{z}) + H_{p,q+1}(z,\bar{z}). \end{aligned}$$

We also have the important generating function for the product of those Hermite polynomials

$$\sum_{m,n=0}^{\infty} H_{m,n}(z_1,\overline{z}_1)H_{n,m}(z_2,\overline{z}_2)\frac{t^m}{m!}\frac{s^n}{n!} = \frac{1}{1-ts}\exp\left\{\frac{-tsz_1\overline{z}_1+tz_1\overline{z}_2+sz_2\overline{z}_1-tsz_2\overline{z}_2}{1-ts}\right\}. \quad (6.12)$$

Let

$$K(z_1,\overline{z}_1,z_2,\overline{z}_2;s,t) = \frac{1}{\pi(1-ts)}\exp\left\{\frac{-tsz_1\overline{z}_1+tz_1\overline{z}_2+sz_2\overline{z}_1-tsz_2\overline{z}_2}{1-ts}\right\}. \quad (6.13)$$

By multiplying Eq. (6.12) by $\overline{H}_{p,q}(z_2,\overline{z}_2)$ and integrating with respect to the measure $d\omega = e^{-|z_2|^2}dz_2$, where $z_1 = u+iv$, $z_2 = x+iy$, and using the orthogonality relation (6.11), we obtain

$$\begin{aligned}\frac{1}{\pi}\sum_{m,n=0}^{\infty} & H_{m,n}(z_1,\overline{z}_1)\frac{t^m}{m!}\frac{s^n}{n!}\int_{\mathbb{R}^2} H_{n,m}(z_2,\overline{z}_2)\overline{H}_{p,q}(z_2,\overline{z}_2)d\omega \\ &= \int_{\mathbb{R}^2} K(z_1,\overline{z}_1,z_2,\overline{z}_2;s,t)\overline{H}_{p,q}(z_2,\overline{z}_2)d\omega \\ &= \frac{t^q}{q!}\frac{s^p}{p!}q!p!\,H_{q,p}(z_1,\overline{z}_1) = t^q s^p H_{q,p}(z_1,\overline{z}_1). \end{aligned} \quad (6.14)$$

We may now write (6.14) as

$$\begin{aligned} & t^q s^p H_{q,p}(z_1,\overline{z}_1) = \\ & \frac{1}{\pi(1-ts)}\int_{\mathbb{R}^2}\exp\left\{\frac{-tsz_1\overline{z}_1+tz_1\overline{z}_2+sz_2\overline{z}_1-tsz_2\overline{z}_2}{1-ts}\right\}\overline{H}_{p,q}(z_2,\overline{z}_2)d\omega \\ = & \frac{1}{\pi(1-ts)}\int_{\mathbb{R}^2}\exp\left\{\frac{-ts(u^2+v^2+x^2+y^2)+(t+s)(ux+vy)+i(t-s)(vx-uy)}{1-ts}\right\} \\ & \times \overline{H}_{p,q}(x+iy,x-iy)e^{-(x^2+y^2)}dxdy. \end{aligned}$$

In particular, if $t = s = i$, we have

$$\begin{aligned} & (i)^{p+q}H_{q,p}(u+iv,u-iv) = \\ & \quad \frac{1}{2\pi}\int_{\mathbb{R}^2}\exp\left\{\frac{(u^2+v^2+x^2+y^2)+2i(ux+vy)}{2}\right\}\overline{H}_{p,q}(x+iy,x-iy) \\ & \times \; e^{-(x^2+y^2)}dxdy, \end{aligned}$$

which yields

$$\begin{aligned}(i)^{p+q}H_{q,p}(u+iv,u-iv)e^{-(u^2+v^2)/2} &= \frac{1}{2\pi}\int_{\mathbb{R}^2} e^{i(ux+vy)}\overline{H}_{p,q}(x+iy,x-iy) \\ & \times \; e^{-(x^2+y^2)/2}dxdy \end{aligned}$$

But from (6.10) it is easy to see that $\overline{H}_{m,n}(z,\overline{z}) = H_{n,m}(z,\overline{z})$, and hence, we have

$$\begin{aligned}(i)^{p+q} H_{q,p}(z_1,\overline{z}_1) e^{-(u^2+v^2)/2} &= \frac{1}{2\pi}\int_{\mathbb{R}^2} H_{q,p}(x+iy, x-iy) e^{-(x^2+y^2)/2} \\ &\times\ e^{i(ux+vy)}dxdy,\end{aligned}$$

or

$$(i)^{p+q} h_{q,p}(z_1,\overline{z}_1) = e^{i(p+q)\pi/2} h_{q,p}(z_1,\overline{z}_1) = \frac{1}{2\pi}\int_{\mathbb{R}^2} h_{q,p}(z_2,\overline{z}_2) e^{i(ux+vy)}dxdy, \quad (6.15)$$

where

$$h_{q,p}(z,\overline{z}) = H_{q,p}(z,\overline{z}) e^{-|z|^2/2},$$

is the Hermite function of two variables. Eq. (6.15) shows that the Hermite functions are the eigenfunctions of the two-dimensional Fourier transform with eigenvalues $(i)^{p+q} = e^{i(p+q)\pi/2}$; see Eqs. (6.4) and (6.9).

6.3.2 Integral Representation of the Two-Dimensional Fractional Fourier Transform

Now we are in a position to introduce the fractional Fourier transform in two variables [334, 335, 337]. Let

$$s = e^{i\alpha},\ t = e^{i\beta} \text{ so that } st = e^{i(\alpha+\beta)},$$

and let $\gamma = (\alpha+\beta)/2$, $\delta = (\alpha-\beta)/2$. Thus,

$$K(z_1,\overline{z}_1,z_2,\overline{z}_2;s,t) = \frac{1}{\pi(1-e^{i(\alpha+\beta)})}\times$$
$$\exp\left\{\frac{-e^{i(\alpha+\beta)}(u^2+v^2+x^2+y^2)+(e^{i\alpha}+e^{i\beta})(ux+vy)+i(e^{i\beta}-e^{i\alpha})(vx-uy)}{1-e^{i(\alpha+\beta)}}\right\}$$

In addition, we have

$$\frac{1}{1-e^{i(\alpha+\beta)}} = \frac{e^{-i(\alpha+\beta)/2}}{e^{-i(\alpha+\beta)/2}-e^{i(\alpha+\beta)/2}} = \frac{ie^{-i\gamma}}{2\sin\gamma}, \quad (6.16)$$

$$\frac{e^{i\alpha}+e^{i\beta}}{1-e^{i(\alpha+\beta)}} = \frac{e^{i(\alpha-\beta)/2}+e^{-i(\alpha-\beta)/2}}{e^{-i(\alpha+\beta)/2}-e^{i(\alpha+\beta)/2}}, \quad (6.17)$$

$$= \frac{e^{i\delta}+e^{-i\delta}}{e^{-i\gamma}-e^{i\gamma}} = \frac{i\cos\delta}{\sin\gamma}, \quad (6.18)$$

$$i\frac{e^{i\beta}-e^{i\alpha}}{1-e^{i(\alpha+\beta)}} = i\frac{-e^{i\delta}-e^{i\delta}}{e^{-i\gamma}-e^{i\gamma}} = i\frac{\sin\delta}{\sin\gamma}. \quad (6.19)$$

Therefore, if we put

$$k(z_1,\overline{z}_1,z_2,\overline{z}_2;s,t) = K(z_1,\overline{z}_1,z_2,\overline{z}_2;s,t)\, e^{-(x^2+y^2+u^2+v^2)/2}, \quad (6.20)$$

and observing that

$$\left[\frac{e^{2i\gamma}}{1-e^{2i\gamma}}+\frac{1}{2}\right] = \frac{i\cos\gamma}{2\sin\gamma} = i\frac{\cot\gamma}{2},$$

we have

$$k(z_1,\bar{z}_1,z_2,\bar{z}_2;s,t) = d(\gamma)\times$$
$$\exp\left\{-a(\gamma)\left(x^2+y^2+u^2+v^2\right)+b(\gamma,\delta)(ux+vy)+c(\gamma,\delta)(vx-uy)\right\}, \quad (6.21)$$

where

$$a(\gamma) = i\frac{\cot\gamma}{2} \quad , \quad b(\gamma,\delta) = \frac{i\cos\delta}{\sin\gamma} \quad (6.22)$$

$$c(\gamma,\delta) = \frac{i\sin\delta}{\sin\gamma} \quad , \quad d(\gamma) = \frac{ie^{-i\gamma}}{2\pi\sin\gamma}. \quad (6.23)$$

From now on we will write a,b,c,d, instead of $a(\gamma),b(\gamma,\delta),c(\gamma,\delta),d(\gamma)$ for short and replace s,t by α,β, respectively. We will also use the notation $\tilde{a},\tilde{b},\tilde{c},\tilde{d}$ to denote the same quantities but without the imaginary number $i=\sqrt{-1}$, that is, $a=i\tilde{a}, b=i\tilde{b},..$ etc.

We are now ready to introduce the definition of the two-dimensional coupled fractional Fourier transform and some of its properties, such as its inversion formula.

Definition 30. *For a function $f\in L^1(\mathbb{R}^2)$, we define the two-dimensional coupled fractional Fourier transform (CFrFT) as*

$$F_{\alpha,\beta}(z_1,\bar{z}_1) = \int_{\mathbb{R}^2} k(z_1,\bar{z}_1,z_2,\bar{z}_2;\alpha,\beta)\, f(z_2,\bar{z}_2)dz_2 \quad \gamma\neq n\pi$$

or

$$F_{\alpha,\beta}(u,v) = \int_{\mathbb{R}^2} k(x,y,u,v;\alpha,\beta)\, f(x,y)dxdy, \quad \gamma\neq n\pi \quad (6.24)$$

where $z_1=u+iv$, $z_2=x+iy$.

The definition may be extended to $f\in L^2(\mathbb{R}^2)$ in the usual way. When $\alpha=\beta$, $\delta=0$ and $\gamma=\alpha$ and the two-dimensional fraction Fourier transform becomes a tensor product of two one-dimensional fractional Fourier transforms, i.e.,

$$\begin{aligned} F_{\alpha,\alpha}(u,v) &= d(\alpha)\int_{\mathbb{R}^2}\exp\left\{-a(\alpha)\left(x^2+y^2+u^2+v^2\right)+b(\alpha)(ux+vy)\right\} \\ &\times f(x,y)dxdy, \end{aligned} \quad (6.25)$$

where $a(\alpha)=i\cot\alpha/2$, $b(\alpha)=i\csc\alpha$, which is the standard fractional Fourier transform defined in (6.7). Furthermore, if $\alpha=\beta=\pi/2$, the two-dimensional coupled fraction Fourier transforms reduces to the standard two-dimensional Fourier transform.

6.3.3 Inversion Formula

We now derive an inversion formula for the two-dimensional coupled fractional Fourier transform.

Proposition 2. *The inversion formula for the two-dimensional coupled fractional Fourier transform given by (6.24) is*

$$f(x,y) = d(-\gamma)\int_{\mathbb{R}^2} F(u,v)$$
$$\times \exp\left\{a\left(x^2+y^2+u^2+v^2\right)-b(xu+yv)-c(xv-yu)\right\}dudv, \quad (6.26)$$

whenever the integral exists. A sufficient condition for the integral to exist is that $F_{\alpha,\beta} \in L^1(\mathbb{R}^2)$.

Proof *Let*

$$g(x,y) = f(x,y)e^{-a(\gamma)(x^2+y^2)} \quad \textit{and } G(u,v) = F_{\alpha,\beta}(u,v)e^{a(\gamma)(u^2+v^2)}.$$

Then from (6.24) and (6.21), we have

$$G(u,v) = d(\gamma)\int_{\mathbb{R}^2} g(x,y)\exp\left\{ix\left(\tilde{b}u+\tilde{c}v\right)+iy\left(\tilde{b}v-\tilde{c}u\right)\right\}dxdy,$$

where $\tilde{b} = b/i, \tilde{c} = c/i$.

Setting

$$U = \tilde{b}u + \tilde{c}v, \quad \textit{and } V = \tilde{b}v - \tilde{c}u,$$

we have

$$G(u,v) = d(\gamma)\int_{\mathbb{R}^2} g(x,y)e^{i(xU+yV)}dxdy.$$

Using the inversion formula for the two-dimensional Fourier transform, we obtain

$$g(x,y) = \frac{1}{(2\pi)^2 d(\gamma)}\int_{\mathbb{R}^2} G(u,v)\exp\{-i(xU+yV)\}dUdV.$$

But since

$$dUdV = \begin{vmatrix} \tilde{b} & \tilde{c} \\ -\tilde{c} & \tilde{b} \end{vmatrix} dudv = \left(\tilde{b}^2+\tilde{c}^2\right)dudv = \csc^2\gamma dudv,$$

we have after some simplification

$$g(x,y) = d(-\gamma)\int_{\mathbb{R}^2} G(u,v)\exp\left\{-ix(\tilde{b}u+\tilde{c}v)-iy(\tilde{b}v-\tilde{c}u)\right\}dudv,$$

or

$$f(x,y) = d(-\gamma)\times$$
$$\int_{\mathbb{R}^2} F_{\alpha,\beta}(u,v)\exp\left\{a\left(x^2+y^2+u^2+v^2\right)-b(xu+yv)-c(xv-yu)\right\}dudv,$$

which is the inversion formula for the two-dimensional coupled Fractional Fourier transform.

Remark. The above result shows that the kernel of the inverse transform is obtained from the kernel of the forward transform by replacing γ by $-\gamma$ in much the same way as in the standard fractional Fourier transform, but instead of replacing α,β by $-\alpha,-\beta$, separately, here we replace $(\alpha+\beta)/2$ by $-(\alpha+\beta)/2$.

6.3.4 Examples

Example 1: If $f(x,y) = e^{a(x^2+y^2)}\chi_\Omega$, where χ_Ω is the characteristic function of the rectangular region $\Omega = [-r,r] \times [-R,R]$, then

$$F_{\alpha,\beta}(u,v) = d(\gamma)e^{-a(u^2+v^2)} \int_\Omega e^{i(xA+yB)}dxdy,$$

where

$$A = \tilde{b}u + \tilde{c}v = \frac{\cos\delta}{\sin\gamma}u + \frac{\sin\delta}{\sin\gamma}v = (u\cos\delta + v\sin\delta)\,/\sin\gamma, \tag{6.27}$$

$$B = \tilde{b}v - \tilde{c}u = \frac{\cos\delta}{\sin\gamma}v - \frac{\sin\delta}{\sin\gamma}u = (v\cos\delta - u\sin\delta)\,/\sin\gamma. \tag{6.28}$$

Therefore, using the fact that

$$AB = \left[uv\cos 2\delta + (v^2 - u^2)(\sin 2\delta)/2\right]/\sin^2\gamma,$$

we have

$$\begin{aligned} F_{\alpha,\beta}(u,v) &= d(\gamma)e^{-a(u^2+v^2)} \int_{-r}^{r} e^{ixA}dx \int_{-R}^{R} e^{iyB}dy \\ &= 4d(\gamma)e^{-a(u^2+v^2)}\frac{\sin Ar}{A}\frac{\sin BR}{B} \\ &= \left(\frac{2ie^{-i\gamma}}{\pi\sin\gamma}\right)\exp\left[-i\cot\gamma(u^2+v^2)/2\right]\frac{\sin Ar}{A}\frac{\sin BR}{B} \\ &= \frac{2i\sin\gamma e^{-i\gamma}}{\pi}e^{-a(u^2+v^2)} \\ &\times \frac{\sin\left(\frac{r}{\sin\gamma}(u\cos\delta+v\sin\delta)\right)\sin\left(\frac{R}{\sin\gamma}(v\cos\delta-u\sin\delta)\right)}{uv\cos 2\delta+\sin 2\delta(v^2-u^2)/2}. \end{aligned} \tag{6.29}$$

As a special case, if $\alpha = \beta = \pi/2$, then $\gamma = \pi/2, \delta = 0$, and it follows that $f(x,y) = \chi_\Omega$. Under these conditions Eq. (6.29) reduces to

$$F(u,v) = \frac{2}{\pi}\frac{\sin ru}{u}\frac{\sin Rv}{v}, \tag{6.30}$$

which is consistent with the classical result that the Fourier transform of the characteristic function of the set Ω is given by (6.30).

Example 2:

If $f(x,y) = 1$, then

$$\begin{aligned} F_{\alpha,\beta}(u,v) &= d(\gamma)e^{-a(u^2+v^2)} \int_{\mathbb{R}^2} \exp\left[-a(x^2+y^2)+ixA+iyB\right]dxdy \\ &= d(\gamma)e^{-a(u^2+v^2)} \int_{\mathbb{R}} e^{-i\tilde{a}x^2+ixA}dx \int_{\mathbb{R}} e^{-i\tilde{a}y^2+iyB}dy, \end{aligned}$$

where A, B are given by Eqs. (6.27, 6.28).

Using the following formulas (see Eq. (4.8))

$$\int_{\mathbb{R}} e^{-p^2x^2+qx}dx = e^{q^2/4p^2}\sqrt{\frac{\pi}{p^2}}, \quad \Im p^2 < 0,$$

and

$$\int_{\mathbb{R}} e^{-ix^2}dx = (1-i)\sqrt{\frac{\pi}{2}},$$

we obtain

$$F_{\alpha,\beta}(u,v) = d(\gamma)e^{-a(u^2+v^2)}\left(\frac{(1-i)\sqrt{\pi}}{\sqrt{2\tilde{a}}}e^{iA^2/4\tilde{a}}\right)\left(\frac{(1-i)\sqrt{\pi}}{\sqrt{2\tilde{a}}}e^{iB^2/4\tilde{a}}\right),$$

or

$$F_{\alpha,\beta}(u,v) = d(\gamma)e^{-a(u^2+v^2)}\frac{\pi(1-i)^2}{2\tilde{a}}\exp\left[i(A^2+B^2)/4\tilde{a}\right].$$

But since

$$\tilde{b}^2+\tilde{c}^2 = \frac{\cos^2\delta}{\sin^2\gamma}+\frac{\sin^2\delta}{\sin^2\gamma} = \csc^2\gamma,$$

one can easily verify that

$$A^2+B^2 = \tilde{b}^2(u^2+v^2)+c^2(u^2+v^2) = (\tilde{b}^2+\tilde{c}^2)(u^2+v^2) = \csc^2\gamma(u^2+v^2).$$

Replacing $\tilde{a}$ by its value $\cot\gamma/2$ and simplifying the expression for $F_{\alpha,\beta}$, we obtain

$$\begin{aligned}
F_{\alpha,\beta}(u,v) &= d(\gamma)e^{-i\cot\gamma(u^2+v^2)/2}(-2\pi i\tan\gamma)\exp\left[\frac{i}{2\sin\gamma\cos\gamma}(u^2+v^2)\right] \\
&= d(\gamma)(-2\pi i\tan\gamma)\exp\left[\frac{i}{2}(u^2+v^2)\left(\frac{1}{\sin\gamma\cos\gamma}-\cot\gamma\right)\right] \\
&= d(\gamma)(-2\pi i\tan\gamma)\exp\left[i\tan\gamma(u^2+v^2)/2\right] \\
&= e^{-i\gamma}\sec\gamma\exp\left[i\tan\gamma(u^2+v^2)/2\right]
\end{aligned}$$

6.4 ADDITIVE PROPERTY

In this section, we prove the additive property of the two-dimensional coupled fractional Fourier transform [154].

Theorem 27. *Let* $\gamma_i = \frac{\alpha_i+\beta_i}{2}$, $i = 1,2$ *and asssume that* $\gamma_1, \gamma_2, \gamma_1+\gamma_2 \neq n\pi$. *Then the following formula holds*

$$\mathcal{F}_{\gamma_2}\left[\mathcal{F}_{\gamma_1}\right] = \mathcal{F}_{\gamma_1+\gamma_2}.$$

In particular

$$\mathcal{F}_{\alpha_2,\beta_2}\left[\mathcal{F}_{\alpha_1,\beta_1}\right] = \mathcal{F}_{\alpha_1+\alpha_2,\beta_1+\beta_2}. \tag{6.31}$$

Proof *Since*

$$F_{\alpha_1,\beta_1}[f](u,v) = \int_{\mathbb{R}^2} f(x,y)k_1(x,y;u,v;\alpha_1,\beta_1)dxdy,$$

we have

$$J = F_{\alpha_2,\beta_2}\left[F_{\alpha_1,\beta_1}\right](z,w) = \int_{\mathbb{R}^2} F_{\alpha_1,\beta_1}(u,v)k_2(u,v;z,w,\alpha_2,\beta_2)dudv$$
$$= \int_{\mathbb{R}^2}\int_{\mathbb{R}^2} f(x,y)k_1(x,y;u,v)k_2(u,v;z,w)dx\,dy\,du\,dv. \tag{6.32}$$

where

$$k_1(x,y;u,v;\alpha_1,\beta_1) = d_1\exp\left\{-a_1(x^2+y^2+u^2+v^2)+b_1(xu+yv)+c_1(vx-uy)\right\},$$

$$k_2(u,v;z,w;\alpha_2,\beta_2) = d_2\exp\left\{-a_2(u^2+v^2+z^2+w^2)+b_2(uz+vw)+c_2(uw-vz)\right\},$$

and

$$a_i = a(\gamma_i), b_i = b(\gamma_i,\delta_i), c_i = c(\gamma_i,\delta_i), d_i = d(\gamma_i),\ i = 1,2.$$

To simplify the notation, we will drop the angles α and β from the kernel and write $k_i(x,y;u,v)$ instead of $k_i(x,y;u,v;\alpha_i,\beta_i)$, $i = 1,2$.

One can verify that

$$\begin{aligned} &k_1(x,y;u,v)k_2(u,v;z,w) = \\ &\quad d_1d_2\exp\left\{-a_1(x^2+y^2)-a_2(z^2+w^2)-(a_1+a_2)(u^2+v^2)\right. \\ &\quad \left.+u(b_1x-c_1y+b_2z+c_2w)+v(b_1y+c_1x+b_2w-c_2z)\right\}. \end{aligned} \tag{6.33}$$

Therefore, we have

$$J = \int_{\mathbb{R}^2} f(x,y)\exp\left\{-a_1(x^2+y^2)-a_2(z^2+w^2)\right\}I_1I_2dxdy, \tag{6.34}$$

where

$$I_1 = d_1\int_{\mathbb{R}}\exp\left\{-(a_1+a_2)u^2+iu(\tilde{b}_1x-\tilde{c}_1y+\tilde{b}_2z+\tilde{c}_2w)\right\}du,$$

$$I_2 = d_2\int_{\mathbb{R}}\exp\left\{-(a_1+a_2)v^2+iv(\tilde{b}_1y+\tilde{c}_1x+\tilde{b}_2w-\tilde{c}_2z)\right\}dv,$$

and $a_1 = i\tilde{a}_1, b_1 = i\tilde{b}_1$, etc.. To evaluate I_1 and I_2, we use the relation

$$\int_{\mathbb{R}} e^{-p^2x^2+iwx}dx = \sqrt{\frac{\pi}{p^2}}e^{-w^2/4p^2},$$

to obtain

$$I_1 = d_1\sqrt{\frac{\pi}{a_1+a_2}}\exp\left\{-\frac{A^2}{4(a_1+a_2)}\right\},$$

$$I_2 = d_2\sqrt{\frac{\pi}{a_1+a_2}}\exp\left\{-\frac{B^2}{4(a_1+a_2)}\right\},$$

where

$$A = (\tilde{b}_1x-\tilde{c}_1y+\tilde{b}_2z+\tilde{c}_2w),$$
$$B = (\tilde{b}_1y+\tilde{c}_1x+\tilde{b}_2w-\tilde{c}_2z).$$

Therefore,

$$I_1 I_2 = d_1 d_2 \left[\frac{\pi}{a_1+a_2}\right] \exp\left\{\frac{-1}{4(a_1+a_2)}\right\} \left[A^2+B^2\right]. \tag{6.35}$$

With some computations, we obtain

$$\begin{aligned} A^2+B^2 &= (\tilde{b}_1^2+\tilde{c}_1^2)(x^2+y^2)+(\tilde{b}_2^2+\tilde{c}_2^2)(z^2+w^2)+2xz(\tilde{b}_1\tilde{b}_2-\tilde{c}_1\tilde{c}_2)\\ &+ 2xw(\tilde{b}_1\tilde{c}_2+\tilde{b}_2\tilde{c}_1)-2yz(\tilde{b}_2\tilde{c}_1+\tilde{b}_1\tilde{c}_2)+2yw(\tilde{b}_1\tilde{b}_2-\tilde{c}_1\tilde{c}_2)\\ &= \frac{(x^2+y^2)}{\sin^2\gamma_1}+\frac{(z^2+w^2)}{\sin^2\gamma_2}+\frac{2xz\cos(\delta_1+\delta_2)}{\sin\gamma_1\sin\gamma_2}\\ &+ \frac{2xw\sin(\delta_1+\delta_2)}{\sin\gamma_1\sin\gamma_2}-\frac{2yz\sin(\delta_1+\delta_2)}{\sin\gamma_1\sin\gamma_2}+\frac{2yw\cos(\delta_1+\delta_2)}{\sin\gamma_1\sin\gamma_2}. \end{aligned} \tag{6.36}$$

Because

$$\frac{-1}{4(a_1+a_2)} = \frac{i\sin\gamma_1\sin\gamma_2}{2\sin(\gamma_1+\gamma_2)},$$

we have after simplifying Eq. (6.35)

$$\begin{aligned} I_1 I_2 &= d_1 d_2 \left[\frac{\pi}{a_1+a_2}\right] \exp i\left\{\frac{\sin\gamma_2(x^2+y^2)}{2\sin\gamma_1\sin(\gamma_1+\gamma_2)}+\frac{\sin\gamma_1(z^2+w^2)}{2\sin\gamma_2\sin(\gamma_1+\gamma_2)}\right.\\ &+ \left.\frac{\cos(\delta_1+\delta_2)(xz+yw)}{\sin(\gamma_1+\gamma_2)}+\frac{\sin(\delta_1+\delta_2)(xw-yz)}{\sin(\gamma_1+\gamma_2)}\right\}. \end{aligned} \tag{6.37}$$

Before we substitute this back into Eq. (6.34), let us observe that

$$\begin{aligned} &-a_1(x^2+y^2)+i\frac{\sin\gamma_2(x^2+y^2)}{2\sin\gamma_1\sin(\gamma_1+\gamma_2)}\\ &= \frac{-i(x^2+y^2)}{2\sin\gamma_1}\left[\cos\gamma_1-\frac{\sin\gamma_2}{\sin(\gamma_1+\gamma_2)}\right]\\ &= \frac{-i(x^2+y^2)}{2\sin\gamma_1}\left[\frac{\sin\gamma_1\cos(\gamma_1+\gamma_2)}{\sin(\gamma_1+\gamma_2)}\right]\\ &= \frac{-i\cot(\gamma_1+\gamma_2)}{2}(x^2+y^2) = a_{1,2}(x^2+y^2) \end{aligned} \tag{6.38}$$

where

$$a_{1,2} = a(\gamma_1+\gamma_2) = \frac{-i\cot(\gamma_1+\gamma_2)}{2}.$$

Similar calculation holds for the term z^2+w^2. Finally, we have

$$\begin{aligned} d_1 d_2\left(\frac{\pi}{a_1+a_2}\right) &= \left(\frac{ie^{-i\gamma_1}}{2\pi\sin\gamma_1}\right)\left(\frac{ie^{-i\gamma_2}}{2\pi\sin\gamma_2}\right)\left(\frac{2\pi\sin\gamma_1\sin\gamma_2}{i\sin(\gamma_1+\gamma_2)}\right)\\ &= \frac{ie^{-i(\gamma_1+\gamma_2)}}{(2\pi)\sin(\gamma_1+\gamma_2)} = d_{1,2} = d(\gamma_1+\gamma_2). \end{aligned} \tag{6.39}$$

Therefore, by substituting (6.39) into (6.37) and then (6.37) into (6.34),we obtain

$$\begin{aligned} J(z,w) &= d(\gamma_1+\gamma_2)\int_{\mathbb{R}^2} f(x,y)\exp\left\{-a_{1,2}\left(x^2+y^2+z^2+w^2\right)\right.\\ &+ \left. b_{1,2}(xz+yw)+c_{1,2}(xw-yz)\right\}dxdy \end{aligned} \tag{6.40}$$

where

$$a_{1,2}=a(\gamma_1+\gamma_2),\ b_{1,2}=b(\gamma_1+\gamma_2,\delta_1+\delta_2),\ c_{1,2}=c(\gamma_1+\gamma_2,\delta_1+\delta_2).$$

Hence, J equals the right-hand side of Eq. (6.31), and this completes the proof.

Under similar conditions to those in the above theorem, one can show that the composition of the transform is associative

$$\mathcal{F}_{\gamma 3}\left(\mathcal{F}_{\gamma_2}\mathcal{F}_{\gamma_1}\right)=\left(\mathcal{F}_{\gamma_3}\mathcal{F}_{\gamma_2}\right)\mathcal{F}_{\gamma_1}.$$

The next theorem completes Theorem 27 by covering the case when $\gamma_1+\gamma_2=n\pi$.

Theorem 28. *Let $\gamma_1.\gamma_2\neq n\pi$ and assume that $\gamma_1+\gamma_2=n\pi$. Then*

$$\begin{aligned}&\mathcal{F}\alpha_2,\beta_2\left(\mathcal{F}\alpha_1,\beta_1\left[f\right]\right)(z,w)\\ =\ & f\left(z\cos(\delta_1+\delta_2)+w\sin(\delta_1+\delta_2),-z\sin(\delta_1+\delta_2)+w\cos(\delta_1+\delta_2)\right), \quad n \text{ is even}\\ =\ & f\left(-z\cos(\delta_1+\delta_2)-w\sin(\delta_1+\delta_2),z\sin(\delta_1+\delta_2)-w\cos(\delta_1+\delta_2)\right), \quad n \text{ is odd.}\end{aligned}$$

Proof *We have as in Theorem 27*

$$\begin{aligned}J(z,w)&=F_{\alpha_2,\beta_2}\left[F_{\alpha_1,\beta_1}\right](z,w)\\ &=\int_{\mathbb{R}^2}\int_{\mathbb{R}^2}f(x,y)k_1(x,y;u,v)k_2(u,v;z,w)dx\,dy\,du\,dv.\end{aligned}\qquad(6.41)$$

But since $\gamma_1=n\pi-\gamma_2$, $\cot\gamma_1=-\cot\gamma_2$; *hence, $a_1=-a_2$, and it follows that*

$$\begin{aligned}&k_1(x,y;u,v)k_2(u,v;z,w)=d_1d_2\exp\left\{-a_1(x^2+y^2)+a_1(z^2+w^2)\right.\\ &\left.+u(b_1x-c_1y+b_2z+c_2w)+v(b_1y+c_1x+b_2w-c_2z)\right\}.\end{aligned}\qquad(6.42)$$

Thus,

$$J(z,w)=d_1d_2\int_{\mathbb{R}^2}f(x,y)\exp\left\{-a_1(x^2+y^2)+a_1(z^2+w^2)\right\}I_1I_2dxdy,\qquad(6.43)$$

where

$$\begin{aligned}I_1&=\int_{\mathbb{R}}\exp\left\{iu(\tilde{b}_1x-\tilde{c}_1y+\tilde{b}_2z+\tilde{c}_2w)\right\}du\\ &=2\pi\delta(U++\tilde{b}_2z+\tilde{c}_2w),\end{aligned}\qquad(6.44)$$

and

$$\begin{aligned}I_2&=\int_{\mathbb{R}}\exp\left\{iv(\tilde{b}_1y+\tilde{c}_1x+\tilde{b}_2w-\tilde{c}_2z)\right\}dv,\\ &=2\pi\delta(V+\tilde{b}_2w-\tilde{c}_2z),\end{aligned}\qquad(6.45)$$

where

$$U=\tilde{b}_1x-\tilde{c}_1y,\quad V=\tilde{b}_1y+\tilde{c}_1x.$$

Solving for x, y in terms of U, V, we obtain

$$x = (\tilde{b}_1 U + \tilde{c}_1 V)\sin^2\gamma_1, \quad y = (-\tilde{c}_1 U + \tilde{b}_1 V)\sin^2\gamma_1,$$

and $dxdy = (\sin^2\gamma_1)dUdV$. Substituting (6.44) and (6.45) into (6.43) leads to

$$\begin{aligned} J &= (2\pi)^2 \sin^2\gamma_1 d_1 d_2 \int_{\mathbb{R}^2} f(A,B) \\ &\times \exp\left[a_1(z^2+w^2)\right] \exp\left\{-a_1 \sin^4\gamma_1 \left[(\tilde{b}_1 U + \tilde{c}_1 V)^2 + (-\tilde{c}_1 U + \tilde{b}_1 V)^2\right]\right\} \\ &\times \delta(U + +\tilde{b}_2 z + \tilde{c}_2 w)\delta(V + \tilde{b}_2 w - \tilde{c}_2 z) dUdV \end{aligned} \tag{6.46}$$

where

$$A = (\tilde{b}_1 U + \tilde{c}_1 V)\sin^2\gamma_1, \quad B = (-\tilde{c}_1 U + \tilde{b}_1 V)\sin^2\gamma_1$$

But because

$$\begin{aligned} &\exp\left\{-a_1 \sin^4\gamma_1 \left[(\tilde{b}_1 U + \tilde{c}_1 V)^2 + (-\tilde{c}_1 U + \tilde{b}_1 V)^2\right]\right\} \\ = \; &\exp\left\{-a_1 \sin^4\gamma_1 \left[\frac{U^2+V^2}{\sin^2\gamma_1}\right]\right\} = \exp\left\{-a_1 \sin^2\gamma_1 (U^2+V^2)\right\}, \end{aligned}$$

we have

$$\begin{aligned} J &= (2\pi)^2 \sin^2\gamma_1 d_1 d_2 \int_{\mathbb{R}^2} f(A,B) \\ &\times \exp\left[a_1(z^2+w^2)\right] \exp\left\{-a_1 \sin^2\gamma_1 (U^2+V^2)\right\} \\ &\times \delta(U + +\tilde{b}_2 z + \tilde{c}_2 w)\delta(V + \tilde{b}_2 w - \tilde{c}_2 z) dUdV. \end{aligned} \tag{6.47}$$

Applying the delta functions to the integrant in (6.47) amounts to replacing U by $-\tilde{b}_2 z - \tilde{c}_2 w$ and V by $-\tilde{b}_2 w + \tilde{c}_2 z$, to obtain

$$\begin{aligned} A &= \left(-\tilde{b}_1\tilde{b}_2 z - \tilde{b}_1\tilde{c}_2 w + \tilde{c}_1\tilde{c}_2 z - \tilde{b}_2\tilde{c}_1 w\right)\sin^2\gamma_1 \\ &= -z\frac{\sin\gamma_1\cos(\delta_1+\delta_2)}{\sin\gamma_2} - w\frac{\sin\gamma_1\sin(\delta_1+\delta_2)}{\sin\gamma_2}, \end{aligned} \tag{6.48}$$

and

$$\begin{aligned} B &= \left(\tilde{c}_1\tilde{b}_2 z + \tilde{c}_1\tilde{c}_2 w + \tilde{b}_1\tilde{c}_2 z - \tilde{b}_1\tilde{b}_2 w\right)\sin^2\gamma_1 \\ &= z\frac{\sin\gamma_1\sin(\delta_1+\delta_2)}{\sin\gamma_2} - w\frac{\sin\gamma_1\cos(\delta_1+\delta_2)}{\sin\gamma_2}. \end{aligned} \tag{6.49}$$

As for the exponential factor in the integral in (6.47), let us first observe that

$$U^2+V^2 = (-\tilde{b}_2 z - \tilde{c}_2 w)^2 + (-\tilde{b}_2 w + \tilde{c}_2 z)^2 = \frac{z^2+w^2}{\sin^2\gamma_2},$$

hence

$$\begin{aligned} & \exp\left\{a_1(z^2+w^2) - a_1\sin^2\gamma_1\frac{z^2+w^2}{\sin^2\gamma_2}\right\} \\ = & \exp\left\{a_1(z^2+w^2)\left[1-\frac{\sin^2\gamma_1}{\sin^2\gamma_2}\right]\right\} = 1. \end{aligned} \tag{6.50}$$

The last step follows from the fact that $\sin\gamma_1 = \sin(n\pi-\gamma_2) = (-1)^{n+1}\sin\gamma_2$. *Moreover, we have*

$$(2\pi)^2(\sin^2\gamma_1)d_1d_2 = \frac{ie^{-i\gamma_1}}{2\pi\sin\gamma_1}\frac{ie^{-i\gamma_2}}{2\pi\sin\gamma_2}(2\pi)^2\sin^2\gamma_1 = 1 \tag{6.51}$$

Substituting (6.51) and (6.50) into (6.47), we finally arrive at

$$J = f\left(-z\eta\cos(\delta_1+\delta_2) - w\eta\sin(\delta_1+\delta_2), z\eta\sin(\delta_1+\delta_2) - w\eta\cos(\delta_1+\delta_2)\right),$$

where $\eta = \frac{\sin\gamma_1}{\sin\gamma_2} = (-1)^{n+1}$. *Hence,*

$$J = f\left(z\cos(\delta_1+\delta_2) + w\sin(\delta_1+\delta_2), -z\sin(\delta_1+\delta_2) + w\cos(\delta_1+\delta_2)\right),$$

for n *even and*

$$J = f\left(-z\cos(\delta_1+\delta_2) - w\sin(\delta_1+\delta_2), z\sin(\delta_1+\delta_2) - w\cos(\delta_1+\delta_2)\right),$$

for n *odd.*

6.5 CONVOLUTION THEOREM

In this section, we derive a convolution theorem for the two-dimensional coupled fractional Fourier transform (CFrFT).

Definition 31. *Let*

$$\tilde{f}(x,y) = e^{-a(x^2+y^2)}f(x,y), \quad \text{and } \tilde{g}(x,y) = e^{-a(x^2+y^2)}g(x,y),$$

and define

$$h(x,y) = \left(f\bigotimes g\right) = d(\gamma)e^{a(x^2+y^2)}\left(\tilde{f}*\tilde{g}\right)(x,y),$$

where $*$ *stands for the standard convolution given by*

$$(f*g)(x,y) = \int_{\mathbb{R}^2} f(x-\eta, y-\zeta)g(\eta,\zeta)d\eta d\zeta.$$

Theorem 29. *Let* $h(x,y) = (f\bigotimes g)$, *and* F, G, H *denote the two-dimensional coupled fraction Fourier transform of* f, g, h *respectively. Then*

$$H(u,v) = e^{a(u^2+v^2)}F(u,v)G(u,v).$$

Proof *Let*

$$A = (bu + cv), \quad \text{and } B = (bv - cu).$$

Hence

$$\begin{aligned} H(u,v) &= d(\gamma)\int_{\mathbb{R}^2} e^{-a(x^2+y^2+u^2+v^2)+Ax+By} h(x,y)dxdy \\ &= d^2(\gamma)e^{-a(u^2+v^2)}\int_{\mathbb{R}^2} e^{Ax+By}dxdy\int_{\mathbb{R}^2} f(x-\eta,y-\zeta)e^{-a(x-\eta)^2-a(y-\zeta)^2} \\ &\times\ g(\eta,\zeta)e^{-a\eta^2-a\zeta^2}d\eta d\zeta \\ &= d^2(\gamma)e^{-a(u^2+v^2)}\int_{\mathbb{R}^2} f(x-\eta,y-\zeta)e^{-a(x-\eta)^2-a(y-\zeta)^2+A(x-\eta)+B(y-\zeta)}dxdy \\ &\times \int_{\mathbb{R}^2} g(\eta,\zeta)e^{-a\eta^2-a\zeta^2+A\eta+B\zeta}d\eta d\zeta. \end{aligned}$$

By making a change of variable in the first integral on the right-hand side, we obtain

$$\begin{aligned} H(u,v) &= d^2(\gamma)e^{-a(u^2+v^2)}\int_{\mathbb{R}^2} f(X,Y)e^{-aX^2-aY^2+AX+BY}dXdY \\ &\times \int_{\mathbb{R}^2} g(\eta,\zeta)e^{-a\eta^2-a\zeta^2+A\eta+B\zeta}d\eta d\zeta \\ &= d^2(\gamma)e^{a(u^2+v^2)}\int_{\mathbb{R}^2} f(X,Y)e^{-aX^2-aY^2-au^2-av^2+AX+BY}dXdY \\ &\times \int_{\mathbb{R}^2} g(\eta,\zeta)e^{-a\eta^2-a\zeta^2-au^2-av^2+A\eta+B\zeta}d\eta d\zeta \\ &= e^{a(u^2+v^2)}F(u,v)G(u,v), \end{aligned}$$

which completes the proof.

6.6 POISSON SUMMATION FORMULA

This section is devoted to the Poisson summation formula for the two-dimensional coupled fractional Fourier transform .

Theorem 30. *The Poisson summation formula for the two-dimensional coupled fractional Fourier transform is given by*

$$\begin{aligned} &\sum_{k,m\in\mathbb{Z}} f(x+2\pi k,y+2\pi m)e^{-a\left[(x+2\pi k)^2+(y+2\pi m)^2\right]} \\ =\ &\frac{1}{(2\pi)^2 d(\gamma)}\sum_{k,m\in\mathbb{Z}} e^{-i(kx+my)}F_{\alpha,\beta}\left(\sin\gamma\left(k\cos\delta-m\sin\delta\right),\sin\gamma\left(k\sin\delta+m\cos\delta\right)\right) \\ \times\ &\exp\left\{i\sin 2\gamma(k^2+m^2)/4\right\}. \end{aligned} \tag{6.52}$$

For a sufficient condition for formula (6.52) to hold, see Eq. (6.3).

Proof *: Let*

$$\tilde{f}(x,y) = \sum_{k,m\in\mathbb{Z}} f(x+2\pi k,y+2\pi m)e^{-a\left[(x+2\pi k)^2+(y+2\pi m)^2\right]}.$$

Clearly, $\tilde{f}$ is periodic with period 2π in both x and y; hence, it has a Fourier series of the form

$$\tilde{f}(x,y) = \sum_{k,m\in\mathbb{Z}} \hat{f}_{k,m} e^{-i(kx+my)},$$

where

$$\hat{f}_{k,m} = \frac{1}{(2\pi)^2} \int_{-\pi}^{\pi}\int_{-\pi}^{\pi} \tilde{f}(x,y) e^{i(kx+my)} dx dy.$$

Thus,

$$\begin{aligned}
\tilde{f}(x,y) &= \sum_{k,m\in\mathbb{Z}} e^{-i(kx+my)} \frac{1}{(2\pi)^2} \int_{-\pi}^{\pi}\int_{-\pi}^{\pi} e^{i(k\tilde{x}+m\tilde{y})} \\
&\times \sum_{r,s\in\mathbb{Z}} f(\tilde{x}+2\pi r, \tilde{y}+2\pi s) e^{-a\left[(\tilde{x}+2\pi r)^2+(\tilde{y}+2\pi s)^2\right]} d\tilde{x} d\tilde{y} \\
&= \sum_{k,m\in\mathbb{Z}} e^{-i(kx+my)} \frac{1}{(2\pi)^2} \sum_{r,s\in\mathbb{Z}} \int_{-\pi}^{\pi}\int_{-\pi}^{\pi} e^{i(k\tilde{x}+m\tilde{y})} \\
&\times f(\tilde{x}+2\pi r, \tilde{y}+2\pi s) e^{-a\left[(\tilde{x}+2\pi r)^2+(\tilde{y}+2\pi s)^2\right]} d\tilde{x} d\tilde{y}.
\end{aligned}$$

By substituting

$$X = \tilde{x} + 2\pi r, \quad Y = \tilde{y} + 2\pi s,$$

in the last integral, we obtain

$$\begin{aligned}
\tilde{f}(x,y) &= \frac{1}{(2\pi)^2} \sum_{k,m\in\mathbb{Z}} e^{-i(kx+my)} \\
&\times \sum_{r,s\in\mathbb{Z}} \int_{\pi(2r-1)}^{\pi(2r+1)} \int_{\pi(2s-1)}^{\pi(2s+1)} f(X,Y) e^{-a(X^2+Y^2)} e^{ik(X-2\pi r)+im(Y-2\pi s)} dX dY \\
&= \frac{1}{(2\pi)^2} \sum_{k,m\in\mathbb{Z}} e^{-i(kx+my)} \int_{-\infty}^{\infty}\int_{-\infty}^{\infty} e^{ikX+imY} f(X,Y) e^{-a(X^2+Y^2)} dX dY \\
&= \frac{1}{(2\pi)^2 d(\gamma)} \sum_{k,m\in\mathbb{Z}} e^{-i(kx+my)} d(\gamma) e^{a(u^2+v^2)} \\
&\times \int_{-\infty}^{\infty}\int_{-\infty}^{\infty} e^{ikX+imY} e^{-a(X^2+Y^2)} e^{-a(u^2+v^2)} f(X,Y) dX dY.
\end{aligned}$$

Interchanging the summation and integration signs is permissible in view of condition Eq. (6.3). As for the last integral to be the two-dimensional coupled fractional Fourier integral transform of f, we have to solve the system of equations

$$k = \tilde{b}u + \tilde{c}v, \quad m = \tilde{b}v - \tilde{c}u$$

for u, v in terms of k, m. It is easy to see that since $\tilde{b}^2 + \tilde{c}^2 = \csc^2\gamma$, the solution is given by

$$u = \sin\gamma\,(k\cos\delta - m\sin\delta), \quad v = \sin\gamma\,(k\sin\delta + m\cos\delta).$$

Therefore, we finally arrive at the Poisson summation formula for the two-dimensional coupled fractional Fourier transform, namely,

$$\begin{aligned}
&\sum_{k,m\in\mathbb{Z}} f(x+2\pi k, y+2\pi m)e^{-a\left[(x+2\pi k)^2+(y+2\pi m)^2\right]} \\
= &\frac{1}{(2\pi)^2 d(\gamma)} \sum_{k,m\in\mathbb{Z}} e^{-i(kx+my)} F_{\alpha,\beta}\left(\sin\gamma\left[k\cos\delta - m\sin\delta\right], \sin\gamma\left[k\sin\delta + m\cos\delta\right]\right) \\
\times &\exp\left\{a\left[\sin\gamma\left(k\cos\delta - m\sin\delta\right)\right]^2 + a\left[\sin\gamma\left(k\sin\delta + m\cos\delta\right)\right]^2\right\} \\
= &\frac{1}{(2\pi)^2 d(\gamma)} \sum_{k,m\in\mathbb{Z}} e^{-i(kx+my)} F_{\alpha,\beta}\left(\sin\gamma\left[k\cos\delta - m\sin\delta\right], \sin\gamma\left[k\sin\delta + m\cos\delta\right]\right) \\
\times &\exp\left\{a\sin^2\gamma\left[\left(k\cos\delta - m\sin\delta\right)^2 + \left(k\sin\delta + m\cos\delta\right)^2\right]\right\} \\
= &\frac{1}{(2\pi)^2 d(\gamma)} \sum_{k,m\in\mathbb{Z}} e^{-i(kx+my)} F_{\alpha,\beta}\left(\sin\gamma\left[k\cos\delta - m\sin\delta\right], \sin\gamma\left[k\sin\delta + m\cos\delta\right]\right) \\
\times &\exp\left\{i\sin 2\gamma(k^2+m^2)/4\right\}.
\end{aligned}$$

Remark. For $\alpha = \beta = \pi/2$, we have $\gamma = \pi/2, \delta = 0$, and hence $a(\gamma) = 0$, and Eq. (6.52) reduces to Eq. (6.2).

Theorem 31. *Let $\mathcal{F}_{\alpha,\beta}$ be the coupled fractional Fourier transformation with angle α, β. Then $\lim_{\alpha,\beta\to 0} \mathcal{F}_{\alpha,\beta} = \mathcal{I}$ is the identity operator, where the convergence is in the strong topology.*

Proof *From the generating function of the complex Hermite polynomials we have*

$$\sum_{m,n=0}^{\infty} H_{m,n}(z_1,\overline{z}_1)H_{n,m}(z_2,\overline{z}_2)\frac{t^m}{m!}\frac{s^n}{n!} = K(z_1,\overline{z}_1,z_2,\overline{z}_2;s,t),\ |ts|<1, \tag{6.53}$$

where

$$K(z_1,\overline{z}_1,z_2,\overline{z}_2;s,t) = \frac{1}{\pi(1-ts)}\exp\left\{\frac{-tsz_1\overline{z}_1 + tz_1\overline{z}_2 + sz_2\overline{z}_1 - tsz_2\overline{z}_2}{1-ts}\right\},$$

which as in (6.14) leads to

$$\begin{aligned}
t^q s^p H_{q,p}(z_1,\overline{z}_1) &= \int_{\mathbb{R}^2} K(z_1,\overline{z}_1,z_2,\overline{z}_2;s,t)\overline{H}_{p,q}(z_2,\overline{z}_2)e^{-|z_2|^2}dz_2 \\
&= \int_{\mathbb{R}^2} K(x,y;u,v)\overline{H}_{p,q}(x+iy,x-iy)e^{-(x^2+y^2)}dxdy. \quad (6.54)
\end{aligned}$$

But since

$$k(z_1,\overline{z}_1,z_2,\overline{z}_2;s,t) = K(z_1,\overline{z}_1,z_2,\overline{z}_2;s,t)\,e^{-(|z_1|^2+|z_2|^2)/2}, \tag{6.55}$$

and after multiplying both sides of (6.54) by $e^{-|z_1|^2/2} = e^{-(u^2+v^2)/2}$, we have

$$t^q s^p h_{q,p}(z_1,\overline{z}_1) = \int_{\mathbb{R}^2} k(z_1,\overline{z}_1,z_2,\overline{z}_2;s,t)\overline{h}_{p,q}(z_2,\overline{z}_2)dz_2, \tag{6.56}$$

where

$$h_{q,p}(z,\overline{z}) = H_{q,p}(z,\overline{z})e^{-|z|^2/2},$$

is the Hermite function of two complex variables.

In view of Eq. (6.56), we have

$$e^{i\beta q}e^{i\alpha p}h_{q,p}(z_1,\overline{z}_1) = \int_{\mathbb{R}^2} k\left(z_1,\overline{z}_1,z_2,\overline{z}_2;\alpha,\beta\right)\overline{h}_{p,q}(z_2,\overline{z}_2)dz_2. \tag{6.57}$$

But since $\overline{h}_{p,q}(z_2,\overline{z}_2) = h_{q,p}(z_2,\overline{z}_2)$, *we obtain*

$$\mathcal{F}_{\alpha,\beta}(h_{q,p})\left(z_1,\overline{z}_1\right) = e^{i\beta q}e^{i\alpha p}h_{q,p}(z_1,\overline{z}_1),$$

or

$$\mathcal{F}_{\alpha,\beta}(h_{q,p})(u,v) = e^{i\beta q}e^{i\alpha p}h_{q,p}(u,v), \tag{6.58}$$

and by taking the limit of Eq. (6.58), as α *and* β *go to zero, we have*

$$\lim_{\alpha,\beta\to 0} \mathcal{F}_{\alpha,\beta}(h_{q,p})(u,v) = h_{q,p}(u,v) \tag{6.59}$$

Finally, let $f \in L^2(\mathbb{R}^2)$, *hence*

$$f(x,y) = \sum_{m,n=0}^{\infty} \hat{f}_{m,n}h_{m,n}(x,y), \quad \text{with} \sum_{m,n=0}^{\infty} \left|\hat{f}_{m,n}\right|^2 < \infty,$$

where $\hat{f}_{m,n}$ *are the generalized Fourier coefficients of* f. *It follows that*

$$\begin{aligned}
\mathcal{F}_{\alpha,\beta}[f](u,v) &= \int_{\mathbb{R}^2} k\left(x,y,u,v;\alpha,\beta\right)f(x,y)dxdy, \\
&= \sum_{m,n=0}^{\infty} \hat{f}_{m,n}\int_{\mathbb{R}^2} k\left(x,y,u,v;\alpha,\beta\right)h_{m,n}(x,y)dxdy \\
&= \sum_{m,n=0}^{\infty} \hat{f}_{m,n}\mathcal{F}_{\alpha,\beta}[h_{m,n}](u,v).
\end{aligned}$$

Therefore, we have, in view of Eq. (6.58),

$$\left\|\mathcal{F}_{\alpha,\beta}[f(x,y)](u,v) - f(u,v)\right\|^2 = \left\|\sum_{m,n=0}^{\infty} \hat{f}_{m,n}\mathcal{F}_{\alpha,\beta}[h_{m,n}] - \sum_{m,n=0}^{\infty} \hat{f}_{m,n}h_{m,n}\right\|^2$$

$$= \left\|\sum_{m,n=0}^{\infty} \left(e^{i(\alpha n+\beta m)} - 1\right)\hat{f}_{m,n}h_{m,n}\right\|^2 = \sum_{m,n=0}^{\infty} \left|e^{i(\alpha n+\beta m)} - 1\right|^2 \left|\hat{f}_{m,n}\right|^2. \tag{6.60}$$

But since

$$\sum_{m,n=0}^{\infty} \left|e^{i(\alpha n+\beta m)} - 1\right|^2 \left|\hat{f}_{m,n}\right|^2 \le 2\sum_{m,n=0}^{\infty} \left|\hat{f}_{m,n}\right|^2 < \infty,$$

taking the limit of Eq. (6.60), yields

$$\lim_{\alpha,\beta\to 0} \mathcal{F}_{\alpha,\beta}[f](u,v) = f(u,v),$$

that is $\lim_{\alpha,\beta\to 0}\mathcal{F}_{\alpha,\beta} = \mathcal{I}$ *is the identity operator.*

6.7 A SPACE OF BANDLIMITED SIGNALS AND ITS SAMPLING THEOREM

In this section, we introduce a space of functions bandlimited to a rectangular region in the coupled fractional Fourier transform (CFrFT) domain, derive its reproducing kernel and then derive its sampling theorem.

6.7.1 Space of Bandlimited Signals

Let $\Omega = [-r, r] \times [-R, R]$, and

$$B_\Omega = \left\{ f : f \in L^2(\mathbb{R}^2) \text{ such that Supp } F_{\alpha,\beta} \subseteq \Omega \right\}.$$

Theorem 32. *The reproducing kernel of the space B_Ω is given by*

$$h(x, y, \eta, \zeta) = \frac{\sin Ar}{\pi A \sin\gamma} \frac{\sin BR}{\pi B \sin\gamma} \exp\left\{ a\left(x^2 + y^2 - \eta^2 - \zeta^2\right)\right\}. \tag{6.61}$$

That is for any $f \in B_\Omega$, we have

$$f(x, y) = \int_{\mathbb{R}^2} f(\eta, \zeta) h(x, y, \eta, \zeta) d\eta d\zeta.$$

Proof *We have*

$$\begin{aligned}
f(x,y) &= d(-\gamma) \int_\Omega F(u,v) \exp\left\{a\left(x^2+y^2+u^2+v^2\right) - b(xu+yv) - c(xv-yu)\right\} dudv \\
&= d(\gamma)d(-\gamma) \int_\Omega \exp\left\{a\left(x^2+y^2+u^2+v^2\right) - b(xu+yv) - c(xv-yu)\right\} dudv \\
&\quad \times \int_{\mathbb{R}^2} f(\eta,\zeta) \exp\left\{-a\left(\eta^2+\zeta^2+u^2+v^2\right) + b(\eta u+\zeta v) + c(\eta v-\zeta u)\right\} d\eta d\zeta \\
&= \frac{\csc^2\gamma}{(2\pi)^2} \int_{\mathbb{R}^2} f(\eta,\zeta) \exp\left\{a\left(x^2+y^2-\eta^2-\zeta^2\right)\right\} d\eta d\zeta \\
&\quad \times \int_\Omega \exp\{iAu + iBv\}\, dudv,
\end{aligned} \tag{6.62}$$

where

$$\begin{aligned}
A &= \tilde{b}(\eta - x) - \tilde{c}(\zeta - y), \\
B &= \tilde{b}(\zeta - y) + \tilde{c}(\eta - x).
\end{aligned}$$

But

$$\int_\Omega e^{iAu+iBv} dudv = \int_{-r}^{r} e^{iAu} du \int_{-R}^{R} e^{iBv} dv = \frac{2\sin Ar}{A} \frac{2\sin BR}{B};$$

therefore, from (6.62), we have

$$f(x, y) = \int_{\mathbb{R}^2} f(\eta, \zeta) h(x, y, \eta, \zeta) d\eta d\zeta,$$

where $h(x, y, \eta, \zeta)$ is the reproducing kernel given by

$$h(x, y, \eta, \zeta) = \frac{\sin Ar}{\pi A \sin\gamma} \frac{\sin BR}{\pi B \sin\gamma} \exp\left\{ a\left(x^2 + y^2 - \eta^2 - \zeta^2\right)\right\}.$$

Remark. It should be noted that the reproducing kernel depends on the angle γ, which is one half of the sum of the transform two angles α, β, but not on δ, which is one half the difference of the two angles.

6.7.2 Sampling Theorems

In this section, we derive the sampling theorem for the two-dimensional coupled FrFT (CFrFT). But first, let us recall that for the fractional Fourier transform given by (6.6), the sampling theorem reads as follows: if f is bandlimited to the N-dimensional parallelepiped, $[-\sigma_1,\sigma_1]\times\cdots\times[-\sigma_N,\sigma_N]$, then f can be reconstructed from its samples according to the following formula

$$\begin{aligned} f(t_1,t_2,\ldots,t_N) &= \prod_{k=1}^{N} e^{-i(\cot\theta_k)t_k^2/2} \sum_{n_1,n_2,\ldots,n_N=-\infty}^{\infty} \\ &\times f\left(n_1\pi\sin\theta_1/\sigma_1,\ldots,n_N\pi\sin\theta_N/\sigma_N\right) \\ &\times \prod_{k=1}^{N} e^{-i(\cot\theta_k)(n_k\pi\sin\theta_k/\sigma_k)^2/2}\frac{\sin\left[\sigma_k\csc\theta_k(t_k-n_k\pi\sin\theta_k/\sigma_k)\right]}{\sigma_k\csc\theta_k(t_k-n_k\pi\sin\theta_k/\sigma_k)}. \end{aligned}$$

In particular, in two dimensions, the formula and notation can be simplified to

$$\begin{aligned} f(x,y) &= e^{-ix^2\cot\alpha/2}e^{-iy^2\cot\beta/2}\sum_{m,n} f(x_m,y_n)e^{-ix_m^2\cot\alpha/2}e^{-iy_n^2\cot\beta/2} \\ &\times \frac{\sin\sigma_1\csc\alpha(x-x_m)}{\sigma_1\csc\alpha(x-x_m)}\frac{\sin\sigma_2\csc\beta(y-y_n)}{\sigma_2\csc\beta(y-y_n)}, \end{aligned} \tag{6.63}$$

where $x_m = m\pi\sin\alpha/\sigma_1$ and $y_n = n\pi\sin\beta/\sigma_2$.

If $f(x,y)$ is bandlimited to $[-\pi r,\pi r]\times[-\pi R,\pi R]$, the sampling points are located at

$$x_m = m\sin\alpha/r,\; y_n = n\sin\beta/R, \tag{6.64}$$

In particular, when $\sigma_1=\sigma_2=\pi$ and $\alpha=\beta=\pi/2$, we obtain the sampling theorem for the two-dimensional Fourier transform with $x_m=m$ and $y_n=n$.

Here it should be emphasized that there is a striking difference between the sampling formula we are about to derive and the sampling formula that is based on the tensor product of two one-dimensional fractional Fourier transforms given by Eq. (6.63). In the latter sampling theorem, the sampling function is a product of two sinc functions, one in each variable, say x and y. In the former, as we will see, the sampling function is a product of two sinc functions but the argument of one of them is a weighted difference, w_1, of the two transform variables and the argument of the second one, w_2, is a weighted sum of the two variables.

Theorem 33. *[334] Let f be bandlimited to $\Omega=[-r,r]\times[-R,R]$, in the CFrFT domain. Then f can be reconstructed from its samples via the formula*

$$\tilde{f}(x,y) = \sum_{m,n=-\infty}^{\infty}\tilde{f}(x_{m,n},y_{m,n})\frac{\sin(m\pi-rw_1)\sin(n\pi-Rw_2)}{(m\pi-rw_1)(n\pi-Rw_2)}, \tag{6.65}$$

where $\tilde{f}(x,y)=e^{-a(x^2+y^2)}f(x,y)$, $\gamma=(\alpha+\beta)/2$ and $\delta=(\alpha-\beta)/2$,

$$w_2 = \tilde{b}y+\tilde{c}x \quad \text{and } w_1 = \tilde{b}x-\tilde{c}y, \tag{6.66}$$

and

$$x_{m,n} = \pi \sin\gamma \left(\frac{m}{r} \cos\delta + \frac{n}{R} \sin\delta \right), \quad y_{m,n} = \pi \sin\gamma \left(\frac{n}{R} \cos\delta - \frac{m}{r} \sin\delta \right), \tag{6.67}$$

The sampling points lie on parallel lines and the slope of the line connecting the points $(x_{m,n}, y_{m,n})$ *and* $(x_{m_1,n_1}, y_{m_1,n_1})$ *is given by*

$$\frac{\frac{n_1-n}{R}\cos\delta - \frac{m_1-m}{r}\sin\delta}{\frac{m_1-m}{r}\cos\delta + \frac{n_1-n}{R}\sin\delta}.$$

Proof *Let* f *be bandlimited to* $\Omega = [-r,r] \times [-R,R]$*. From the inversion formula (6.26), we have*

$$f(x,y) = d(-\gamma) \int_\Omega F(u,v) \exp\left\{ a\left(x^2+y^2+u^2+v^2\right) - b(xu+yv) - c(xv-yu) \right\} dudv.$$

By setting $\tilde{f}(x,y) = e^{-a(x^2+y^2)} f(x,y)$ *and* $\tilde{F}(u,v) = e^{a(u^2+v^2)} F(u,v)$*, we obtain*

$$\tilde{f}(x,y) = d(-\gamma) \int_\Omega \tilde{F}(u,v) \exp\left\{ -iu(\tilde{b}x - \tilde{c}y) - iv(\tilde{b}y + \tilde{c}x) \right\} dudv.$$

By putting

$$w_2 = \tilde{b}y + \tilde{c}x \quad \text{and } w_1 = \tilde{b}x - \tilde{c}y,$$

we obtain

$$\tilde{f}(x,y) = d(-\gamma) \int_\Omega \tilde{F}(u,v) \exp\{-iuw_2 - ivw_1\} dudv \tag{6.68}$$

Since the support of $F(u,v)$ *is* $\Omega = [-r,r] \times [-R,R]$*, then clearly, the support of* $\tilde{F}$ *is also* Ω*. Therefore, we can expand it in a double Fourier series as*

$$\tilde{F}(u,v) = \sum_{m,n=-\infty}^{\infty} \hat{F}_{m,n} \exp\{im\pi u/r + in\pi v/R\}, \tag{6.69}$$

where

$$\hat{F}_{m,n} = \frac{1}{4rR} \int_\Omega \tilde{F}(u,v) \exp\{-im\pi u/r - in\pi v/R\} dudv. \tag{6.70}$$

By substituting (6.69) and (6.70) into (6.68), we have

$$\tilde{f}(x,y) = d(-\gamma) \sum_{m,n=-\infty}^{\infty} \hat{F}_{m,n} \int_\Omega \exp\{iu(m\pi/r - w_1) + iv(n\pi/R - w_2\} dudv.$$

But it is easy to see that

$$\int_{-r}^{r} e^{iu(m\pi/r - w_1)} du = \frac{2\sin r(m\pi/r - w_1)}{(m\pi/r - w_1)}$$

and similar expression for the integral over v*. Therefore, we have*

$$\begin{aligned} \tilde{f}(x,y) &= d(-\gamma) \sum_{m,n=-\infty}^{\infty} \hat{F}_{m,n} \frac{4\sin r(m\pi/r - w_1)\sin R(n\pi/R - w_2)}{(m\pi/r - w_1)(n\pi/R - w_2)} \\ &= 4rRd(-\gamma) \sum_{m,n=-\infty}^{\infty} \hat{F}_{m,n} \frac{\sin(m\pi - rw_1)\sin(n\pi - Rw_2)}{(m\pi - rw_1)(n\pi - Rw_2)}. \end{aligned} \tag{6.71}$$

To express $\hat{F}_{m,n}$ *in terms of the values of* $\tilde{f}(x,y)$, *we have*

$$\begin{aligned}\tilde{f}(x,y) &= d(-\gamma)\int_{\Omega}\tilde{F}(u,v)\exp\left\{-iu(\tilde{b}x-\tilde{c}y)-iv(\tilde{b}y+\tilde{c}x)\right\}dudv\\ &= d(-\gamma)\int_{\Omega}\tilde{F}(u,v)\exp\left\{-i\left[x(\tilde{b}u+\tilde{c}v)+y(\tilde{b}v-\tilde{c}u)\right]\right\}dudv;\end{aligned}$$

hence

$$\begin{aligned}\tilde{f}(x_{m,n},y_{m,n}) &= d(-\gamma)\int_{\Omega}\tilde{F}(u,v)\exp\left\{-ix_{m,n}(\tilde{b}u+\tilde{c}v)-iy_{m,n}(\tilde{b}v-\tilde{c}u)\right\}dudv\\ &= d(-\gamma)\int_{\Omega}\tilde{F}(u,v)\exp\left\{-iu\left(\tilde{b}x_{m,n}-\tilde{c}y_{m,n}\right)-iv\left(\tilde{c}x_{m,n}+\tilde{b}y_{m,n}\right)\right\}dudv\end{aligned}$$

Putting

$$\tilde{b}x_{m,n}-\tilde{c}y_{m,n}=\frac{m\pi}{r},\quad \textit{and}\quad \tilde{c}x_{m,n}+\tilde{b}y_{m,n}=\frac{n\pi}{R},\tag{6.72}$$

we obtain

$$\tilde{f}(x_{m,n},y_{m,n})=4rRd(-\gamma)\hat{F}_{m,n}.$$

Therefore, Eq. (6.71) may be written as

$$\tilde{f}(x,y)=\sum_{m,n=-\infty}^{\infty}\tilde{f}(x_{m,n},y_{m,n})\frac{\sin(m\pi-rw_1)\sin(n\pi-Rw_2)}{(m\pi-rw_1)(n\pi-Rw_2)}.\tag{6.73}$$

Now we solve for $(x_{m,n},y_{m,n})$ *in terms of the transformation parameters. Since*

$$\tilde{b}=\frac{\cos\delta}{\sin\gamma},\quad \textit{and}\ \tilde{c}=\frac{\sin\delta}{\sin\gamma},$$

it follows that

$$\tilde{b}^2+\tilde{c}^2=\csc^2\gamma.$$

By solving the system of equations (6.72), we get

$$\begin{aligned}\tilde{b}^2x_{m,n}-\tilde{b}\tilde{c}y_{m,n} &= \tilde{b}\frac{m\pi}{r}\\ \tilde{c}^2x_{m,n}+\tilde{c}\tilde{b}y_{m,n} &= \tilde{c}\frac{n\pi}{R},\end{aligned}$$

which leads to

$$\begin{aligned}x_{m,n} &= \pi\sin\gamma\left(\frac{m}{r}\cos\delta+\frac{n}{R}\sin\delta\right),\\ y_{m,n} &= \pi\sin\gamma\left(\frac{n}{R}\cos\delta-\frac{m}{r}\sin\delta\right).\end{aligned}$$

By substituting the last two equations into (6.73), we obtain (6.65).

To find the slope of the lines connecting the sampling points, we write

$$\begin{aligned}\textit{slope} &= \frac{y_{m_1,n_1}-y_{m,n}}{x_{m_1,n_1}-x_{m,n}}\\ &= \frac{\pi\sin\gamma\left[\left(\frac{n_1}{R}\cos\delta-\frac{m_1}{r}\sin\delta\right)-\left(\frac{n}{R}\cos\delta-\frac{m}{r}\sin\delta\right)\right]}{\pi\sin\gamma\left[\left(\frac{m_1}{r}\cos\delta+\frac{n_1}{R}\sin\delta\right)-\left(\frac{m}{r}\cos\delta+\frac{n}{R}\sin\delta\right)\right]}\\ &= \frac{\frac{n_1-n}{R}\cos\delta-\frac{m_1-m}{r}\sin\delta}{\frac{m_1-m}{r}\cos\delta+\frac{n_1-n}{R}\sin\delta}\end{aligned}$$

In particular, for fixed m, i.e., $m_1 = m$, we have

$$\textit{slope} = \frac{\cos\delta}{\sin\delta} = \cot\delta,$$

and for fixed n, i.e., $n_1 = n$, we have

$$\textit{slope} = \frac{-\sin\delta}{\cos\delta} = -\tan\delta.$$

6.7.3 Examples and Sampling Points Configuration

In this subsection, we discuss some special cases of the sampling points. For simplicity, we assume that the signal is bandlimited to $[-\pi r, \pi r] \times [-\pi R, \pi R]$.

Case I:

If $\alpha = \beta$, then $\gamma = \alpha$ and $\delta = 0$, hence,

$$x_{m,n} = \frac{m}{r}\sin\alpha, \quad y_{m,n} = \frac{n}{R}\sin\alpha,$$

and the sampling points are

$$\left(\frac{m}{r}\sin\alpha, \frac{n}{R}\sin\alpha\right),$$

which is consistent with (6.64). The slope of the lines are

$$\text{slope} = \frac{r(n_1 - n_2)}{R(m_1 - m_2)}.$$

When $r = R = 1$, and $\alpha = \beta = \pi/2$, we get the standard sampling points (m, n) for the standard two-dimensional Fourier transform. A graph displaying sampling points in this case can be seen in Fig. 6.1.

Case II:

If $\alpha = 2\beta$, then $\gamma = 3\beta/2$, and $\delta = \beta/2$. Hence,

$$\begin{aligned} x_{m,n} &= \sin(3\beta/2)\left(\frac{m}{r}\cos(\beta/2) + \frac{n}{R}\sin(\beta/2)\right) \\ y_{m,n} &= \sin(3\beta/2)\left(\frac{n}{R}\cos(\beta/2) - \frac{m}{r}\sin(\beta/2)\right). \end{aligned}$$

In particular, if $\beta = \pi/3$, we have

$$x_{m,n} = \frac{1}{2}\left(\frac{m\sqrt{3}}{r} + \frac{n}{R}\right) \quad \text{and } y_{m,n} = \frac{1}{2}\left(\frac{n\sqrt{3}}{R} - \frac{m}{r}\right).$$

A graph displaying the sampling points in this case can be seen in Fig. 6.2.

Case III:

If $\alpha = 3\beta$, then $\gamma = 2\beta$ and $\delta = \beta$. Hence,

$$\begin{aligned} x_{m,n} &= \sin(2\beta)\left(\frac{m}{r}\cos(\beta) + \frac{n}{R}\sin(\beta)\right) \\ y_{m,n} &= \sin(2\beta)\left(\frac{n}{R}\cos(\beta) - \frac{m}{r}\sin(\beta)\right), \end{aligned}$$

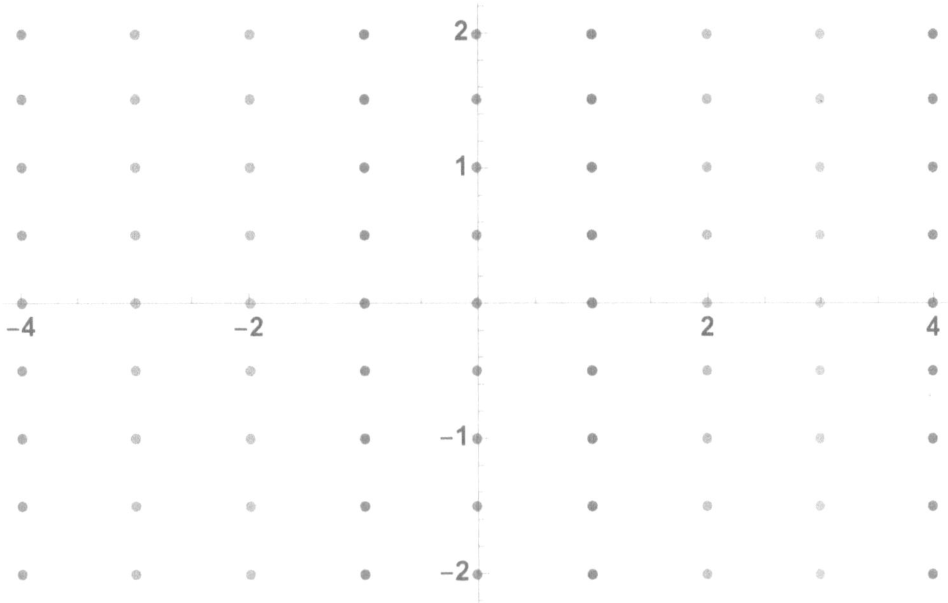

Figure 6.1 Graphs of the sampling points $(x_{m,n}, y_{m,n})$ when $\alpha = \beta = \pi/2$; or $\gamma = \pi/2, \delta = 0, r = 1, R = 2$

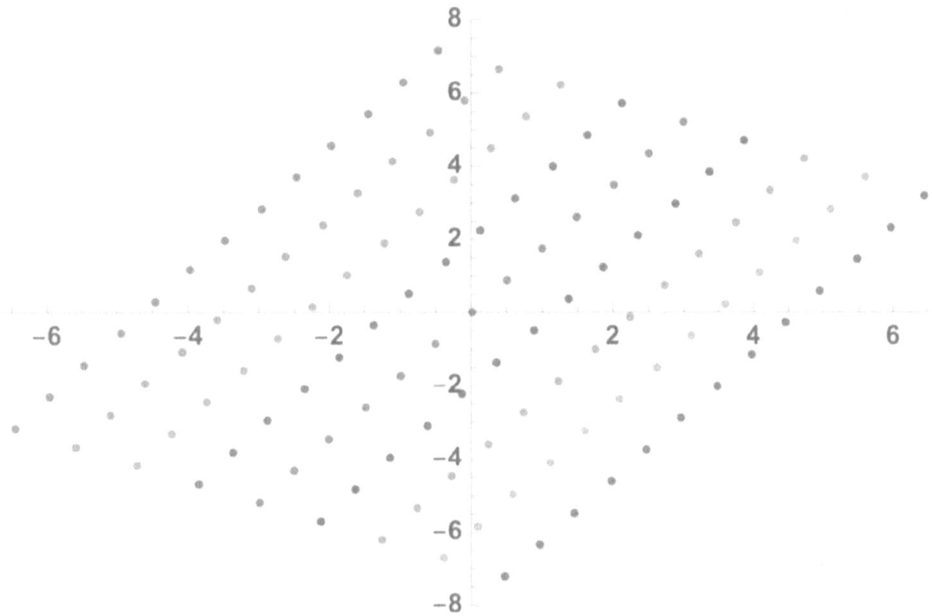

Figure 6.2 Graphs of the sampling points $(x_{m,n}, y_{m,n})$ when $\alpha = 2\beta, \beta = \pi/3$; or $\gamma = \pi/2, \delta = \pi/6, r = R = 1$

and when $\beta = \pi/4$, we have

$$x_{m,n} = \frac{1}{\sqrt{2}}\left(\frac{m}{r} + \frac{n}{R}\right) \quad \text{and } y_{m,n} = \frac{1}{\sqrt{2}}\left(\frac{n}{R} - \frac{m}{r}\right).$$

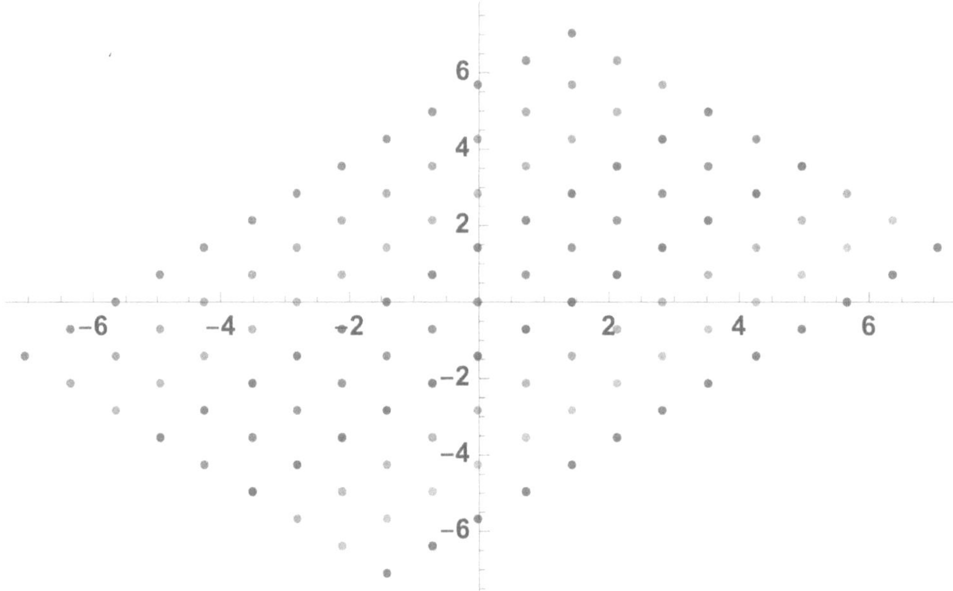

Figure 6.3 Graphs of the sampling points $(x_{m,n}, y_{m,n})$ when $\alpha = 3\beta, \beta = \pi/4$; or $\gamma = \pi/2, \delta = \pi/4, r = R = 1$

In particular, when $r = R = 1$, we have the sampling points

$$\frac{1}{\sqrt{2}}(m+n, n-m).$$

A graph displaying the sampling points in this case can be seen in Fig. 6.3. Other configurations of the sample points are shown in Figs. 6.4 and 6.5.

6.8 THE COUPLED FRACTIONAL FOURIER TRANSFORM OF GENERALIZED FUNCTIONS

The coupled fractional Fourier transform is defined for $f \in L^p(\mathbb{R}^2)$, $1 \le p \le 2$. In this section, we extend the transform to a space of generalized functions. The following lemma will be needed.

Lemma 3. *To simplify the notation, let $k = k_{\alpha,\beta}(x, y; u, v)$ be the kernel of the coupled fractional Fourier transform given by Eq. (6.21). We have*

$$\frac{\partial^{m+n} k}{\partial u^m \partial v^n} = k\left[(-2au)^m + P_m(x, y; u)\right]\left[(-2av)^n + \tilde{P}_n(x, y; v)\right],$$

where $P_m(x, y; u)$ and $\tilde{P}_n(x, y; v)$ are polynomials of degree m and n in x and y and of degree $m-1$ in u and $n-1$ in v, respectively.

Proof *It is easy to see that*

$$\frac{\partial k}{\partial u} = kH, \quad \frac{\partial k}{\partial v} = k\tilde{H},$$

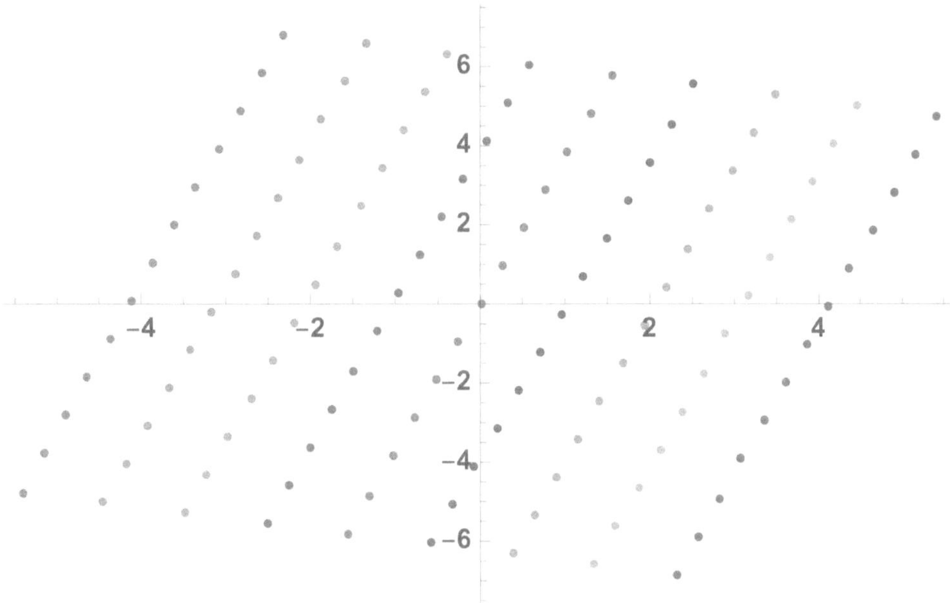

Figure 6.4 Graphs of the sampling points $(x_{m,n}, y_{m,n})$ when $\alpha = 7\pi/12, \beta = 5\pi/12$; or $\gamma = \pi/2, \delta = \pi/12, r = R = 1$

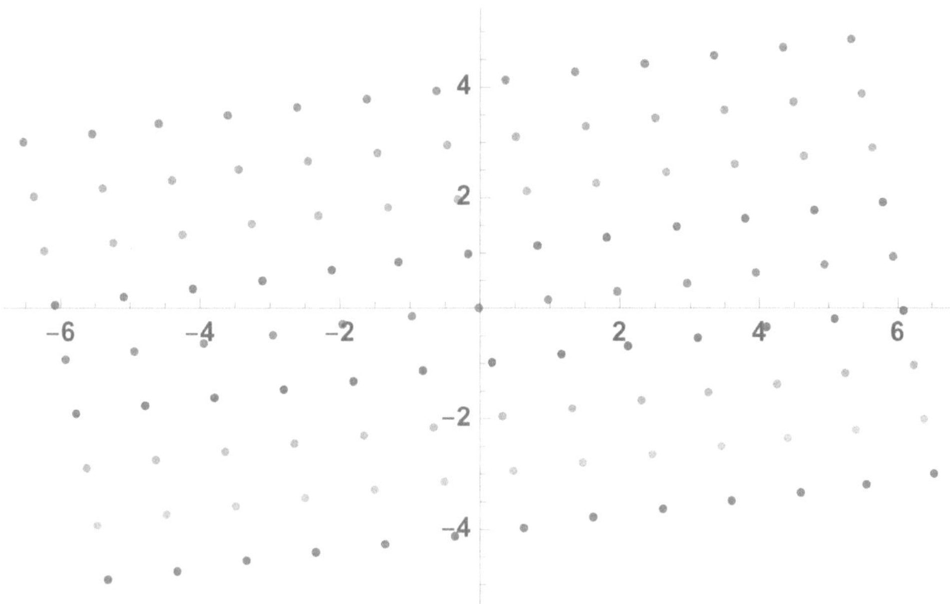

Figure 6.5 Graphs of the sampling points $(x_{m,n}, y_{m,n})$ when $\alpha = 19\pi/20, \beta = \pi/20$; or $\gamma = \pi/2, \delta = 9\pi/20, r = R = 1$

where

$$H = H(x,y;u) = -2au + \eta(x,y),\ \eta(x,y) = bx - cy,$$

and

$$\tilde{H} = -2av + \tilde{\eta}(x,y), \quad \tilde{\eta}(x,y) = by + cx.$$

Hence

$$\frac{\partial^2 k}{\partial u^2} - 2ak + kH^2 = k\left(4a^2u^2 + \eta^2 - 4au\eta - 2a\right) = k\left[(-2au)^2 + P_2(x,y;u)\right],$$

where $P_2(x,y;u) = \eta^2 - 4au\eta - 2a$ *is a polynomial of degree 2 in* x *and* y *and of degree 1 in* u. *In general,*

$$\frac{\partial^n k}{\partial u^n} = k\left[(-2au)^n + P_n(x,y;u)\right], \tag{6.74}$$

where $P_n(x,y;u)$ *is a polynomial of degree* n *in* x *and* y *and of degree* $n-1$ *in* u.

We prove (6.74) by induction. By differentiating the last equation with respect to u, *we have*

$$\begin{aligned}
\frac{\partial^{n+1}k}{\partial u^{n+1}} &= \frac{\partial}{\partial u} k\left[(-2au)^n + P_n(x,y;u)\right] = k\left[n(-2a)^n u^{n-1} + \frac{\partial}{\partial u}P_n(x,y;u)\right] \\
&+ \left[(-2au)^n + P_n(x,y;u)\right] Hk \\
&= k\left[n(-2a)^n u^{n-1} + \frac{\partial}{\partial u}P_n(x,y;u) + \{(-2au)^n + P_n(x,y;u)\}(-2au + \eta(x,y))\right] \\
&= k\left[(-2au)^{n+1} + P_{n+1}(x,y;u)\right],
\end{aligned}$$

where

$$\begin{aligned}
P_{n+1}(x,y;u) &= n(-2a)^n u^{n-1} + \frac{\partial}{\partial u}P_n(x,y;u) \\
&+ (-2au)^n \eta + (-2au)P_n(x,y;u) + P_n(x,y;u)\eta
\end{aligned}$$

is a polynomial of degree $n+1$ *in* x *and* y *and of degree* n *in* u *because* $P_n(x,y;u)$ *is of degree* n *in* x *and* y *and* $n-1$ *in* u. *Similarly,*

$$\frac{\partial^n k}{\partial v^n} = k\left[(-2av)^n + \tilde{P}_n(x,y;v)\right],$$

where $\tilde{P}_n(x,y;u)$ *is a polynomial of degree* n *in* x *and* y *and of degree* $n-1$ *in* v. *Since*

$$\frac{\partial^2 k}{\partial u \partial v} = kH\tilde{H},$$

it can be easily shown as above that

$$\frac{\partial^{m+n}k}{\partial u^m \partial v^n} = k\left[(-2au)^m + P_m(x,y;u)\right]\left[(-2av)^n + \tilde{P}_n(x,y;v)\right],$$

where $P_m(x,y;u)$ *and* $\tilde{P}_n(x,y;v)$ *are polynomials of degree* m *and* n *in* x *and* y *and of degree* $m-1$ *in* u *and* $n-1$ *in* v, *respectively.*

Now we extend the transform to a space of generalized functions. Let $\mathcal{E}(\mathbb{R}^2)$ be the testing-function space of all infinitely differentiable functions on $\mathbb{R}^2$ and $\mathcal{E}^*$ be its dual space which is the space of all generalized functions with compact support. It is known that $\mathcal{E}^*$ is a subspace of the space $\mathfrak{D}^*$ of Schwartz distributions. Since the kernel of the coupled fractional Fourier transform $k_{\alpha,\beta}(x,y;u,v)$ is in the space $\mathcal{E}(\mathbb{R}^2)$, we have the following definition

Definition 32. *Let $f \in \mathcal{E}^*$. We define the coupled fractional Fourier transform of f as*

$$F(u,v) = F_{\alpha,\beta}(u,v) = \langle f(x,y), k_{\alpha,\beta}(x,y;u,v)\rangle.$$

Theorem 34. *Let $f \in \mathcal{E}^*$. Then its coupled fractional Fourier transform $F(u,v)$ is in the space $\mathcal{E}(\mathbb{R}^2)$ and satisfies*

$$\left|\frac{\partial^{m+n}F}{\partial u^m \partial v^n}\right| \leq C|u|^m|v|^n$$

for sufficiently large u and v. That is F is a $C^\infty(\mathbb{R}^2)$ function in both u and v and does not grow faster than a polynomial as $|u|$ and $|v|$ go to infinity, i.e., F is a tempered function.

Proof *By taking the derivatives of F with respect to u and v, we have by Lemma 3*

$$\begin{aligned}
\frac{\partial^{m+n}F}{\partial u^m \partial v^n} &= \langle f, \frac{\partial^{m+n}k}{\partial u^m \partial v^n}\rangle \\
&= \langle f, k\left[(-2au)^m + P_m(x,y;u)\right]\left[(-2av)^n + \tilde{P}_n(x,y;u)\right]\rangle \\
&= \langle f, (-2au)^m(-2av)^n k\left[1 + \frac{\tilde{P}_n}{(-2av)^n} + \frac{P_m}{(-2au)^m} + \frac{P_m\tilde{P}_n}{(-2au)^m(-2av)^n}\right]\rangle \\
&= (-2au)^m(-2av)^n\langle f, k\left[1 + G(x,y,;u,v)\right]\rangle \\
&= (-2au)^m(-2av)^n\left\{\langle f,k\rangle + \langle f,kG\rangle\right\}
\end{aligned}$$

where

$$G(x,y;u,v) = \frac{\tilde{P}_n}{(-2av)^n} + \frac{P_m}{(-2au)^m} + \frac{P_m\tilde{P}_n}{(-2au)^m(-2av)^n}.$$

Since G is a polynomial in x and y, it is a multiplier of the space $\mathcal{E}(\mathbb{R}^2)$ and; hence, $\langle f,kG\rangle$ is well defined and consequently so is $\frac{\partial^{m+n}F}{\partial u^m \partial v^n}$. Finally, since $|G(x,y;u,v)|$ can be made arbitrary small for u and v sufficiently large, it will follow that

$$\left|\frac{\partial^{m+n}F}{\partial u^m \partial v^n}\right| \leq C|u|^m|v|^n,$$

for some constant C.

Remark. The coupled fractional Fourier transform maps the space $\mathcal{E}^*$ into a subspace of the space of tempered distribution. Next we derive the inversion formula for the coupled fractional Fourier transform of the generalized function f.

Theorem 35. *Let $\mathcal{F}_{\alpha,\beta}[f](u,v) = F_{\alpha,\beta}(u,v)$ be the coupled FrFT of a generalized function f with compact support. Then*

$$f(x,y) = \lim_{r\to\infty}\lim_{R\to\infty}\int_{-r}^{r}\int_{-R}^{R} F_{\alpha,\beta}(u,v)k_{-\alpha,-\beta}(x,y;u,v)\,dudv, \qquad (6.75)$$

where the limits are taken in the space S^* of tempered distributions.

Proof *We need to show that*

$$\langle f(x,y),\phi(x,y)\rangle = \lim_{r\to\infty}\lim_{R\to\infty}\langle \int_{-r}^{r}\int_{-R}^{R} F_{\alpha,\beta}(u,v)k_{-\alpha,-\beta}(x,y,u,v)\,dudv,\phi(x,y)\rangle,$$

for all $\phi \in \mathcal{S}(\mathbb{R}^2)$, where $\mathcal{S}(\mathbb{R}^2)$ is the Schwartz space of all infinitely differentiable functions with rapid decay . We have

$$\begin{aligned}
&\langle \int_{-r}^{r}\int_{-R}^{R} F_{\alpha,\beta}(u,v)k_{-\alpha,-\beta}(x,y,u,v)\,dudv,\phi(x,y)\rangle \\
&= \int_{\mathbb{R}^2}\phi(x,y)\,dxdy\int_{-r}^{r}\int_{-R}^{R} F_{\alpha,\beta}(u,v)k_{-\alpha,-\beta}(x,y,u,v)\,dudv \\
&= \int_{\mathbb{R}^2}\phi(x,y)\,dxdy\int_{-r}^{r}\int_{-R}^{R}\langle f(w,z),k_{\alpha,\beta}(w,z,u,v)\rangle k_{-\alpha,-\beta}(x,y,u,v)\,dudv \\
&= \langle f(w,z),\int_{\mathbb{R}^2}\phi(x,y)\,dxdy\int_{-r}^{r}\int_{-R}^{R} k_{\alpha,\beta}(w,z,u,v)k_{-\alpha,-\beta}(x,y,u,v)\,dudv\rangle
\end{aligned}$$

Changing the order of integration is possible because ϕ is infinitely differentiable with rapid decay and the integrant is a continuous function of x,y and u,v. Therefore,

$$\begin{aligned}
&\lim_{r\to\infty}\lim_{R\to\infty}\langle \int_{-r}^{r}\int_{-R}^{R} F_{\alpha,\beta}(u,v)k_{-\alpha,-\beta}(x,y;u,v)\,dudv,\phi(x,y)\rangle \\
&= \lim_{r\to\infty}\lim_{R\to\infty}\langle f(w,z),\int_{R^2}\phi(x,y)\,dxdy\int_{-r}^{r}\int_{-R}^{R} k_{\alpha,\beta}(w,z;u,v) \\
&\times\ k_{-\alpha,-\beta}(x,y;u,v)\,dudv\rangle \\
&= \langle f(w,z),\lim_{r\to\infty}\lim_{R\to\infty}\int_{\mathbb{R}^2}\phi(x,y)\,dxdy\int_{-r}^{r}\int_{-R}^{R} k_{\alpha,\beta}(w,z;u,v) \\
&\times\ k_{-\alpha,-\beta}(x,y;u,v)\,dudv\rangle \\
&= \langle f(w,z),\int_{\mathbb{R}^2}\phi(x,y)\,dxdy\int_{\mathbb{R}^2} k_{\alpha,\beta}(w,z;u,v) \\
&\times\ k_{-\alpha,-\beta}(x,y;u,v)\,dudv\rangle \\
&= \langle f(w,z),\int_{\mathbb{R}^2}\phi(x,y)\,\delta(x,y;w,z)dxdy\rangle = \langle f(w,z),\phi(w,z)\rangle,
\end{aligned}$$

which is equivalent to (6.75).

6.9 THE GYRATOR TRANSFORM

6.9.1 Motivation and Definitions

The Gyrator transform was introduced in optics in the early 2000's by R. Simon and K. B. Wolf [278, 279] and Wolf [320], but its mathematical properties and optical applications were developed by J. A. Rodrigo, T. Alieva, M. Calvo in [250, 251, 252]. Both the Gyrator and the fractional Fourier transforms are special cases of the general class of linear canonical integral transforms which represents rotations in the time-frequency plane and in the position-spatial frequency plane of phase space. More properties and extensions of the Gyrator transform can be found in [152].

The Gyrator transform of order α is a two-dimensional integral transform defined for a function $f(t_1,t_2)$ as

$$(R_\alpha f)(\omega_1,\omega_2) = \int_{\mathbb{R}}\int_{\mathbb{R}} f(t_1,t_2)G_\alpha(t_1,t_2,\omega_1,\omega_2)\,dt_1dt_2, \quad \alpha \neq n\pi \tag{6.76}$$

where

$$G_\alpha(t_1,t_2,\omega_1,\omega_2) = \frac{1}{2\pi|\sin\alpha|}\exp\left\{\frac{i}{\sin\alpha}[(t_1t_2+\omega_1\omega_2)\cos\alpha-(t_2\omega_1+t_1\omega_2)]\right\}.$$

Because G_α is bounded, the transform is defined for all $f \in L^1(\mathbb{R}^2)\bigcap L^2(\mathbb{R}^2)$, and as was done for the fractional Fourier transform, the definition can be extended to all $f \in L^2(\mathbb{R}^2)$. It is evident that:

1. R_α is periodic with period 2π, in α
2. R_0 is the identity transformation, and
3. $R_{\pi/2}$ is the Fourier transform with rotation of the coordinates by $\pi/2$.

We will show that $R_\alpha R_\beta = R_{\alpha+\beta}$, and in addition if $f,g \in L^2(\mathbb{R}^2)$, then

$$\langle R_\alpha f, R_\alpha g\rangle = \langle f,g\rangle.$$

We begin by discussing the relationship between the Gyrator and the fractional Fourier transforms. Consider the two-dimensional fractional Fourier transform of the form

$$F_{\alpha,\beta}(w_1,w_2) = \int_{\mathbb{R}}\int_{\mathbb{R}} f(x_1,x_2)K_\alpha(x_1,w_1)K_\beta(x_2,w_2)dx_1dx_2,$$

which reduces to

$$\begin{aligned} F_{-\alpha,\alpha}(w_1,w_2) &= \frac{1}{2\pi|\sin\alpha|}\int_{\mathbb{R}}\int_{\mathbb{R}} f(x_1,x_2) \\ \times\ & \exp\left\{\frac{i}{2\sin\alpha}\Big[(x_1^2-x_2^2+w_1^2-w_2^2)\cos\alpha-2(x_1w_1-x_2w_2)\Big]\right\}dx_1dx_2. \\ &= \int_{\mathbb{R}}\int_{\mathbb{R}} f(x_1,x_2)H_\alpha(x_1,x_2,w_1,w_2)dx_1dx_2, \end{aligned} \tag{6.77}$$

where

$$\begin{aligned} H_\alpha(x_1,x_2,w_1,w_2) &= K_{-\alpha}(x_1,w_1)K_\alpha(x_2,w_2) \\ &= \frac{1}{2\pi|\sin\alpha|}\exp\left\{\frac{i}{2\sin\alpha}\Big[(x_1^2-x_2^2+w_1^2-w_2^2)\cos\alpha-2(x_1w_1-x_2w_2)\Big]\right\} \end{aligned} \tag{6.78}$$

Let

$$A = \frac{1}{\sqrt{2}}\begin{pmatrix} 1 & 1 \\ -1 & 1 \end{pmatrix}. \tag{6.79}$$

and

$$\mathbf{x} = (x_1,x_2)^T, \mathbf{w} = (w_1,w_2)^T, \ldots, \text{ etc.},$$

where T stands for the transpose operation. The matrix A corresponds to a plane rotation by angle $\pi/4$. By making the substitution

$$\mathbf{x} = A\mathbf{t}, \quad \mathbf{w} = A\boldsymbol{\omega},$$

that is

$$x_1 = \frac{(t_1+t_2)}{\sqrt{2}}, \quad x_2 = \frac{(-t_1+t_2)}{\sqrt{2}}, \quad w_1 = \frac{(\omega_1+\omega_2)}{\sqrt{2}}, \quad w_2 = \frac{(-\omega_1+\omega_2)}{\sqrt{2}},$$

in Equation (6.77), we obtain with some calculations,

$$\begin{aligned} F_{-\alpha,\alpha}(\omega_1,\omega_2) &= \frac{1}{2\pi|\sin\alpha|}\int_{\mathbb{R}}\int_{\mathbb{R}} f(t_1,t_2) \\ &\quad \times \exp\left\{\frac{i}{\sin\alpha}\left[(t_1t_2+\omega_1\omega_2)\cos\alpha-(t_1\omega_2+t_2\omega_1)\right]\right\}dt_1d_2 \\ &= \int_{\mathbb{R}}\int_{\mathbb{R}} f(t_1,t_2)H_\alpha(A\mathbf{t},A\boldsymbol{\omega})dt_1dt_2 \\ &= \int_{\mathbb{R}}\int_{\mathbb{R}} f(t_1,t_2)G_\alpha(t_1,t_2,\omega_1,\omega_2)dt_1dt_2 \\ &= \int_{\mathbb{R}}\int_{\mathbb{R}} f(\mathbf{t})G_\alpha(\mathbf{t},\boldsymbol{\omega})d\mathbf{t} = R_\alpha[f](\boldsymbol{\omega}), \end{aligned} \tag{6.80}$$

where

$$G_\alpha(\mathbf{t},\boldsymbol{\omega}) = H_\alpha(A\mathbf{t},A\boldsymbol{\omega}).$$

We may write the last equation as

$$R_\alpha[f](\boldsymbol{\omega}) = F_{-\alpha,\alpha}[f(A\mathbf{t})](A\boldsymbol{\omega}).$$

The inverse of the Gyrator transform is easily seen to be $G_{-\alpha}$ which could be derived from that of the fractional Fourier transform or directly from the relation

$$\begin{aligned} R_{-\alpha}(R_\alpha[f])(\mathbf{t}) &= \int_{\mathbb{R}^2} R_\alpha[f](\mathbf{w})G_{-\alpha}(\mathbf{w},\mathbf{t})d\mathbf{w} \\ &= \int_{\mathbb{R}^2} G_{-\alpha}(\mathbf{w},\mathbf{t})d\mathbf{w}\int_{\mathbb{R}^2} f(\mathbf{x})G_\alpha(\mathbf{x.w})d\mathbf{x} \\ &= \int_{\mathbb{R}^2} f(\mathbf{x})d\mathbf{x}\int_{\mathbb{R}^2} G_{-\alpha}(\mathbf{w},\mathbf{t})G_\alpha(\mathbf{x},\mathbf{w})d\mathbf{w}. \end{aligned} \tag{6.81}$$

But

$$\begin{aligned} &\int_{\mathbb{R}^2} G_{-\alpha}(\mathbf{w},\mathbf{t})G_\alpha(\mathbf{x.w})d\mathbf{w} \\ =\ & \frac{1}{(2\pi\sin\alpha)^2}\int_{\mathbb{R}}\exp\left\{\frac{i}{\sin\alpha}\left[(x_1x_2-t_1t_2)\cos\alpha-w_1(x_2-t_2)-w_2(x_1-t_1)\right]\right\}dw_1dw_2 \\ =\ & \delta(x_1-t_1)\delta(x_2-t_2), \end{aligned} \tag{6.82}$$

and by substituting this into (6.81), we have

$$R_{-\alpha}(R_\alpha[f])(\mathbf{t}) = f(\mathbf{t}) = f(t_1,t_2).$$

6.9.2 Elementary Properties of the Gyrator Transform

1. **Linearity:**
$$R_\alpha[af+bg] = aR_\alpha[f]+bR_\alpha[g], \quad \text{where } a,b\in\mathbb{C}.$$

2. **Translation:**
$$\begin{aligned} R_\alpha\left[f(t_1-a,t_2-b)\right](w_1,w_2) = \\ \exp\left\{\frac{i}{2}ab\sin 2\alpha - i(bw_1+aw_2)\sin\alpha\right\} \\ \times \quad R_\alpha\left[f(t_1,t_2)\right](w_1-a\cos\alpha, w_2-b\cos\alpha). \end{aligned} \tag{6.83}$$

3. **Frequency Shift:**
$$\begin{aligned} R_\alpha\left[e^{-i(at_1+bt_2)}f(t_1,t_2)\right](w_1,w_2) = \\ \exp\left\{-i\left[\frac{1}{2}ab\sin 2\alpha+(aw_1+bw_2)\cos\alpha\right]\right\} \\ \times \quad R_\alpha\left[f(t_1,t_2)\right](w_1+b\sin\alpha, w_2+a\sin\alpha). \end{aligned} \tag{6.84}$$

4. **Differentiation:**
$$\begin{aligned} & e^{-i\alpha}R_\alpha\left[\left\{(t_1+t_2)+\frac{\partial}{\partial t_1}+\frac{\partial}{\partial t_2}\right\}f(t_1,t_2)\right](w_1,w_2) \\ = & \left\{(w_1+w_2)+\left(\frac{\partial}{\partial w_1}+\frac{\partial}{\partial w_2}\right)\right\}R_\alpha\left[f(t_1,t_2)\right](w_1,w_2), \end{aligned} \tag{6.85}$$

and

$$\begin{aligned} & e^{i\alpha}R_\alpha\left[\left\{(t_1+t_2)-\left(\frac{\partial}{\partial t_1}+\frac{\partial}{\partial t_2}\right)\right\}f(t_1,t_2)\right](w_1,w_2) \\ = & \left\{(w_1+w_2)-\left(\frac{\partial}{\partial w_1}+\frac{\partial}{\partial w_2}\right)\right\}R_\alpha\left[f(t_1,t_2)\right](w_1,w_2). \end{aligned} \tag{6.86}$$

5. **Eigenfunctions and Eigenvalues:** The eigenfunctions of the Gyrator transform are $\Psi_{m,n}(t_1,t_2)$ and the corresponding eigenvalues are $e^{i(n-m)\alpha}$, that is
$$R_\alpha\left[\Psi_{m,n}(t_1,t_2)\right] = e^{i(n-m)\alpha}\Psi_{m,n}(\omega_1,\omega_2), \tag{6.87}$$
where
$$\Psi_{m,n}(t_1,t_2) = H_{m,n}\left(\frac{(t_1+t_2)}{\sqrt{2}},\frac{(-t_1+t_2)}{\sqrt{2}}\right) = H_{m,n}(\mathbf{A}\mathbf{t}),$$
$H_{m,n}(x_1,x_2) = h_m(x_1)h_n(x_2)$, and h_m is the Hermite function of order m.

6. **Isometry:** The Gyrator transform is an isometry from $L^2(\mathbb{R}^2)$ into itself, and if $f,g\in L^2(\mathbb{R}^2)$, then
$$\langle R_\alpha[f], R_\alpha[g]\rangle = \langle f,g\rangle. \tag{6.88}$$
In particular $\|R_\alpha[f]\|_2 = \|f\|_2$.

7. **Additive Property:**
$$R_\beta(R_\alpha(f)) = R_{\alpha+\beta}(f).$$

The proof of these properties will follow. Linearity is obvious. To prove (6.83), we have

$$\begin{aligned} R_\alpha[f(t_1-a,t_2-b)](w_1,w_2) &= \int_{\mathbb{R}}\int_{\mathbb{R}} f(t_1-a,t_2-b)G_\alpha(t_1,t_2,w_1,w_2)dt_1dt_2 \\ &= \int_{\mathbb{R}}\int_{\mathbb{R}} f(x_1,x_2)G_\alpha(x_1+a,x_2+b,w_1,w_2)dx_1dx_2 \end{aligned}$$

But with some calculations, one can show that

$$\begin{aligned} G_\alpha(x_1+a,x_2+b,w_1,w_2) &= \exp\left\{i\left[\frac{1}{2}ab\sin 2\alpha-(bw_1+aw_2)\sin\alpha\right]\right\} \\ &\times \; G_\alpha(x_1,x_2,w_1-a\cos\alpha,w_2-b\cos\alpha) \end{aligned}$$

which completes the proof.
The proof of (6.84) follows from the relation

$$R_\alpha\left[e^{-i(at_1+bt_2)}f(t_1,t_2)\right](w_1,w_2) = \int_{\mathbb{R}}\int_{\mathbb{R}} e^{-i(at_1+bt_2)}f(t_1,t_2)G_\alpha(t_1,t_2,w_1,w_2)dt_1dt_2$$
$$= \frac{1}{2\pi|\sin\alpha|}\int_{\mathbb{R}}\int_{\mathbb{R}} f(t_1,t_2)dt_1dt_2$$
$$\times \exp\left\{\frac{i}{\sin\alpha}[(t_1t_2+w_1w_2)\cos\alpha-(t_1w_2+t_2w_1)-(at_1+bt_2)\sin\alpha]\right\}.$$

With some computations one can verify that

$$\begin{aligned} e^{-i(at_1+bt_2)}G_\alpha(t_1,t_2,w_1,w_2) &= \exp\left\{-i\left[\frac{1}{2}ab\sin 2\alpha+(aw_1+bw_2)\cos\alpha\right]\right\} \\ &\times \; G_\alpha(t_1,t_2,w_1+b\sin\alpha,w_2+a\sin\alpha), \end{aligned}$$

which is (6.84) .

The proof of (6.89) and (6.86) goes as follows.

Proof

$$R_\alpha\left[\frac{\partial f}{\partial t_1}+\frac{\partial f}{\partial t_2}\right](w_1,w_2) =$$
$$-\frac{i}{\sin\alpha}\int_{\mathbb{R}}\int_{\mathbb{R}} f(t_1,t_2)G(t_1,t_2,w_1,w_2)\left[(t_1+t_2)\cos\alpha-(w_1+w_2)\right]dt_1dt_2,$$

hence

$$\begin{aligned} &(w_1+w_2)R_\alpha[f](w_1,w_2) = \cos\alpha R_\alpha[(t_1+t_2)f(t_1,t_2)] \\ &-i\sin\alpha R_\alpha\left[\frac{\partial f}{\partial t_1}+\frac{\partial f}{\partial t_2}\right]. \end{aligned} \tag{6.89}$$

On the other hand, we have

$$\left(\frac{\partial}{\partial w_1}+\frac{\partial}{\partial w_2}\right)R_\alpha[f](w_1,w_2)$$
$$=\frac{i}{\sin\alpha}\{(w_1+w_2)\cos\alpha R_\alpha[f](w_1,w_2)-R_\alpha\left[(t_1+t_2)f(t_1,t_2)\right](w_1,w_2)\}. \tag{6.90}$$

By combining (6.89) and (6.90), we obtain

$$\begin{aligned}&\left(\frac{\partial}{\partial w_1}+\frac{\partial}{\partial w_2}\right)R_\alpha[f](w_1,w_2)\\ =\;& \frac{i}{\sin\alpha}\Big\{-\sin^2\alpha R_\alpha[(t_1+t_2)f(t_1,t_2)](w_1,w_2)\\ -\;& i\sin\alpha\cos\alpha R_\alpha\left[\frac{\partial f}{\partial t_1}+\frac{\partial f}{\partial t_2}\right](w_1,w_2)\Big\}\\ =\;& -i\sin\alpha R_\alpha[(t_1+t_2)f(t_1,t_2)](w_1,w_2)+\cos\alpha R_\alpha\left[\frac{\partial f}{\partial t_1}+\frac{\partial f}{\partial t_2}\right](w_1,w_2)\end{aligned} \tag{6.91}$$

Adding (6.89) and (6.91), leads to

$$\begin{aligned}&\left\{(w_1+w_2)+\frac{\partial}{\partial w_1}+\frac{\partial}{\partial w_2}\right\}R_\alpha[f](w_1,w_2)\\ =\;& e^{-i\alpha}R_\alpha\left\{(t_1+t_2)f(t_1,t_2)+\frac{\partial f}{\partial t_1}+\frac{\partial f}{\partial t_2}\right\}(w_1,w_2),\end{aligned}$$

which is (6.85). Similarly, by subtracting (6.91) from (6.89), we obtain (6.86).

To find the eigenfunctions and eigenvalues of the Gyrator transform, let us recall that the eigenvalues and eigenfunctions of the fractional Fourier transform are given by the equation

$$\mathcal{F}_\alpha[h_m(x)](w)=e^{im\alpha}h_m(w),$$

where $h_m(x)$ is the Hermite function, see (6.9). Hence,

$$\int_{\mathbb{R}}\int_{\mathbb{R}}H_{m,n}(x_1,x_2)K_{-\alpha}(x_1,w_1)K_\alpha(x_2,w_2)dx_1dx_2=e^{i(n-m)\alpha}H_{m,n}(w_1,w_2), \tag{6.92}$$

where $H_{m,n}(x_1,x_2)=h_m(x_1)h_n(x_2)$. Therefore, if we set $\mathbf{x}=A\mathbf{t}$, $\mathbf{w}=A\boldsymbol{\omega}$, and

$$\Psi_{m,n}(t_1,t_2)=H_{m,n}\left(\frac{(t_1+t_2)}{\sqrt{2}},\frac{(-t_1+t_2)}{\sqrt{2}}\right)=H_{m,n}\left(A\mathbf{t}\right),$$

in Equation (6.92), we obtain

$$\begin{aligned}&\int_{\mathbb{R}}\int_{\mathbb{R}}H_{m,n}(A\mathbf{t})H_\alpha\left(A\mathbf{t},A\boldsymbol{\omega}\right)d\mathbf{t}\\ =\;&\int_{\mathbb{R}}\int_{\mathbb{R}}H_{m,n}(A\mathbf{t})G_\alpha\left(\mathbf{t},\boldsymbol{\omega}\right)d\mathbf{t}=e^{i(n-m)\alpha}H_{m,n}(A\boldsymbol{\omega})\end{aligned}$$

or

$$R_\alpha\left[\Psi_{m,n}(t_1,t_2)\right](\omega_1,\omega_2)=e^{i(n-m)\alpha}\Psi_{m,n}(\omega_1,\omega_2).$$

The proof of (6.88) goes as follows:

$$\begin{aligned}
\langle R_\alpha[f], R_\alpha[g]\rangle &= \int_{\mathbb{R}}\int_{\mathbb{R}} R_\alpha[f](\mathbf{w})\overline{R_\alpha[g]}(\mathbf{w})\, d\mathbf{w} \\
&= \int_{\mathbb{R}}\int_{\mathbb{R}} d\mathbf{w} \int_{\mathbb{R}}\int_{\mathbb{R}} f(\mathbf{x})G_\alpha(\mathbf{x},\mathbf{w})d\mathbf{x} \int_{\mathbb{R}}\int_{\mathbb{R}} \overline{g}(\mathbf{t})G_{-\alpha}(\mathbf{t},\mathbf{w})d\mathbf{t} \\
&= \int_{\mathbb{R}}\int_{\mathbb{R}} f(\mathbf{x})d\mathbf{x} \int_{\mathbb{R}}\int_{\mathbb{R}} \overline{g}(\mathbf{t})d\mathbf{t} \int_{\mathbb{R}}\int_{\mathbb{R}} G_\alpha(\mathbf{x},\mathbf{w})G_{-\alpha}(\mathbf{t},\mathbf{w})d\mathbf{w} \\
&= \int_{\mathbb{R}}\int_{\mathbb{R}} f(\mathbf{x})d\mathbf{x} \int_{\mathbb{R}}\int_{\mathbb{R}} \overline{g}(\mathbf{t})d\mathbf{t}\, \delta(x_1 - t_1)\,\delta(x_2 - t_2) \qquad (6.93) \\
&= \int_{\mathbb{R}}\int_{\mathbb{R}} f(\mathbf{x})\overline{g}(\mathbf{x})d\mathbf{x} = \langle f, g\rangle,
\end{aligned}$$

where (6.93) follows from Eq. (6.82).

We now present the additive property as a theorem whose proof follows from the additive property of the fractional Fourier transform

Theorem 36. *The Gyrator transform is additive, that is*

$$R_\beta(R_\alpha(f)) = R_{\alpha+\beta}(f).$$

In particular, the inverse R_α^{-1} of R_α is $R_{-\alpha}$.

Proof *Let*

$$R_\alpha(f)(\boldsymbol{\omega}) = \int_{\mathbb{R}}\int_{\mathbb{R}} f(\mathbf{t})G_\alpha(\mathbf{t},\boldsymbol{\omega})d\mathbf{t} = g(\boldsymbol{\omega}),$$

and $J = R_\beta(g)$. Hence,

$$\begin{aligned}
J(\mathbf{z}) &= R_\beta(g)(\mathbf{z}) = \int_{\mathbb{R}}\int_{\mathbb{R}} g(\boldsymbol{\omega})G_\beta(\boldsymbol{\omega},\mathbf{z})d\boldsymbol{\omega} \\
&= \int_{\mathbb{R}}\int_{\mathbb{R}} G_\beta(\boldsymbol{\omega},\mathbf{z})d\boldsymbol{\omega} \int_{\mathbb{R}}\int_{\mathbb{R}} f(\mathbf{t})G_\alpha(\mathbf{t},\boldsymbol{\omega})d\mathbf{t} \\
&= \int_{\mathbb{R}}\int_{\mathbb{R}} f(\mathbf{t})d\mathbf{t} \int_{\mathbb{R}}\int_{\mathbb{R}} G_\alpha(\mathbf{t},\boldsymbol{\omega})G_\beta(\boldsymbol{\omega},\mathbf{z})d\boldsymbol{\omega} \\
&= \int_{\mathbb{R}}\int_{\mathbb{R}} f(\mathbf{t})I(\mathbf{t},\mathbf{z})d\mathbf{t}, \qquad (6.94)
\end{aligned}$$

where

$$\begin{aligned}
I(\mathbf{t},\mathbf{z}) &= \int_{\mathbb{R}}\int_{\mathbb{R}} G_\alpha(\mathbf{t},\boldsymbol{\omega})G_\beta(\boldsymbol{\omega},\mathbf{z})d\boldsymbol{\omega} \\
&= \int_{\mathbb{R}}\int_{\mathbb{R}} H_\alpha(A\mathbf{t}, A\boldsymbol{\omega})H_\beta(A\boldsymbol{\omega}, A\mathbf{z})d\boldsymbol{\omega}. \qquad (6.95)
\end{aligned}$$

In view of the fact that

$$\mathbf{x} = A\mathbf{t},\ \mathbf{w} = A\boldsymbol{\omega},\ \mathbf{v} = A\mathbf{z},$$

we have

$$
\begin{aligned}
I(\mathbf{t},\mathbf{z}) &= \int_{\mathbb{R}}\int_{\mathbb{R}} H_\alpha(\mathbf{x},\mathbf{w})H_\beta(\mathbf{w},\mathbf{v})d\mathbf{w} \\
&= \int_{\mathbb{R}} K_{-\alpha}(x_1,w_1)K_\alpha(x_2,w_2)dw_1 \int_{\mathbb{R}} K_{-\beta}(w_1,v_1)K_\beta(w_2,v_2)dw_2 \\
&= \int_{\mathbb{R}} K_{-\alpha}(x_1,w_1)K_{-\beta}(w_1,v_1)dw_1 \int_{\mathbb{R}} K_\alpha(x_2,w_2)K_\beta(w_2,v_2)dw_2 \\
&= K_{-(\alpha+\beta)}(x_1,v_1)K_{(\alpha+\beta)}(x_2,v_2) \qquad (6.96)\\
&= H_{\alpha+\beta}(\mathbf{x},\mathbf{v}) = H_{\alpha+\beta}(A\mathbf{t},A\mathbf{z}). \qquad (6.97)
\end{aligned}
$$

where (6.96) follows from the additive property of the fractional Fourier transform. By substituting (6.97) into (6.94), we obtain

$$
J(\mathbf{z}) = \int_{\mathbb{R}}\int_{\mathbb{R}} f(\mathbf{t})H_{\alpha+\beta}(A\mathbf{t},A\mathbf{z})d\mathbf{t} = \int_{\mathbb{R}}\int_{\mathbb{R}} f(\mathbf{t})G_{\alpha+\beta}(\mathbf{t},\mathbf{z})d\mathbf{t} = R_{\alpha+\beta}(f)(\mathbf{z}),
$$

which completes the proof.

CHAPTER 7

The Two-Dimensional Fractional Fourier Transform and The Wigner Distribution

7.1 INTRODUCTION

As we have seen in Chapter 4, the fractional Fourier transform and the Wigner distribution are closely related. It was shown that the Wigner distribution $W_{F_\theta}(u,v)$ of the fractional Fourier transform F_θ may be obtained from the Wigner distribution $W_f(u,v)$ of f by a two-dimensional rotation with the angle θ in the time-frequency, $u-v$, plane.

When the fractional Fourier transform is extended to two dimensions by taking the tensor product of two one-dimensional transforms, or in other words, when the transform is taken in two separable variables, the Wigner distribution $W_{F_{\theta,\phi}}(u_1,u_2;v_1,v_2)$ of the two-dimensional fractional Fourier transform $F_{\theta,\phi}(v,w)$ may be obtained from the Wigner distribution $W_f(u_1,u_2;v_1,v_2)$ of $f(x,y)$ by a simple four-dimensional rotation with the angle θ in the u_1-v_1 plane and with the angle ϕ in the u_2-v_2 plane.

The aim of this chapter is to investigate the relationship between the non-separable two-dimensional fractional Fourier transform (CFrFT) introduced in Chapter 6 and the four-dimensional Wigner distribution. We shall give an explicit matrix representation of a four-dimensional rotation that verifies that the Wigner distribution of the two-dimensional coupled fractional Fourier transform $F_{\theta,\phi}(v,w)$ may be obtained from the Wigner distribution of $f(x,y)$ by a four-dimensional rotation. The matrix representation is more general than the one for the tensor product case and it corresponds to a four-dimensional rotation with two planes of rotations, one with the angle $(\theta+\phi)/2$ and the other with the angle $(\theta-\phi)/2$.

7.2 THE WIGNER DISTRIBUTION

Let us recall that the Wigner distribution of a function f is defined as

$$W_f(u,v) = \frac{1}{\sqrt{2\pi}} \int_{\mathbb{R}} f\left(u+\frac{x}{2}\right) \overline{f}\left(u-\frac{x}{2}\right) e^{ivx}\, dx$$

DOI: 10.1201/9781003089353-7

We have seen in Chapter 4 that the Wigner distribution of the Fourier transform $\hat{f}$ of f is related to the Wigner distribution of f by the relation

$$W_{\hat{f}}(u,v) = W_f(-v,u).$$

Hence, it may be said that $W_{\hat{f}}(u,v)$ is obtained from $W_f(u,v)$ by a rotation in the $u-v$ plane with an angle $\pi/2$ in the clockwise direction.

More generally [186, 187], the fractional Fourier transform with angle θ corresponds to the rotation of the Wigner distribution of f by the angle θ. More precisely,

$$W_{F_\theta}(u,v) = W_f(u\cos\theta - v\sin\theta, u\sin\theta + v\cos\theta).$$

This rotation matrix is

$$\begin{pmatrix} \cos\theta & -\sin\theta \\ \sin\theta & \cos\theta \end{pmatrix}.$$

In a more general setting, the Wigner distribution of a function of two variables is a function of four variables. It was shown [220, P. 115, 176] that the Wigner distribution of the fractional Fourier transform given by (6.7) is obtained from the Wigner distribution of f by a four-dimensional rotation, namely,

$$\begin{aligned} &W_{F_{\theta_1,\theta_2}}(u_1,u_2;v_1,v_2) = \\ &W_f\left(u_1\cos\theta_1 - v_1\sin\theta_1, u_2\cos\theta_2 - v_2\sin\theta_2; u_1\sin\theta_1 + v_1\cos\theta_1, u_2\sin\theta_2 + v_2\cos\theta_2\right), \end{aligned}$$

which corresponds to the four-dimensional rotation matrix

$$\begin{pmatrix} \cos\theta_1 & 0 & -\sin\theta_1 & 0 \\ 0 & \cos\theta_2 & 0 & -\sin\theta_2 \\ \sin\theta_1 & 0 & \cos\theta_1 & 0 \\ 0 & \sin\theta_2 & 0 & \cos\theta_2 \end{pmatrix}. \tag{7.1}$$

This amounts to a rotation by θ_1 in the $u_1 - v_1$ plane and a rotation by θ_2 in the $u_2 - v_2$ plane. For other related work, see [7, 278]. This simple result will be generalized later in this chapter.

7.3 FOUR-DIMENSIONAL ROTATIONS

In this subsection, we give an introduction to four-dimensional rotations that will be used later in the proof of one of our main theorems. For more details, see [195, 313].

A counterclockwise rotation in the $x-y$ plane by an angle θ maybe described by the equation

$$\begin{bmatrix} \tilde{u} \\ \tilde{v} \end{bmatrix} = \begin{pmatrix} \cos\theta & \sin\theta \\ -\sin\theta & \cos\theta \end{pmatrix} \begin{bmatrix} u \\ v \end{bmatrix}.$$

A clockwise rotation is obtained by replacing θ by $-\theta$ to get

$$\begin{bmatrix} \tilde{u} \\ \tilde{v} \end{bmatrix} = \begin{pmatrix} \cos\theta & -\sin\theta \\ \sin\theta & \cos\theta \end{pmatrix} \begin{bmatrix} u \\ v \end{bmatrix},$$

which when viewed as a rotation in three dimensions about the $z-$axis, can be represented by the matrix

$$\begin{pmatrix} 1 & 0 & 0 \\ 0 & \cos\theta & -\sin\theta \\ 0 & \sin\theta & \cos\theta \end{pmatrix}.$$

Let $\mathbb{R}^4$ denote the four-dimensional Euclidean space provided with the inner product

$$\langle \mathbf{u}, \mathbf{v} \rangle = x_1 x_2 + y_1 y_2 + z_1 z_2 + w_1 w_2,$$

where $\mathbf{u} = (x_1, y_1, z_1, w_1)$ and $\mathbf{v} = (x_2, y_2, z_2, w_2)$, with $x_i, y_i, z_i, w_i \in \mathbb{R}$, $i = 1, 2$. Let us denote the set of all 4×4 matrices over the real numbers by $M_{4\times 4}(\mathbb{R})$. A matrix $A \in M_{4\times 4}(\mathbb{R})$ is said to be orthogonal if $AA^T = A^T A = I$, where I is the 4×4 identity matrix and A^T denotes the transpose of A.

In four dimensions, instead of axis of rotation, there is a plane of rotation which describes rotations in $\mathbb{R}^4$. A plane of rotation for a four-dimensional rotation is a plane that is mapped into itself by the rotation. The plane is not fixed but all vectors in the plane are mapped to other vectors in the same plane by the rotation [98].

Rotations in $\mathbb{R}^4$ have at most two planes of rotations. Thus, a simple rotation is a rotation with one fixed plane and the rotation is about this fixed plane. The plane of rotation is orthogonal to the plane which is being rotated. For example, a rotation about the $x-y$ plane in which the points in $z-w$ plane are rotated by an angle θ is given by the matrix

$$\begin{pmatrix} 1 & 0 & 0 & 0 \\ 0 & 1 & 0 & 0 \\ 0 & 0 & \cos\theta & -\sin\theta \\ 0 & 0 & \sin\theta & \cos\theta \end{pmatrix}.$$

A double rotation is a rotation with two planes of rotations. The two planes are orthogonal and the rotation depends on two angles, one for each plane. For examples if the $x-y$ plane is rotated by an angle α and the $z-w$ plane is rotated by an angle β, the matrix of rotation is given by

$$\begin{pmatrix} \cos\alpha & -\sin\alpha & 0 & 0 \\ \sin\alpha & \cos\alpha & 0 & 0 \\ 0 & 0 & \cos\beta & -\sin\beta \\ 0 & 0 & \sin\beta & \cos\beta \end{pmatrix}.$$

It is known [195] that each four-dimensional rotation can be decomposed in two ways into a matrix representing left-multiplication by a unit quaternion and a matrix representing right-multiplication by a unit quaternion . These decompositions differ only in the signs of the component matrices. In fact, let

$$M_L = \begin{pmatrix} a & -b & -c & -d \\ b & a & -d & c \\ c & d & a & -b \\ d & -c & b & a \end{pmatrix}, \quad M_R = \begin{pmatrix} p & -q & -r & -s \\ q & p & s & -r \\ r & -s & p & q \\ s & r & -q & p \end{pmatrix}. \tag{7.2}$$

where

$$a^2 + b^2 + c^2 + d^2 = 1, \quad p^2 + q^2 + r^2 + s^2 = 1,$$

be matrices representing the left-and right- multiplication by a unit quaternion, then their product

$$A = M_L M_R = \begin{pmatrix} a_{0,0} & a_{0,1} & a_{0,2} & a_{0,3} \\ a_{1,0} & a_{1,1} & a_{1,2} & a_{1,3} \\ a_{2,0} & a_{2,1} & a_{2,2} & a_{2,3} \\ a_{3,0} & a_{3,1} & a_{3,2} & a_{3,3} \end{pmatrix} \tag{7.3}$$

is a four-dimensional rotation matrix. The entries $a_{i,j}, i, j = 0,1,2,3$ can be expressed in terms of a, b, c, d and p, q, r, s. The general three-dimensional rotation matrix can be obtained from the general four-dimensional rotation matrix by putting

$$a_{0,0} = 1, a_{0,1} = a_{0,2} = a_{0,3} = a_{1,0} = a_{2,0} = a_{3,0} = 0,$$

to yield

$$A = \begin{pmatrix} 1 & 0 & 0 & 0 \\ 0 & a_{1,1} & a_{1,2} & a_{1,3} \\ 0 & a_{2,1} & a_{2,2} & a_{2,3} \\ 0 & a_{3,1} & a_{3,2} & a_{3,3} \end{pmatrix}.$$

7.4 THE FOUR-DIMENSIONAL WIGNER DISTRIBUTION

We begin this section by introducing the four-dimensional Wigner distribution and some of its elementary properties. For more general forms, see [101] and [114, P. 63].

Definition 33. *The cross-Wigner distribution function of two functions $f(x,y), g(x,y) \in L^2(\mathbb{R}^2)$ is defined as*

$$W_{f,g}(u_1,u_2;v_1,v_2) = \frac{1}{2\pi}\int_{\mathbb{R}^2} f(u_1+\frac{x}{2}, u_2+\frac{y}{2})\overline{g}(u_1-\frac{x}{2}, u_2-\frac{y}{2})e^{i(xv_1+yv_2)}dxdy,$$

and the auto-Wigner distribution function of f (or the Wigner distribution function of f for short) is defined as

$$W_f(u_1,u_2;v_1,v_2) = \frac{1}{2\pi}\int_{\mathbb{R}^2} f(u_1+\frac{x}{2}, u_2+\frac{y}{2})\overline{f}(u_1-\frac{x}{2}, u_2-\frac{y}{2})e^{i(xv_1+yv_2)}dxdy.$$

Next we derive some properties of the Wigner distribution. One of the salient properties of the Wigner distribution is that it satisfies Moyal's identity which we will state in the following proposition. Although the result may be obtained from a more general result in harmonic analysis [101, p. 56], [115], for the reader's convenience we will sketch a more direct proof for it, see Section 4.3.4.

Proposition 3. *The Wigner distribution satisfies Moyal's identity , namely*

$$\langle W_{f_1,g_1}, W_{f_2,g_2}\rangle_{L^2(\mathbb{R}^4)} = \langle f_1, f_2\rangle_{L^2(\mathbb{R}^2)}\langle g_1, g_2\rangle_{L^2(\mathbb{R}^2)}.$$

Moreover, we have the inversion formula

$$f(\tau_1,\tau_2)\overline{g}(\tau_1,\tau_2) = \frac{1}{2\pi}\int_{\mathbb{R}^2} W_{f,g}(\tau_1,\tau_2;v_1,v_2)dv_1dv_2.$$

Proof *From the definition of the Wigner distribution, we have*

$$\begin{aligned}
\langle W_{f_1,g_1}, W_{f_2,g_2}\rangle &= \int_{\mathbb{R}^4} W_{f_1,g_1}(u_1,u_2;v_1,v_2)\overline{W}_{f_2,g_2}(u_1,u_2;v_1,v_2)du_1du_2dv_1dv_2 \\
&= \frac{1}{(2\pi)^2}\int_{\mathbb{R}^4} du_1du_2dv_1dv_2 \int_{\mathbb{R}^2} f_1(u_1+x/2,u_2+y/2) \\
&\times \overline{g}_1(u_1-x/2,u_2-y/2)e^{i(xv_1+yv_2)}dxdy \\
&\times \int_{\mathbb{R}^2} \overline{f}_2(u_1+\tilde{x}/2,u_2+\tilde{y}/2)g_2(u_1-\tilde{x}/2,u_2-\tilde{y}/2)e^{-i(\tilde{x}v_1+\tilde{y}v_2)}d\tilde{x}d\tilde{y} \\
&= \frac{1}{(2\pi)^2}\int_{\mathbb{R}^2} e^{i(x-\tilde{x})v_1-i(y-\tilde{y})v_2}dv_1dv_2\int_{\mathbb{R}^2} du_1du_2 \\
&\times \int_{\mathbb{R}^2} f_1(u_1+x/2,u_2+y/2)\overline{g}_1(u_1-x/2,u_2-y/2)dxdy \\
&\times \int_{\mathbb{R}^2} \overline{f}_2(u_1+\tilde{x}/2,u_2+\tilde{y}/2)g_2(u_1-\tilde{x}/2,u_2-\tilde{y}/2)d\tilde{x}d\tilde{y} \\
&= \int_{\mathbb{R}^2} du_1du_2\int_{\mathbb{R}^2} f_1(u_1+x/2,u_2+y/2)\overline{g}_1(u_1-x/2,u_2-y/2) \\
&\times \int_{\mathbb{R}^2} \overline{f}_2(u_1+x/2,u_2+y/2)g_2(u_1-x/2,u_2-y/2)dxdy.
\end{aligned}$$

Here we have used the relation

$$\int_{\mathbb{R}} e^{ixw}dx = 2\pi\boldsymbol{\delta}(w),$$

where $\boldsymbol{\delta}$ is the Dirac-delta function. If we set

$$x_1 = u_1+x/2,\ x_2 = u_2+y/2,\ x_3 = u_1-x/2,\ x_4 = u_2-y/2,$$

we obtain after some calculations

$$\begin{aligned}
\langle W_{f_1,g_1}, W_{f_2,g_2}\rangle &= \int_{\mathbb{R}^2} f_1(x_1,x_2)\overline{f}_2(x_1,x_2)dx_1dx_2\int_{\mathbb{R}^2}\overline{g}_1(x_3,x_4)g_2(x_3,x_4)dx_3dx_4 \\
&= \langle f_1,f_2\rangle\langle g_1,g_2\rangle.
\end{aligned}$$

Finally, using the inversion formula for the two-dimensional Fourier transform, we have

$$f(u_1+\frac{x}{2},u_2+\frac{y}{2})\overline{g}(u_1-\frac{x}{2},u_2-\frac{y}{2}) = \frac{1}{(2\pi)}\int_{\mathbb{R}^2} W_{f,g}(u_1,u_2;v_1,v_2)e^{-i(xv_1+yv_2)}dv_1dv_2,$$

and if we set $\tau_1 = u_1+x/2$, $\tau_2 = u_2+y/2$, $\sigma_1 = u_1-x/2$, $\sigma_2 = u_2-y/2$, we obtain

$$f(\tau_1,\tau_2)\overline{g}(\sigma_1,\sigma_2) = \frac{1}{(2\pi)}\int_{\mathbb{R}^2} W_{f,g}(\frac{\tau_1+\sigma_1}{2},\frac{\tau_2+\sigma_2}{2};v_1,v_2)e^{-i[v_1(\tau_1-\sigma_1)+v_2(\tau_2-\sigma_2)]}dv_1dv_2,$$

from which we obtain if $\tau_1=\sigma_1, \tau_2=\sigma_2$

$$f(\tau_1,\tau_2)\overline{g}(\tau_1,\tau_2) = \frac{1}{(2\pi)}\int_{\mathbb{R}^2} W_{f,g}(\tau_1,\tau_2;v_1,v_2)dv_1dv_2.$$

Corollary 3. *If $f_2 = f_1 = f$ and $g_2 = g_1 = g$, we have*

$$\|W_{f,g}\|_{L^2(\mathbb{R}^4)} = \|f\|_{L^2(\mathbb{R}^2)}\|g\|_{L^2(\mathbb{R}^2)};$$

hence $\|W_f\|_{L^2(\mathbb{R}^4)} = \|f\|^2_{L^2(\mathbb{R}^2)}$.

7.5 THE FOUR-DIMENSIONAL WIGNER DISTRIBUTION AND THE COUPLED FRACTIONAL FOURIER TRANSFORM

In this section, we present one of the main results of this chapter which is the relationship between the two-dimensional coupled fractional Fourier transform and the four-dimensional Wigner distribution. Recall that for a function of one variable, the Wigner distribution of the fractional Fourier transform W_{F_θ} is obtained from the Wigner distribution W_f of f by a two-dimensional rotation with the angle θ. Likewise, we will show that the Wigner distribution of the two-dimensional fractional Fourier transform $W_{F_{\alpha,\beta}}$ is obtained from the Wigner distribution W_f of f by a four-dimensional rotation that is more general and profound than the one given by (7.1). It will be shown that the four-dimensional rotation has two planes of rotation, one with angle $\gamma = (\alpha+\beta)/2$ and another with an angle $\delta = (\alpha-\beta)/2$.

Remark. In this section, we adopt a slightly different version of the definition of the Wigner distribution in which the factor $1/2\pi$ in front of the integral is absent, the reason being that factor is already embedded in the constant $d(\gamma)$ that appears in the definition of the two-dimensional factional Fourier transform.

Theorem 37. *[335] The Wigner distribution $W_{F_{\alpha,\beta}}$ of the two-dimensional coupled fractional Fourier transform $F_{\alpha,\beta}$ of a function $f(x,y)$ is obtained from the Wigner distribution W_f of f by a four-dimensional rotation through the matrix*

$$\mathcal{A} = \begin{pmatrix} \cos\gamma\cos\delta & \cos\gamma\sin\delta & -\sin\gamma\cos\delta & -\sin\gamma\sin\delta \\ -\cos\gamma\sin\delta & \cos\gamma\cos\delta & \sin\gamma\sin\delta & -\sin\gamma\cos\delta \\ \sin\gamma\cos\delta & \sin\gamma\sin\delta & \cos\gamma\cos\delta & \cos\gamma\sin\delta \\ -\sin\gamma\sin\delta & \sin\gamma\cos\delta & -\cos\gamma\sin\delta & \cos\gamma\cos\delta \end{pmatrix},$$

where $\gamma = (\alpha+\beta)/2$, $\delta(\alpha-\beta)/2$. The matrix $\mathcal{A}$ is a genuine four-dimensional rotation matrix.

Proof *Because the proof is relatively long, we will break it into 3 parts to make it easier for the reader to follow. In part one, we express $W_{F_{\alpha,\beta}}$ in terms of f rather than $F_{\alpha,\beta}$. In part two, we relate $W_{F_{\alpha,\beta}}$ to W_f, and finally in part three, we show that $W_{F_{\alpha,\beta}}$ is obtained from W_f, by a four-dimensional rotation.*

Part One: *The Wigner distribution of the two-dimensional fractional Fourier transform $F_{\alpha,\beta}$ (or F for short) of $f(x,y)$ is given by*

$$\begin{aligned} W_{F_{\alpha,\beta}}(u_1,u_2;v_1,v_2) &= \int_{\mathbb{R}^2} F_{\alpha,\beta}(u_1+\frac{x}{2},u_2+\frac{y}{2})\overline{F_{\alpha,\beta}}(u_1-\frac{x}{2},u_2-\frac{y}{2})e^{i(xv_1+yv_2)}dxdy \\ &= \int_{\mathbb{R}^2} e^{i(xv_1+yv_2)}dxdy \int_{\mathbb{R}^2} d(\gamma)f(x_1,y_1) \\ &\times \exp\left\{-a(x_1^2+y_1^2+U^2+V^2)+b(Ux_1+Vy_1)+c(Vx_1-Uy_1)\right\}dx_1dy_1 \\ &\times \int_{\mathbb{R}^2} \overline{d}(\gamma)\overline{f}(x_2,y_2) \\ &\times \exp\left\{a(x_2^2+y_2^2+W^2+Z^2)-b(Wx_2+Zy_2)-c(Zx_2-Wy_2)\right\}dx_2dy_2, \end{aligned} \tag{7.4}$$

where

$$U = u_1 + \frac{x}{2} \quad , \qquad V = u_2 + \frac{y}{2},$$
$$W = u_1 - \frac{x}{2} \quad , \qquad Z = u_2 - \frac{y}{2}.$$

Thus,

$$\begin{aligned} W_{F_{\alpha,\beta}}(u_1,u_2;v_1,v_2) &= \frac{1}{(2\pi\sin\gamma)^2}\int_{\mathbb{R}^2} e^{i(xv_1+yv_2)}dxdy \\ \times & \int_{\mathbb{R}^2} f(x_1,y_1)\exp\left\{-a(x_1^2+y_1^2+U^2+V^2)+x_1(bU+cV)+y_1(bV-cU)\right\}dx_1dy_1 \\ \times & \int_{\mathbb{R}^2} \overline{f}(x_2,y_2)\exp\left\{a(x_2^2+y_2^2+W^2+Z^2)-x_2(bW+cZ)-y_2(bZ-cW)\right\}dx_2dy_2. \end{aligned} \tag{7.5}$$

But since

$$\begin{aligned} U^2+V^2 &= u_1^2+u_2^2+u_1x+u_2y+x^2/4+y^2/4, \\ bU+cV &= bu_1+cu_2+bx/2+cy/2, \\ bV-cU &= bu_2-cu_1+by/2-cx/2, \end{aligned}$$

and

$$\begin{aligned} W^2+Z^2 &= u_1^2+u_2^2-u_1x-u_2y+x^2/4+y^2/4, \\ bW+cZ &= bu_1+cu_2-bx/2-cy/2, \\ bZ-cW &= bu_2-cu_1+cx/2-by/2, \end{aligned}$$

we have, by substituting these expressions into (7.5) and simplifying, that

$$\begin{aligned} W_{F_{\alpha,\beta}}(u_1,u_2;v_1,v_2) &= \frac{1}{(2\pi\sin\gamma)^2}\int_{\mathbb{R}^2} e^{i(xv_1+yv_2)}e^{-2a(u_1x+u_2y)} \\ \times & \; e^{x(bx_1/2+bx_2/2-cy_1/2-cy_2/2)}e^{y(by_1/2+by_2/2+cx_1/2+cx_2/2)}dxdy \\ \times & \int_{\mathbb{R}^2} f(x_1,y_1)\exp\left\{-a(x_1^2+y_1)+x_1(bu_1+cu_2)+y_1(bu_2-cu_1)\right\}dx_1dy_1 \\ \times & \int_{\mathbb{R}^2} \overline{f}(x_2,y_2)\exp\left\{a(x_2^2+y_2^2)-x_2(bu_1+cu_2)-y_2(bu_2-cu_1)\right\}dx_2dy_2. \end{aligned} \tag{7.6}$$

The first integral becomes

$$I_1 = \int_{\mathbb{R}^2}\exp\left\{ix\left[v_1-2\tilde{a}u_1+\tilde{b}\eta-\tilde{c}\zeta\right]+iy\left[v_2-2\tilde{a}u_2+\tilde{b}\eta+\tilde{c}\zeta\right]\right\}dxdy,$$

where

$$\eta = (x_1+x_2)/2\,, \zeta = (y_1+y_2)/2.$$

It is easy to show that

$$I_1 = (2\pi)^2\boldsymbol{\delta}(v_1-2\tilde{a}u_1+\tilde{b}\eta-\tilde{c}\zeta)\boldsymbol{\delta}(v_2-2\tilde{a}u_2+\tilde{b}\eta+\tilde{c}\zeta). \tag{7.7}$$

where δ *denotes the delta function. Changing the variables* $x_2 = 2\eta - x_1,\ y_2 = 2\zeta - y_1$ *in the third integral yields*

$$\begin{aligned} I_3 &= 4\int_{\mathbb{R}^2} \overline{f}(2\eta - x_1, 2\zeta - y_1)\exp\left\{a\left[(2\eta - x_1)^2 + (2\zeta - y_1)^2\right]\right\} \\ &\times \exp\{-(2\eta - x_1)(bu_1 + cu_2) - (2\zeta - y_1)(bu_2 - cu_1)\}\,d\eta d\zeta. \end{aligned}$$

Let

$$U = \tilde{b}\eta - \tilde{c}\zeta, \quad V = \tilde{c}\eta + \tilde{b}\zeta,$$

so that

$$\eta = \sin^2\gamma\left(\tilde{b}U + \tilde{c}V\right), \quad \text{and } \zeta = \sin^2\gamma\left(\tilde{b}V - \tilde{c}U\right),$$

and hence

$$d\eta d\zeta = \sin^2\gamma dU dV.$$

Therefore, I_3 *may take the form*

$$I_3 = 4\int_{\mathbb{R}^2} \overline{f}\left(2\sin^2\gamma\left(\tilde{b}U + \tilde{c}V\right) - x_1, 2\sin^2\gamma\left(\tilde{b}V - \tilde{c}U\right) - y_1\right)\exp\{\ldots\}\sin^2\gamma dU dV,$$

where the exponential factor $\exp\{\cdots\}$ *in* I_3*, after some calculations, can be written as a product of two factors*

$$h_1(x_1, y_1, u_1, u_2), \ \text{and } h_2(u_1, u_2, U, V),$$

where

$$h_1(x_1, y_1, u_1, u_2) = \exp\left\{a(x_1^2 + y_1^2) + x_1(bu_1 + cu_2) + y_1(bu_2 - cu_1)\right\},$$

and

$$h_2(u_1, u_2, U, V) = \exp\left\{4a(\eta^2 + \zeta^2) - 4ax_1\eta - 4ay_1\zeta - 2\eta(bu_1 + cu_2) - 2\zeta(bu_2 - cu_1)\right\}.$$

After some calculations, we have

$$\begin{aligned} h_2(u_1, u_2, U, V) &= \exp\left\{4a(\eta^2 + \zeta^2) - 2\eta(bu_1 + cu_2 + 2ax_1) - 2\zeta(bu_2 - cu_1 + 2ay_1)\right\} \\ &= \exp\left\{4a\sin^2\gamma(U^2 + V^2) - 2\sin^2\gamma(\tilde{b}U + \tilde{c}V)(bu_1 + cu_2 + 2ax_1)\right\} \\ &\times \exp\left\{-2\sin^2\gamma(\tilde{b}V - \tilde{c}U)(bu_2 - cu_1 + 2ay_1)\right\}. \end{aligned}$$

Therefore; I_3 *can now be written as*

$$\begin{aligned} I_3 &= 4\sin^2\gamma h_1(x_1, y_1, u_1, u_2)\int_{\mathbb{R}^2} h_2(u_1, u_2, U, V) \\ &\times \overline{f}\left(2\sin^2\gamma\left(\tilde{b}U + \tilde{c}V\right) - x_1, 2\sin^2\gamma\left(\tilde{b}V - \tilde{c}U\right) - y_1\right) dU dV. \end{aligned}$$

Let us denote by A *and* B *the expressions*

$$A = 2\tilde{a}u_1 - v_1, \quad B = 2\tilde{a}u_2 - v_2, \tag{7.8}$$

so that the integral I_1 in Eq. (7.7) becomes

$$I_1 = (2\pi)^2 \boldsymbol{\delta}(U-A)\boldsymbol{\delta}(V-B).$$

By combining the integrals I_1, I_3 we have

$$\begin{aligned} & 4(2\pi)^2 \sin^2\gamma h_1(x_1,y_1,u_1,u_2) \int_{\mathbb{R}^2} h_2(u_1,u_2,U,V) \\ \times\ & \overline{f}\left(2\sin^2\gamma\left(\tilde{b}U+\tilde{c}V\right)-x_1, 2\sin^2\gamma\left(\tilde{b}V-\tilde{c}U\right)-y_1\right) \\ \times\ & \boldsymbol{\delta}(U-A)\boldsymbol{\delta}(V-B)dUdV, \end{aligned}$$

which yields

$$\begin{aligned} & 4(2\pi)^2 \sin^2\gamma h_1(x_1,y_1,u_1,u_2) h_2(u_1,u_2,A,B) \\ \times\ & \overline{f}\left(2\sin^2\gamma\left(\tilde{b}A+\tilde{c}B\right)-x_1, 2\sin^2\gamma\left(\tilde{b}B-\tilde{c}A\right)-y_1\right). \end{aligned}$$

Similarly, $h_2(u_1,u_2,A,B)$ can be simplified to

$$\begin{aligned} & h_2(u_1,u_2,A,B) = \\ & \exp\left\{4a\sin^2\gamma\left(A^2+B^2\right)-2\sin^2\gamma(\tilde{b}A+\tilde{c}B)(bu_1+cu_2+2ax_1)\right\} \\ \times\ & \exp\left\{-2\sin^2\gamma(\tilde{b}B-\tilde{c}A)(bu_2-cu_1+2ay_1)\right\} \\ =\ & h_3(x_1,y_1,A,B)h_4(u_1,u_2,A,B), \end{aligned}$$

where

$$h_3(x_1,y_1,A,B) = \exp\left\{-4ax_1\sin^2\gamma(\tilde{b}A+\tilde{c}B) - -4ay_1\sin^2\gamma(\tilde{b}B-\tilde{c}A)\right\}$$

and

$$\begin{aligned} h_4(u_1,u_2,A,B) &= \exp\left\{4a\sin^2\gamma\left(A^2+B^2\right)-2\sin^2\gamma(\tilde{b}A+\tilde{c}B)(bu_1+cu_2)\right\} \\ &\times \exp\left\{-2\sin^2\gamma(\tilde{b}B-\tilde{c}A)(bu_2-cu_1)\right\}. \end{aligned}$$

By combining all the three integrals in (7.6), we have

$$\begin{aligned} & W_{F_{\alpha,\beta}}(u_1,u_2,v_1,v_2) = \frac{1}{(2\pi\sin\gamma)^2} \times \\ & \int_{\mathbb{R}^2} f(x_1,y_1)\exp\left\{-a(x_1^2+y_1^2)+x_1(bu_1+cu_2)+y_1(bu_2-cu_1)\right\} \\ \times\ & 4(2\pi\sin\gamma)^2 h_1(x_1,y_1,u_1,u_2)h_3(x_1,y_1,A,B)h_4(u_1,u_2,A,B) \\ \times\ & \overline{f}\left(2\sin^2\gamma\left(\tilde{b}A+\tilde{c}B\right)-x_1, 2\sin^2\gamma\left(\tilde{b}B-\tilde{c}A\right)-y_1\right)dx_1dy_1 \\ =\ & 4\int_{\mathbb{R}^2} f(x_1,y_1)\exp\left\{2x_1(bu_1+cu_2)+2y_1(bu_2-cu_1)\right\} \\ \times\ & \exp\left\{-4ax_1\sin^2\gamma(\tilde{b}A+\tilde{c}B)-4ay_1\sin^2\gamma(\tilde{b}B-\tilde{c}A)\right\} \\ \times\ & \exp\left\{4a\sin^2\gamma(A^2+B^2)-2\sin^2\gamma(\tilde{b}A+\tilde{c}B)(bu_1+cu_2)\right\} \\ \times\ & \exp\left\{-2\sin^2\gamma(\tilde{b}B-\tilde{c}A)(bu_2-cu_1)\right\} \\ \times\ & \overline{f}\left(2\sin^2\gamma\left(\tilde{b}A+\tilde{c}B\right)-x_1, 2\sin^2\gamma\left(\tilde{b}B-\tilde{c}A\right)-y_1\right)dx_1dy_1. \end{aligned}$$

Part Two: *To relate $W_{F_{\alpha,\beta}}$ to W_f, we make the following change of variables in the last equation:*

$$x_1 = \tilde{u}_1 + x/2, \quad y_1 = \tilde{u}_2 + y/2,$$

where

$$\tilde{u}_1 = \sin^2\gamma(\tilde{b}A + \tilde{c}B), \quad \tilde{u}_2 = \sin^2\gamma(\tilde{b}B - \tilde{c}A), \tag{7.9}$$

to obtain

$$W_{F_{\alpha,\beta}}(u_1,u_2;v_1,v_2) = \int_{\mathbb{R}^2} f\left(\tilde{u}_1 + \frac{x}{2}, \tilde{u}_2 + \frac{y}{2}\right)\overline{f}\left(\tilde{u}_1 - \frac{x}{2}, \tilde{u}_2 - \frac{y}{2}\right)\exp\{\cdots\}\,dxdy,$$

which is close to the Wigner distribution of f, except for the exponential factor. We now focus on simplifying the exponential factor $\exp\{\cdots\}$ *inside the integral. To this end, let*

$$\sigma = bu_1 + cu_2, \quad \tau = bu_2 - cu_1$$

with $\tilde{\sigma}$ and $\tilde{\tau}$ have the same meaning as before, i.e., expressed in terms of $\tilde{b}$ and $\tilde{c}$. We have after some simplification

$$\begin{aligned}
\exp\{\cdots\} &= \exp\{2\sigma(\tilde{u}_1 + x/2) + 2\tau(\tilde{u}_2 + y/2) - 4a\tilde{u}_1x_1 - 4a\tilde{u}_2y_1\} \\
&\times \exp\left\{4a\sin^2\gamma(A^2+B^2) - 2\tilde{u}_1\sigma - 2\tilde{u}_2\tau\right\} \\
&= \exp\left\{\sigma x + \tau y - 4a\tilde{u}_1(\tilde{u}_1 + x/2) - 4a\tilde{u}_2(\tilde{u}_2 + y/2) + 4a\sin^2\gamma(A^2+B^2)\right\} \\
&= \exp\left\{(\sigma - 2a\tilde{u}_1)x + (\tau - 2a\tilde{u}_2)y - 4a(\tilde{u}_1^2 + \tilde{u}_2^2) + 4a\sin^2\gamma(A^2+B^2)\right\}.
\end{aligned}$$

But in view of the fact that

$$\tilde{u}_1^2 + \tilde{u}_2^2 = \sin^4\gamma\left(\tilde{b}^2A^2 + \tilde{c}^2B^2 + \tilde{b}^2B^2 + \tilde{c}^2A^2\right),$$

and that $\tilde{b}^2 + \tilde{c}^2 = \csc^2\gamma$, we have

$$\tilde{u}_1^2 + \tilde{u}_2^2 = \sin^2\gamma(A^2+B^2).$$

Thus, the exponential factor is reduced to

$$\begin{aligned}
\exp\{\cdots\} &= \exp\{(\sigma - 2a\tilde{u}_1)x + (\tau - 2a\tilde{u}_2)y\} \\
&= e^{ix(\tilde{\sigma} - 2\tilde{a}\tilde{u}_1) + iy(\tilde{\tau} - 2\tilde{a}\tilde{u}_2)},
\end{aligned}$$

and, hence, we finally have

$$\begin{aligned}
W_{F_{\alpha,\beta}}(u_1,u_2;v_1,v_2) &= \int_{\mathbb{R}^2} f\left(\tilde{u}_1 + \frac{x}{2}, \tilde{u}_2 + \frac{y}{2}\right)\overline{f}\left(\tilde{u}_1\frac{x}{2}, \tilde{u}_2 - \frac{y}{2}\right) \\
&\times\ e^{ix(\tilde{\sigma} - 2\tilde{a}\tilde{u}_1) + iy(\tilde{\tau} - 2\tilde{a}\tilde{u}_2)}dxdy,
\end{aligned}$$

or equivalently

$$\begin{aligned}
W_{F_{\alpha,\beta}}(u_1,u_2;v_1,v_2) &= \int_{\mathbb{R}^2} f\left(\tilde{u}_1 + \frac{x}{2}, \tilde{u}_2 + \frac{y}{2}\right)\overline{f}\left(\tilde{u}_1\frac{x}{2}, \tilde{u}_2 - \frac{y}{2}\right)e^{i\tilde{v}_1x + i\tilde{v}_2y}dxdy \\
&= W_f(\tilde{u}_1, \tilde{u}_2; \tilde{v}_1, \tilde{v}_2),
\end{aligned} \tag{7.10}$$

where

$$\tilde{v}_1 = \tilde{\sigma} - 2\tilde{a}\tilde{u}_1, \text{ and } \tilde{v}_2 = \tilde{\tau} - 2\tilde{a}\tilde{u}_2.$$

Part Three: *We now relate the new variables* $\tilde{u}_1, \tilde{u}_2, \tilde{v}_1, \tilde{v}_2$ *to the original variables* u_1, u_2, v_1, v_2. *Using the values of* a, b, c, d, *and* A, B *given by Eqs. (6.22), (6.23) and (7.8), we have from (7.9)*

$$\begin{aligned}
\tilde{u}_1 &= \sin^2\gamma(\tilde{b}A + \tilde{c}B) \\
&= \sin^2\gamma\left[\frac{\cos\delta}{\sin\gamma}(2\tilde{a}u_1 - v_1) + \frac{\sin\delta}{\sin\gamma}(2\tilde{a}u_2 - v_2)\right] \\
&= \sin\gamma\cos\delta(u_1\cot\gamma - v_1) + \sin\gamma\sin\delta(u_2\cot\gamma - v_2) \\
&= (\cos\gamma\cos\delta)u_1 + (\cos\gamma\sin\delta)u_2 - (\sin\gamma\cos\delta)v_1 - (\sin\gamma\sin\delta)v_2. \qquad (7.11)
\end{aligned}$$

Similarly, we have

$$\tilde{u}_2 = (-\cos\gamma\sin\delta)u_1 + (\cos\gamma\cos\delta)u_2 + (\sin\gamma\sin\delta)v_1 - (\sin\gamma\cos\delta)v_2.$$

As for $\tilde{v}_1$, *we have, in view of the relation*

$$2\tilde{a}\tilde{u}_1 = \cot\gamma(\sin^2\gamma)(\tilde{b}A + \tilde{c}B) = \cos\gamma\sin\gamma(\tilde{b}A + \tilde{c}B),$$

that

$$\begin{aligned}
\tilde{v}_1 &= \tilde{\sigma} - 2\tilde{a}\tilde{u}_1 = \tilde{b}u_1 + \tilde{c}u_2 - \cos\gamma\sin\gamma\left[\tilde{b}(2\tilde{a}u_1 - v_1) + \tilde{c}(2\tilde{a}u_2 - v_2)\right] \\
&= u_1\tilde{b}[1 - \cos\gamma\sin\gamma\cot\gamma] + u_2\tilde{c}[1 - \cos\gamma\sin\gamma\cot\gamma] \\
&+ v_1\left[\tilde{b}\cos\gamma\sin\gamma\right] + v_2[\tilde{c}\cos\gamma\sin\gamma] \\
&= u_1\frac{\cos\delta}{\sin\gamma}\sin^2\gamma + u_2\frac{\sin\delta}{\sin\gamma}\sin^2\gamma + v_1\cos\gamma\cos\delta + v_2\cos\gamma\sin\delta \\
&= (\sin\gamma\cos\delta)u_1 + (\sin\gamma\sin\delta)u_2 + (\cos\gamma\cos\delta)v_1 + (\cos\gamma\sin\delta)v_2.
\end{aligned}$$

Similarly, one can show that

$$\tilde{v}_2 = (-\sin\gamma\sin\delta)u_1 + (\sin\gamma\cos\delta)u_2 - (\cos\gamma\sin\delta)v_1 + (\cos\gamma\cos\delta)v_2.$$

Now if we denote by U^T *and* $\tilde{U}^T$ *the row vectors* (u_1, u_2, v_1, v_2) *and* $(\tilde{u}_1, \tilde{u}_2, \tilde{v}_1\,\tilde{v}_2)$, *respectively, we can write the transformation relation in the form* $\tilde{U} = \mathcal{A}U$, *where*

$$\mathcal{A} = \begin{pmatrix}
\cos\gamma\cos\delta & \cos\gamma\sin\delta & -\sin\gamma\cos\delta & -\sin\gamma\sin\delta \\
-\cos\gamma\sin\delta & \cos\gamma\cos\delta & \sin\gamma\sin\delta & -\sin\gamma\cos\delta \\
\sin\gamma\cos\delta & \sin\gamma\sin\delta & \cos\gamma\cos\delta & \cos\gamma\sin\delta \\
-\sin\gamma\sin\delta & \sin\gamma\cos\delta & -\cos\gamma\sin\delta & \cos\gamma\cos\delta
\end{pmatrix},$$

which is in the form

$$\begin{pmatrix}
A & B & -C & -D \\
-B & A & D & -C \\
C & D & A & B \\
-D & C & -B & A
\end{pmatrix},$$

with

$$A = \cos\gamma\cos\delta,\ B = \cos\gamma\sin\delta,\ C = \sin\gamma\cos\delta,\ D = \sin\gamma\sin\delta.$$

In addition, one can verify that

$$\begin{aligned} A^2+B^2+C^2+D^2 &= cos^2\gamma\cos^2\delta + cos^2\gamma\sin^2\delta + \sin^2\gamma\cos^2\delta + \sin^2\gamma\sin^2\delta \\ &= \cos^2\gamma + \sin^2\gamma = 1, \end{aligned}$$

that is, all the rows and columns are unit vectors. The matrix $\mathcal{A}$ may be written as

$$\mathcal{A} = \begin{pmatrix} \cos\gamma\begin{pmatrix} \cos\delta & \sin\delta \\ -\sin\delta & \cos\delta \end{pmatrix} & -\sin\gamma\begin{pmatrix} \cos\delta & \sin\delta \\ -\sin\delta & \cos\delta \end{pmatrix} \\ \sin\gamma\begin{pmatrix} \cos\delta & \sin\delta \\ -\sin\delta & \cos\delta \end{pmatrix} & \cos\delta\begin{pmatrix} \cos\delta & \sin\delta \\ -\sin\delta & \cos\delta \end{pmatrix} \end{pmatrix},$$

which shows that the Matrix $\mathcal{A}$ comprises two rotations in two planes, one counterclockwise with angle δ in one plane and one clockwise with angle γ in the second plane.

The matrix $\mathcal{A}$ is an orthogonal matrix because

$$\mathcal{A}^T = \begin{pmatrix} A & -B & C & -D \\ B & A & D & C \\ -C & D & A & -B \\ -D & -C & B & A \end{pmatrix},$$

and hence

$$\mathcal{A}^T\mathcal{A} = \begin{pmatrix} \mathbb{K} & 0 & 0 & \mathbb{H} \\ 0 & \mathbb{K} & \mathbb{H} & 0 \\ 0 & \mathbb{H} & \mathbb{K} & 0 \\ \mathbb{H} & 0 & 0 & \mathbb{K} \end{pmatrix},$$

where $\mathbb{K} = A^2+B^2+C^2+D^2 = 1$, and

$$\mathbb{H} = 2(BC - AD) = 2\left[(\cos\gamma\sin\delta)(\sin\gamma\cos\delta) - (\cos\gamma\cos\delta)(\sin\gamma\sin\delta)\right] = 0,$$

that is $\mathcal{A}^T\mathcal{A} = I$ and similarly $\mathcal{A}\mathcal{A}^T = I$.

More importantly, the matrix $\mathcal{A}$ can be written as a product of two orthogonal matrices

$$\mathcal{A} = M_L M_R,$$

where

$$M_R = \begin{pmatrix} \cos\delta & \sin\delta & 0 & 0 \\ -\sin\delta & \cos\delta & 0 & 0 \\ 0 & 0 & \cos\delta & \sin\delta \\ 0 & 0 & -\sin\delta & \cos\delta \end{pmatrix}$$

and

$$M_L = \begin{pmatrix} \cos\gamma & 0 & -\sin\gamma & 0 \\ 0 & \cos\gamma & 0 & -\sin\gamma \\ \sin\gamma & 0 & \cos\gamma & 0 \\ 0 & \sin\gamma & 0 & \cos\gamma \end{pmatrix}.$$

which verifies that $\mathcal{A}$ is indeed a four-dimensional rotation matrix; see (7.2) and (7.3).

CHAPTER 8

Short-Time Coupled Fractional Fourier Transform and Uncertainty Relations

8.1 INTRODUCTION AND NOTATION

In this chapter, we derive more properties and extensions of the coupled fractional Fourier transform (CFrFT), and then introduce the short-time coupled fractional Fourier transform (SCFrFT) and some of its properties. The chapter is concluded by a brief discussion of an uncertainty priciple for the CFrFT. We begin by introducing some of the notation, definitions, and properties that will be used in the sequel.

Throughout this chapter, we use the notation $\mathbf{x} = (x_1, x_2) \in \mathbb{R}^2$, and for any $\mathbf{x}, \mathbf{y} \in \mathbb{R}^2$, the scalar product of $\mathbf{x}$ and $\mathbf{y}$ is denoted by $\mathbf{x} \cdot \mathbf{y}$. The standard Fourier transform of $f \in L^1(\mathbb{R}^2)$ is defined by $\hat{f}(\mathbf{x}) = \frac{1}{2\pi} \int_{\mathbb{R}^2} f(\mathbf{t}) e^{-i\mathbf{t}\cdot\mathbf{x}} d\mathbf{t}$, $\forall \mathbf{x} \in \mathbb{R}^2$ and the norm of $f \in L^p(\mathbb{R}^{2k})$ is defined by

$$\|f\|_p = \left(\frac{1}{(2\pi)^k} \int_{\mathbb{R}^{2k}} |f(\mathbf{x})|^p d\mathbf{x} \right)^{1/p}, \quad \text{where } p, k \in \{1, 2\}.$$

Definition 34 ([335]). *Let $\alpha, \beta \in \mathbb{R}$, such that $\alpha + \beta \notin 2\pi\mathbb{Z}$. The coupled fractional Fourier transform CFrFT of $f \in L^1(\mathbb{R}^2)$ is defined by*

$$\mathcal{F}_{\alpha,\beta}(f)(\mathbf{u}) = \int_{\mathbb{R}^2} f(\mathbf{x}) K_{\alpha,\beta}(\mathbf{x}, \mathbf{u})\, d\mathbf{x}, \tag{8.1}$$

where $K_{\alpha,\beta}(\mathbf{x}, \mathbf{u})$ is given by Eq. (6.21). The CFrFT of $f \in L^2(\mathbb{R}^2)$ is defined by

$$\mathcal{F}_{\alpha,\beta}(f) = L^2\text{-} \lim_{n\to\infty} \mathcal{F}_{\alpha,\beta}(f_n),$$

where (f_n) is a sequence in $L^1(\mathbb{R}^2) \cap L^2(\mathbb{R}^2)$ such that $f_n \to f$ in $L^2(\mathbb{R}^2)$ as $n \to \infty$.

Using the notations

$$E(\mathbf{x}) = e^{a\|\mathbf{x}\|^2} \text{ and } E^{-1}(\mathbf{x}) = e^{-a\|\mathbf{x}\|^2}, \ \forall \mathbf{x} \in \mathbb{R}^2, \tag{8.2}$$

DOI: 10.1201/9781003089353-8

the CFrFT (8.1) can be written in terms of the Fourier transform as follows.

$$\mathcal{F}_{\alpha,\beta}(f)(\mathbf{u}) = 2\pi\, d(\gamma) E^{-1}(\mathbf{u})(fE^{-1}\hat{)}(-(\tilde{b}u_1+\tilde{c}u_2), -(\tilde{b}u_2-\tilde{c}u_1)), \tag{8.3}$$

where $\tilde{b} = b/i,\ \tilde{c} = c/i$.

8.2 PROPERTIES OF THE COUPLED FRACTIONAL FOURIER TRANSFORM

Let

$$\mathbf{A} = \begin{pmatrix} \cos\delta & -\sin\delta \\ \sin\delta & \cos\delta \end{pmatrix}, \quad \text{and } \mathbf{T} = \begin{pmatrix} t_1 \\ t_2 \end{pmatrix},$$

and let us denote the transpose of a matrix B by B^T.

We derive some elementary properties of CFrFT.

Theorem 38. *[155] The CFrFT satisfies the following properties.*

1. *Time delay: Let $f \in L^1(\mathbb{R}^2)$ and*

$$\mathbf{v} = \cos\gamma(\mathbf{AT})^T = \cos\gamma(t_1\cos\delta - t_2\sin\delta, t_1\sin\delta + t_2\cos\delta),$$

then

$$\begin{aligned}\mathcal{F}_{\alpha,\beta}(T_{\mathbf{t}}f)(\mathbf{u}) &= E^{-1}(\mathbf{u})E^{-1}(\mathbf{t})E(\mathbf{u}-\mathbf{v}.)\\ &\times \exp\{(bu_1+cu_2)t_1 + (bu_2-cu_1)t_2\}\mathcal{F}_{\alpha,\beta}(f)(\mathbf{u}-\mathbf{v}).\end{aligned}$$

2. *Frequency shift: If $f \in L^1(\mathbb{R}^2)$ and $h(\mathbf{x}) = f(\mathbf{x})e^{-i\mathbf{x}\cdot\mathbf{t}}$, then*

$$\mathcal{F}_{\alpha,\beta}(h)(\mathbf{u}) = E^{-1}(\mathbf{u})E(\mathbf{u}-\mathbf{w})\mathcal{F}_{\alpha,\beta}(f)(\mathbf{u}-\mathbf{w}), \tag{8.4}$$

where $\mathbf{w} = \sin\gamma(\mathbf{AT})^T = \sin\gamma(t_1\cos\delta - t_2\sin\delta, t_1\sin\delta + t_2\cos\delta)$.

Proof

1. *From the definition of $\mathcal{F}_{\alpha,\beta}$, we have*

$$\begin{aligned}\mathcal{F}_{\alpha,\beta}(T_{\mathbf{t}}f)(\mathbf{u}) &= \int_{\mathbb{R}^2} f(\mathbf{y}-\mathbf{t})K_{\alpha,\beta}(\mathbf{y},\mathbf{u})d\mathbf{y}\\ &= \int_{\mathbb{R}^2} f(\mathbf{x})K_{\alpha,\beta}(\mathbf{x}+\mathbf{t},\mathbf{u})d\mathbf{x}.\end{aligned}$$

But

$$\begin{aligned}&K_{\alpha,\beta}(\mathbf{x}+\mathbf{t},\mathbf{u}) = d(\gamma)E^{-1}(\mathbf{u})E^{-1}(\mathbf{t})E^{-1}(\mathbf{x})E^{-1}(\mathbf{t})\exp\{-2a(\mathbf{x}\cdot\mathbf{t})\}\\ &\times \exp\{b[u_1(x_1+t_1)+u_2(x_2+t_2)] + c[u_2(x_1+t_1) - u_1(x_2+t_2)]\}\\ &= d(\gamma)E^{-1}(\mathbf{u})E^{-1}(\mathbf{t})E^{-1}(\mathbf{x})\exp\{-2a(\mathbf{x}\cdot\mathbf{t})\}\\ &\times \exp\{b(u_1t_1+u_2t_2) + c(t_1u_2 - t_2u_1)\}\\ &\times \exp\{b(u_1x_1+u_2x_2) + c(u_2x_1 - u_1x_2)\}.\end{aligned} \tag{8.5}$$

The last term can be rewritten as

$$\begin{aligned} & b(x_1u_1+x_2u_2)+c(x_1u_2-x_2u_1) \\ = & \; b\left[x_1(u_1-v_1)+x_2(u_2-v_2)\right]+c\left[x_1(u_2-v_2)-x_2(u_1-v_1)\right] \\ \times & \; x_1\left[bv_1+cv_2\right]+x_2\left[bv_2-cv_1\right] \end{aligned} \tag{8.6}$$

and in view of the fact that

$$[bv_1+cv_2]=t_1\cot\gamma=2at_1, \quad \text{and } [bv_2-cv_1]=t_2\cot\gamma=2at_2,$$

we have

$$\begin{aligned} & b(x_1u_1+x_2u_2)+c(x_1u_2-x_2u_1) \\ = & \; b\left[x_1(u_1-v_1)+x_2(u_2-v_2)\right]+c\left[x_1(u_2-v_2)-x_2(u_1-v_1)\right] \\ \times & \; 2ax_1t_1+2ax_2t_2=2a(\mathbf{x}\cdot\mathbf{t}). \end{aligned}$$

Hence, by substituting this last expression into (8.5), we obtain

$$\begin{aligned} & K_{\alpha,\beta}(\mathbf{x}+\mathbf{t},\mathbf{u})=d(\gamma)E^{-1}(\mathbf{u})E^{-1}(\mathbf{t})E^{-1}(\mathbf{x}) \\ \times & \; \exp\left\{b(u_1t_1+u_2t_2)+c(t_1u_2-t_2u_1)\right\} \\ \times & \; \exp\left\{b\left[x_1(u_1-v_1)+x_2(u_2-v_2)\right]+c\left[x_1(u_2-v_2)-x_2(u_1-v_1)\right]\right\} \\ = & \; E^{-1}(\mathbf{u})E^{-1}(\mathbf{t})E(\mathbf{u}-\mathbf{v})\exp\left\{b(u_1t_1+u_2t_2)+c(t_1u_2-t_2u_1)\right\}K_{\alpha,\beta}(\mathbf{x},\mathbf{u}-\mathbf{v}) \end{aligned}$$

which completes the proof of part (1).

2. *Let* $f\in L^1(\mathbb{R}^2)$ *and*

$$\mathbf{w}=\sin\gamma(t_1\cos\delta-t_2\sin\delta,t_1\sin\delta+t_2\cos\delta),$$

and $h(\mathbf{x})=f(\mathbf{x})e^{-i\mathbf{x}\cdot\mathbf{t}}$*. Then*

$$\mathcal{F}_{\alpha,\beta}(h)(\mathbf{u})=\int_{\mathbb{R}^2}f(\mathbf{x})e^{-i\mathbf{x}\cdot\mathbf{t}}K_{\alpha,\beta}(\mathbf{x},\mathbf{u})d\mathbf{x}.$$

We have

$$\begin{aligned} & K_{\alpha,\beta}(\mathbf{x},\mathbf{u})e^{-i\mathbf{x}\cdot\mathbf{t}}=d(\gamma)E^{-1}(\mathbf{u})E^{-1}E(\mathbf{x}) \\ \times & \; \exp\left\{b(x_1u_1+x_2u_2)+c(x_1u_2-x_2u_1)-i(\mathbf{x}\cdot\mathbf{t})\right\}. \end{aligned} \tag{8.7}$$

But

$$\begin{aligned} & \left\{b(x_1u_1+x_2u_2)+c(x_1u_2-x_2u_1)-i(\mathbf{x}\cdot\mathbf{t})\right\} \\ = & \; \left\{b(x_1(u_1-w_1)+x_2(u_2-w_2))+c(x_1(u_2-w_2)-x_2(u_1-w_1)-i(\mathbf{x}\cdot\mathbf{t})\right\} \\ \times & \; x_1(bw_1+cw_2)+x_2(bw_2-cw_1), \end{aligned}$$

and

$$(bw_1+cw_2)=it_1, \quad \text{and } (bw_2-cw_1)=it_2,$$

hence

$$\begin{aligned} &\{b(x_1u_1+x_2u_2)+c(x_1u_2-x_2u_1)-i(\mathbf{x}\cdot\mathbf{t})\} \\ = \; & b\left[x_1(u_1-w_1)+x_2(u_2-w_2)\right]+c\left[x_1(u_2-w_2)-x_2(u_1-w_1)\right]. \end{aligned}$$

By substituting this last expression into (8.7), we obtain

$$K_{\alpha,\beta}(\mathbf{x},\mathbf{u})=E^{-1}(\mathbf{u})E(\mathbf{u}-\mathbf{w})K_{\alpha,\beta}(\mathbf{x},\mathbf{u}-\mathbf{w}),$$

which implies (8.4).

The following theorem and its corollary will be used in establishing the uncertainty principle for the CFrFT.

Theorem 39. *Let P_k be defined by $P_k(\mathbf{x})=x_k$ for $\mathbf{x}=(x_1,x_2)\in\mathbb{R}^2$. If $f,P_1f,P_2f\in L^1(\mathbb{R}^2)$, and $\gamma\neq n\pi$, then*

$$\frac{\partial}{\partial u_1}[(\mathcal{F}_{\alpha,\beta}(f))(\mathbf{u})]=(-i\cot\gamma)u_1\mathcal{F}_{\alpha,\beta}(f)+b\mathcal{F}_{\alpha,\beta}(P_1f)-c\mathcal{F}_{\alpha,\beta}(P_2f) \qquad (8.8)$$

and

$$\frac{\partial}{\partial u_2}[(\mathcal{F}_{\alpha,\beta}(f))(\mathbf{u})]=(-i\cot\gamma)u_2\mathcal{F}_{\alpha,\beta}(f)+b\mathcal{F}_{\alpha,\beta}(P_2f)+c\mathcal{F}_{\alpha,\beta}(P_1f). \qquad (8.9)$$

Proof *The proof follows easily from the observation that taking the partial derivatives of the kernel in Eq. (6.21) leads to*

$$\begin{aligned} \frac{\partial}{\partial u_1}K_{\alpha,\beta}(\mathbf{x},\mathbf{u}) &= (-2au_1+bx_1-cx_2)\,K_{\alpha,\beta}(\mathbf{x},\mathbf{u}), \\ \frac{\partial}{\partial u_2}K_{\alpha,\beta}(\mathbf{x},\mathbf{u}) &= (-2au_2+bx_2+cx_1)\,K_{\alpha,\beta}(\mathbf{x},\mathbf{u}), \end{aligned}$$

where a,b,c, are given by Eqs. (6.22) and (6.23).

By solving the system of linear equations (8.8) and (8.9) for the variables $\mathcal{F}_{\alpha,\beta}(P_1f)$ and $\mathcal{F}_{\alpha,\beta}(P_2f)$, we obtain the following corollary.

Corollary 4. *Let D_k denote the partial derivative with respect to k^{th} variable, $k=1,2$. Under the hypothesis of Theorem 39, we have*

$$\begin{aligned} \mathcal{F}_{\alpha,\beta}(P_1f) &= -i\sin\gamma\left[\sin\delta D_2(\mathcal{F}_{\alpha,\beta}(f))+\cos\delta D_1(\mathcal{F}_{\alpha,\beta}(f))\right] \\ &\quad +\cos\gamma\left[\sin\delta P_2+(\cos\delta)P_1\right]\mathcal{F}_{\alpha,\beta}(f), \qquad (8.10) \\ \mathcal{F}_{\alpha,\beta}(P_2f) &= -i\sin\gamma\left[\cos\delta D_2(\mathcal{F}_{\alpha,\beta}(f))-\sin\delta D_1(\mathcal{F}_{\alpha,\beta}(f))\right] \\ &\quad +\cos\gamma\left[\cos\delta P_2-(\sin\delta)P_1\right]\mathcal{F}_{\alpha,\beta}(f). \qquad (8.11) \end{aligned}$$

For the interested reader, several other properties of the coupled fractional Fourier transform can be found in [335, 337].

8.3 CONVOLUTION AND EXTENSION OF THE COUPLED FRACTIONAL FOURIER TRANSFORM

We now extend the coupled fractional Fourier transform to $L^2(\mathbb{R}^2)$, and then show that it is a unitary operator on $L^2(\mathbb{R}^2)$. Furthermore, we will derive its convolution and inversion formulas.

Theorem 40. *If $f,g \in L^1(\mathbb{R}^2)\cap L^2(\mathbb{R}^2)$, then $\langle \mathcal{F}_{\alpha,\beta}(f), \mathcal{F}_{\alpha,\beta}(g)\rangle = \langle f,g\rangle$.*

Proof *Let $f,g \in L^1(\mathbb{R}^2)\cap L^2(\mathbb{R}^2)$. Using (8.3), we get*

$$\begin{aligned}
\langle \mathcal{F}_{\alpha,\beta}(f), \mathcal{F}_{\alpha,\beta}(g)\rangle &= \frac{1}{2\pi}\int_{\mathbb{R}^2}(\mathcal{F}_{\alpha,\beta}(f))(u,v)\overline{(\mathcal{F}_{\alpha,\beta}(g))}(u,v)dudv \\
&= \frac{1}{2\pi}\int_{\mathbb{R}^2}\left[2\pi\, d(\gamma)E^{-1}(u,v)(fE^{-1}\hat{)}(U,V)\right] \\
&\qquad \left[2\pi\,\overline{d(\gamma)E^{-1}}(u,v)\overline{(gE^{-1}\hat{)}}(U,V)\right] dudv \\
&\qquad (\text{where } U = -(\tilde{b}u+\tilde{c}v) \text{ and } V = -(\tilde{b}v-\tilde{c}u)\ ) \\
&= \frac{1}{2\pi\sin^2\gamma}\int_{\mathbb{R}^2}(fE^{-1}\hat{)}(U,V)\overline{(gE^{-1}\hat{)}}(U,V)dudv.
\end{aligned}$$

Using the change of variables $u = -sin^2\gamma(\tilde{b}w_1 - \tilde{c}w_2)$ and $v = -\sin^2\gamma(\tilde{c}w_1+\tilde{b}w_2)$, the above equation can be rewritten as

$$\begin{aligned}
\langle \mathcal{F}_{\alpha,\beta}(f), \mathcal{F}_{\alpha,\beta}(g)\rangle &= \frac{1}{2\pi\sin^2\gamma}\int_{\mathbb{R}^2}(fE^{-1}\hat{)}(w_1,w_2)\overline{(gE^{-1}\hat{)}}(w_1,w_2)\sin^2\gamma dw_1dw_2 \\
&\quad (\text{by Parseval's identity for classical Fourier transform}) \\
&= \frac{1}{2\pi}\int_{\mathbb{R}^2}(fE^{-1})(z_1,z_2)\overline{(gE^{-1})}(z_1,z_2)dz_1dz_2 \\
&= \frac{1}{2\pi}\int_{\mathbb{R}^2}f(z_1,z_2)\overline{g}(z_1,z_2)dz_1dz_2 = \langle f,g\rangle.
\end{aligned}$$

Hence, the theorem follows.

From the previous theorem, we conclude that $\mathcal{F}_{\alpha,\beta}$ is an isometry from $L^1(\mathbb{R}^2)\cap L^2(\mathbb{R}^2)$ into $L^2(\mathbb{R}^2)$.

Definition 35. *The CFrFT of $f \in L^2(\mathbb{R}^2)$ is defined by*

$$\mathcal{F}_{\alpha,\beta}(f) = L^2\text{-}\lim_{n\to\infty}\mathcal{F}_{\alpha,\beta}(f_n),$$

where (f_n) is a sequence in $L^1(\mathbb{R}^2)\cap L^2(\mathbb{R}^2)$ such that $f_n \to f$ in $L^2(\mathbb{R}^2)$ as $n\to\infty$. Here $L^2\text{-}\lim_{n\to\infty}\mathcal{F}_{\alpha,\beta}(f_n)$ means the limit of the sequence $(\mathcal{F}_{\alpha,\beta}(f_n))$ in $L^2(\mathbb{R}^2)$ with respect to the metric induced by the norm $\|\cdot\|_2$.

The existence and uniqueness of $L^2\text{-}\lim_{n\to\infty}\mathcal{F}_{\alpha,\beta}(f_n)$ follows from the identity

$$\|\mathcal{F}_{\alpha,\beta}(f)\|_2 = \|f\|_2,\ \forall f\in L^1(\mathbb{R}^2)\cap L^2(\mathbb{R}^2), \tag{8.12}$$

which follows from Theorem 40. The result (8.12) holds for every $f\in L^2(\mathbb{R}^2)$ by the density of $L^1(\mathbb{R}^2)\cap L^2(\mathbb{R}^2)$ in $L^2(\mathbb{R}^2)$.

Definition 36. *[337] For* $f \in L^p(\mathbb{R}^2)$, $p = 1,2$, $g \in L^1(\mathbb{R}^2)$, *we define*

$$(f \circledast_{\alpha,\beta} g)(\mathbf{x}) = 2\pi d(\gamma)E(\mathbf{x})[fE^{-1} * gE^{-1}](\mathbf{x}),$$

where $*$ *is the convolution defined by*

$$(f*g)(x_1,x_2) = \frac{1}{2\pi}\int_{\mathbb{R}^2} f(x_1-y_1,x_2-y_2)g(y_1,y_2)dy_1dy_2.$$

Lemma 4. *If* $f \in L^p(\mathbb{R}^2)$ *and* $g \in L^1(\mathbb{R}^2)$, *then* $f \circledast_{\alpha,\beta} g \in L^p(\mathbb{R}^2)$ *and*

$$\|f \circledast_{\alpha,\beta} g\|_p \leq |\csc\gamma|\,\|f\|_p\|g\|_1,$$

where $p = 1,2$.

Proof *Using the well known result* $\|f*g\|_p \leq \|f\|_p\,\|g\|_1$, *the proof of the lemma follows.*

Theorem 41 (Convolution theorem for CFrFT). *If* $f \in L^p(\mathbb{R}^2)$, $p = 1,2$ *and* $g \in L^1(\mathbb{R}^2)$ *then*

$$\mathcal{F}_{\alpha,\beta}(f \circledast_{\alpha,\beta} g)(\mathbf{u}) = E(\mathbf{u})\mathcal{F}_{\alpha,\beta}(f)(\mathbf{u})\mathcal{F}_{\alpha,\beta}(g)(\mathbf{u}).$$

Proof *For the case* $f,g \in L^1(\mathbb{R}^2)$, *the convolution theorem is proved in [337]. Suppose* $f \in L^2(\mathbb{R}^2)$ *and* $g \in L^1(\mathbb{R}^2)$, *we choose a sequence* (f_n) *in* $L^1(\mathbb{R}^2) \cap L^2(\mathbb{R}^2)$ *such that* $f_n \to f$ *in* $L^2(\mathbb{R}^2)$. *Applying Lemma 4, we have* $f_n * g \in L^1(\mathbb{R}^2) \cap L^2(\mathbb{R}^2)$ *and* $f_n * g \to f*g$ *in* $L^2(\mathbb{R}^2)$, *and therefore;*

$$\begin{aligned}
\mathcal{F}_{\alpha,\beta}(f \circledast_{\alpha,\beta} g)(\mathbf{u}) &= L^2\text{-}\lim_{n\to\infty} \mathcal{F}_{\alpha,\beta}(f_n \circledast_{\alpha,\beta} g)(\mathbf{u}) \\
&= L^2\text{-}\lim_{n\to\infty} \left[E(\mathbf{u})\mathcal{F}_{\alpha,\beta}(f_n)(\mathbf{u})\mathcal{F}_{\alpha,\beta}(g)(\mathbf{u})\right] \\
&= E(\mathbf{u})(\mathcal{F}_{\alpha,\beta}(f))(\mathbf{u})(\mathcal{F}_{\alpha,\beta}(g))(\mathbf{u}).
\end{aligned}$$

Hence, the theorem follows.

Now we state the inversion formula for CFrFT.

Theorem 42 (L^1- inversion theorem for CFrFT). *([337, p.563])*
If $f, \mathcal{F}_{\alpha,\beta}(f) \in L^1(\mathbb{R}^2)$ *then*

$$\begin{aligned}
f(x,y) &= d(-\gamma)\int_{\mathbb{R}^2} \mathcal{F}_{\alpha,\beta}(f)(u,v) \\
&\times \exp\{a(\gamma)(x^2+y^2+u^2+v^2) - b(xu+yv) - c(xv-yu)\}dudv \\
&= \int_{\mathbb{R}^2} \mathcal{F}_{\alpha,\beta}(f)(u,v)K_{-\alpha,-\beta}(u,v,x,y)dudv, \text{ a.e.}
\end{aligned}$$

The inversion formula follows from the relation $(\mathcal{F}_{-\alpha,-\beta} \circ \mathcal{F}_{\alpha,\beta})(f) = f$.

8.4 SHORT-TIME COUPLED FRACTIONAL FOURIER TRANSFORM

In this section, we introduce the short-time coupled fractional Fourier transform, also called windowed coupled fractional Fourier transform, derive its inversion formula, and characterize its range [155].

Definition 37. *Let $0 \neq g \in L^2(\mathbb{R}^2)$ be fixed, and $\alpha, \beta \in \mathbb{R}$ such that $\alpha+\beta \neq 2n\pi$, where $n \in \mathbb{Z}$. For $f \in L^2(\mathbb{R}^2)$, the short-time coupled fractional Fourier transform (SCFrFT) is defined by*

$$SF_{g,\alpha,\beta}(f)(\mathbf{z},\boldsymbol{\nu}) = \int_{\mathbb{R}^2} f(\mathbf{x})\overline{g(\mathbf{x}-\mathbf{z})}K_{\alpha,\beta}(\mathbf{x},\boldsymbol{\nu})d\mathbf{x}.$$

The SCFrFT is also represented as follows:

$$SF_{g,\alpha,\beta}(f)(\mathbf{z},\boldsymbol{\nu}) = \mathcal{F}_{\alpha,\beta}(fT_{\mathbf{z}}\overline{g})(\boldsymbol{\nu}), \forall \mathbf{z} \in \mathbb{R}^2, \tag{8.13}$$

where $T_{\mathbf{z}}\overline{g}(\mathbf{x}) = \overline{g(\mathbf{x}-\mathbf{z})}, \forall \mathbf{x} \in \mathbb{R}^2$ and

$$SF_{g,\alpha,\beta}(f)(\mathbf{z},\boldsymbol{\nu}) = 2\pi\langle f, M_{\boldsymbol{\nu},\alpha,\beta}T_{\mathbf{z}}(g)\rangle,$$

where $M_{\boldsymbol{\nu},\alpha,\beta}(h)(\mathbf{x}) = h(\mathbf{x})\overline{K_{\alpha,\beta}(\mathbf{x},\boldsymbol{\nu})}, \forall \mathbf{x} \in \mathbb{R}^2$, for any $h \in L^2(\mathbb{R}^2)$.

Lemma 5 (The density principle). *([114, p.329]) Let B_1, B_2 be Banach spaces and X be a dense subspace of B_1. If $A : X \to B_2$ is a linear operator that satisfies the inequality*

$$\|Af\| \leq C\|f\|, \forall f \in X \tag{8.14}$$

then (8.14) *holds for all $f \in B_1$ and consequently A extends to a bounded operator from B_1 to B_2.*

Theorem 43 (Parseval's formula). *For $f_1, f_2 \in L^2(\mathbb{R}^2)$, and $0 \neq g \in L^2(\mathbb{R}^2)$, we have*

$$\langle SF_{g,\alpha,\beta}(f_1), SF_{g,\alpha,\beta}(f_2)\rangle_{L^2(\mathbb{R}^4)} = \|g\|_2^2\langle f_1, f_2\rangle_{L^2(\mathbb{R}^2)}.$$

Proof *First assume that $g \in L^2(\mathbb{R}^2) \cap L^\infty(\mathbb{R}^2)$, so that $f_1T_{\mathbf{z}}\overline{g}, f_2T_{\mathbf{z}}\overline{g} \in L^2(\mathbb{R}^2)$, $\forall \mathbf{z} \in \mathbb{R}^2$. The CFrFT Parseval's formula gives*

$$\langle \mathcal{F}_{\alpha,\beta}(f_1T_{\mathbf{z}}\overline{g}), \mathcal{F}_{\alpha,\beta}(f_2T_{\mathbf{z}}\overline{g})\rangle_{L^2(\mathbb{R}^2)} = \langle f_1T_{\mathbf{z}}\overline{g}, f_2T_{\mathbf{z}}\overline{g}\rangle_{L^2(\mathbb{R}^2)}, \forall \mathbf{z} \in \mathbb{R}^2,$$

for all $f_1, f_2 \in L^2(\mathbb{R}^2)$. By direct computation of the inner product, we get $\langle SF_{g,\alpha,\beta}(f_1), SF_{g,\alpha,\beta}(f_2)\rangle_{L^2(\mathbb{R}^4)}$

$$\begin{aligned}
&= \frac{1}{4\pi^2}\int_{\mathbb{R}^2}\int_{\mathbb{R}^2} SF_{g,\alpha,\beta}(f_1)(\mathbf{z},\boldsymbol{\nu})\overline{SF_{g,\alpha,\beta}(f_2)(\mathbf{z},\boldsymbol{\nu})}d\boldsymbol{\nu}d\mathbf{z} \\
&= \frac{1}{4\pi^2}\int_{\mathbb{R}^2}\left(\int_{\mathbb{R}^2} \mathcal{F}_{\alpha,\beta}(f_1T_{\mathbf{z}}\overline{g})(\boldsymbol{\nu})\overline{\mathcal{F}_{\alpha,\beta}(f_2T_{\mathbf{z}}\overline{g})(\boldsymbol{\nu})}d\boldsymbol{\nu}\right)d\mathbf{z} \\
&= \frac{1}{4\pi^2}\int_{\mathbb{R}^2}\left(\int_{\mathbb{R}^2} (f_1T_{\mathbf{z}}\overline{g})(\mathbf{x})\overline{(f_2T_{\mathbf{z}}\overline{g})(\mathbf{x})}d\mathbf{x}\right)d\mathbf{z} \\
&= \frac{1}{4\pi^2}\int_{\mathbb{R}^2}\left(\int_{\mathbb{R}^2} f_1(\mathbf{x})\overline{g(\mathbf{x}-\mathbf{z})f_2(\mathbf{x})}g(\mathbf{x}-\mathbf{z})d\mathbf{x}\right)d\mathbf{z} \\
&= \frac{1}{4\pi^2}\int_{\mathbb{R}^2} f_1(\mathbf{x})\overline{f_2(\mathbf{x})}\left(\int_{\mathbb{R}^2} \overline{g(\mathbf{x}-\mathbf{z})}g(\mathbf{x}-\mathbf{z})d\mathbf{z}\right)d\mathbf{x} \\
&\qquad \textit{(by Fubini's theorem)} \\
&= \|g\|_2^2\langle f_1, f_2\rangle_{L^2(\mathbb{R}^2)}.
\end{aligned}$$

Hence this theorem follows for the case $g \in L^2(\mathbb{R}^2) \cap L^\infty(\mathbb{R}^2)$. Using the density principle ([114, p. 329]), we extend the result to the case $g \in L^2(\mathbb{R}^2)$.

Theorem 44 (Inversion formula for SCFrFT). *Let $g \in L^2(\mathbb{R}^2)$ with $\|g\|_2 \neq 0$ and $f \in L^2(\mathbb{R}^2)$, then*

$$f = \frac{1}{2\pi\|g\|_2^2} \int_{\mathbb{R}^2} \int_{\mathbb{R}^2} SF_{g,\alpha,\beta}(f)(\mathbf{z},\boldsymbol{\nu}) M_{\boldsymbol{\nu},\alpha,\beta}(T_{\mathbf{z}} g) d\mathbf{z}\, d\boldsymbol{\nu}, \tag{8.15}$$

weakly in $L^2(\mathbb{R}^2)$.

Also, if $SF_{g,\alpha,\beta}(f) \in L^1(\mathbb{R}^4)$ then

$$f(\mathbf{x}) = \frac{1}{2\pi\|g\|_2^2} \int_{\mathbb{R}^2} \int_{\mathbb{R}^2} SF_{g,\alpha,\beta}(f)(\mathbf{z},\boldsymbol{\nu}) g(\mathbf{x}-\boldsymbol{\nu}) K_{-\alpha,-\beta}(\boldsymbol{\nu},\mathbf{x}) d\mathbf{z}\, d\boldsymbol{\nu},\ a.e \tag{8.16}$$

Proof *By Theorem 43, $\|SF_{g,\alpha,\beta} f\|_2 = \|f\|_2 \|g\|_2$ and hence $SF_{g,\alpha,\beta}(f) \in L^2(\mathbb{R}^4)$. Therefore, the vector valued integral*

$$\varphi = \frac{1}{2\pi\|g\|_2^2} \int_{\mathbb{R}^2} \int_{\mathbb{R}^2} SF_{g,\alpha,\beta}(f)(\mathbf{z},\boldsymbol{\nu}) M_{\boldsymbol{\nu},\alpha,\beta}(T_{\mathbf{z}} g) d\mathbf{z}\, d\boldsymbol{\nu}$$

is well defined in $L^2(\mathbb{R}^2)$. Hence, for any $h \in L^2(\mathbb{R}^2)$ and by using Theorem 43, we get

$$\begin{aligned}
\langle \varphi, h \rangle &= \frac{1}{4\pi^2\|g\|_2^2} \int_{\mathbb{R}^2} \left(\int_{\mathbb{R}^2} \int_{\mathbb{R}^2} SF_{g,\alpha,\beta}(f)(\mathbf{z},\boldsymbol{\nu}) M_{\boldsymbol{\nu},\alpha,\beta}(T_{\mathbf{z}} g)(\mathbf{x}) d\mathbf{z}\, d\boldsymbol{\nu} \right) \overline{h(\mathbf{x})} d\mathbf{x} \\
&= \frac{1}{2\pi\|g\|_2^2} \int_{\mathbb{R}^2} \int_{\mathbb{R}^2} SF_{g,\alpha,\beta}(f)(\mathbf{z},\boldsymbol{\nu}) \langle M_{\boldsymbol{\nu},\alpha,\beta}(T_{\mathbf{z}} g), h \rangle d\mathbf{z} d\boldsymbol{\nu} \quad \textit{(using [114, p.44])} \\
&= \frac{1}{4\pi^2\|g\|_2^2} \int_{\mathbb{R}^2} \int_{\mathbb{R}^2} SF_{g,\alpha,\beta}(f)(\mathbf{z},\boldsymbol{\nu}) \overline{SF_{g,\alpha,\beta}(h)(\mathbf{z},\boldsymbol{\nu})} d\mathbf{z}\, d\boldsymbol{\nu} \\
&= \frac{1}{\|g\|_2^2} \langle SF_{g,\alpha,\beta}(f), SF_{g,\alpha,\beta}(h) \rangle_{L^2(\mathbb{R}^4)} = \langle f, h \rangle,
\end{aligned}$$

which implies $\varphi = f$, a.e.

If $SF_{g,\alpha,\beta}(f) \in L^1(\mathbb{R}^4)$, then using the fact that $\overline{K_{\alpha,\beta}(\mathbf{x},\boldsymbol{\nu})} = K_{-\alpha,-\beta}(\boldsymbol{\nu},\mathbf{x})$, we get that (8.16) *is the integral representation of* (8.15).

Motivated by the characterization of range of a linear transformation by the reproducing kernel Hilbert space [21, 256, 257], we obtain the following theorem.

Theorem 45 (Characterization of range of $SF_{g,\alpha,\beta}$). *Let $\Phi \in L^2(\mathbb{R}^4)$. Then $\Phi \in SF_{g,\alpha,\beta}(L^2(\mathbb{R}^2))$ if and only if*

$$\Phi(\mathbf{z}',\boldsymbol{\nu}') = \frac{1}{\|g\|_2^2} \int_{\mathbb{R}^2} \int_{\mathbb{R}^2} \Phi(\mathbf{z},\boldsymbol{\nu}) \langle M_{\boldsymbol{\nu},\alpha,\beta} T_{\mathbf{z}} g, M_{\boldsymbol{\nu}',\alpha,\beta} T_{\mathbf{z}'} g \rangle\, d\mathbf{z}\, d\boldsymbol{\nu}. \tag{8.17}$$

Proof *If $\Phi \in SF_{g,\alpha,\beta}\left(L^2(\mathbb{R}^2)\right)$, then there exists $f \in L^2(\mathbb{R}^2)$ such that $SF_{g,\alpha,\beta} f = \Phi$.*

Using the inversion formula for SCFRFT, *(see Eq. (8.15)), we have*

$$\begin{aligned}
\Phi(\mathbf{z}',\boldsymbol{\nu}') &= SF_{g,\alpha,\beta}f(\mathbf{z}',\boldsymbol{\nu}')\\
&= 2\pi\langle f, M_{\boldsymbol{\nu}',\alpha,\beta}T_{\mathbf{z}'}g\rangle\\
&= \int_{\mathbb{R}^2} f(\mathbf{x})\overline{M_{\boldsymbol{\nu}',\alpha,\beta}T_{\mathbf{z}'}g(\mathbf{x})}d\mathbf{x}\\
&= \int_{\mathbb{R}^2}\left(\frac{1}{2\pi\|g\|_2^2}\int_{\mathbb{R}^2}\int_{\mathbb{R}^2} SF_{g,\alpha,\beta}(f)(\mathbf{z},\boldsymbol{\nu})M_{\boldsymbol{\nu},\alpha,\beta}(T_{\mathbf{z}}g)d\mathbf{z}\,d\boldsymbol{\nu}\right)\\
&\times \overline{M_{\boldsymbol{\nu}',\alpha,\beta}T_{\mathbf{z}'}g(\mathbf{x})}d\mathbf{x}\\
&= \frac{1}{\|g\|_2^2}\int_{\mathbb{R}^2}\int_{\mathbb{R}^2} SF_{g,\alpha,\beta}f(\mathbf{z},\boldsymbol{\nu})\langle M_{\boldsymbol{\nu},\alpha,\beta}T_{\mathbf{z}}g, M_{\boldsymbol{\nu}',\alpha,\beta}T_{\mathbf{z}'}g\rangle\, d\mathbf{z}\,d\boldsymbol{\nu}\\
&\quad \text{(Using [114, p.43])}\\
&= \frac{1}{\|g\|_2^2}\int_{\mathbb{R}^2}\int_{\mathbb{R}^2} \Phi(\mathbf{z},\boldsymbol{\nu})\langle M_{\boldsymbol{\nu},\alpha,\beta}T_{\mathbf{z}}g, M_{\boldsymbol{\nu}',\alpha,\beta}T_{\mathbf{z}'}g\rangle\, d\mathbf{z}\,d\boldsymbol{\nu}.
\end{aligned}$$

Conversely, we assume that Φ satisfies (8.17). *We define*

$$f = \frac{1}{2\pi\|g\|_2^2}\int_{\mathbb{R}^2}\int_{\mathbb{R}^2}\Phi(\mathbf{z},\boldsymbol{\nu})M_{\boldsymbol{\nu},\alpha,\beta}(T_{\mathbf{z}}g)d\mathbf{z}\,d\boldsymbol{\nu}.$$

From [114, p.43], it follows that f is well defined and $f \in L^2(\mathbb{R}^2)$. Then

$$\begin{aligned}
SF_{g,\alpha,\beta}f(\mathbf{z}',\boldsymbol{\nu}') &= \int_{\mathbb{R}^2} f(\mathbf{x})\overline{M_{\boldsymbol{\nu}',\alpha,\beta}T_{\mathbf{z}'}g(\mathbf{x})}d\mathbf{x}\\
&= \int_{\mathbb{R}^2}\left(\frac{1}{2\pi\|g\|_2^2}\int_{\mathbb{R}^2}\int_{\mathbb{R}^2}\Phi(\mathbf{z},\boldsymbol{\nu})M_{\boldsymbol{\nu},\alpha,\beta}(T_{\mathbf{z}}g)d\mathbf{z}\,d\boldsymbol{\nu}\right)\overline{M_{\boldsymbol{\nu}',\alpha,\beta}T_{\mathbf{z}'}g(\mathbf{x})}d\mathbf{x}\\
&= \frac{1}{\|g\|_2^2}\int_{\mathbb{R}^2}\int_{\mathbb{R}^2}\Phi(\mathbf{z},\boldsymbol{\nu})\langle M_{\boldsymbol{\nu},\alpha,\beta}T_{\mathbf{z}}g, M_{\boldsymbol{\nu}',\alpha,\beta}T_{\mathbf{z}'}g\rangle\, d\mathbf{z}\,d\boldsymbol{\nu} = \Phi(\mathbf{z}',\boldsymbol{\nu}').
\end{aligned}$$

This completes the proof.

8.5 PROPERTIES OF THE SHORT-TIME COUPLED FRACTIONAL FOURIER TRANSFORM

In this section, we derive some elementary properties of the short-time coupled fractional Fourier transform (SCFrFT).

Theorem 46. *The SCFrFT satisfies the following properties: Let $f, f_1, f_2 \in L^2(\mathbb{R}^2)$,*

1. *Linearity:*

$$SF_{g,\alpha,\beta}(af_1+bf_2)(\mathbf{z},\boldsymbol{\nu}) = aSF_{g,\alpha,\beta}(f_1)(\mathbf{z},\boldsymbol{\nu}) + bSF_{g,\alpha,\beta}(f_2)(\mathbf{z},\boldsymbol{\nu}),\ a,b\in\mathbb{C}.$$

2. *Time delay:*

$$\begin{aligned}
&SF_{g,\alpha,\beta}(T_{\mathbf{t}}f)(\mathbf{z},\boldsymbol{\nu})\\
&\quad = E^{-1}(\boldsymbol{\nu})E^{-1}(\mathbf{t})E(\boldsymbol{\nu}-\mathbf{v})\\
&\qquad \times \exp\{(b\nu_1+c\nu_2)t_1+(b\nu_2-c\nu_1)t_2\}SF_{g,\alpha,\beta}(f)(\mathbf{z}-\mathbf{t},\boldsymbol{\nu}-\mathbf{v}),
\end{aligned}$$

where $T_{\mathbf{t}}f(\mathbf{x}) = f(\mathbf{x}-\mathbf{t})$ and $\mathbf{v} = \cos\gamma(t_1\cos\delta - t_2\sin\delta, t_1\sin\delta + t_2\cos\delta)$.

3. *Frequency shift: If* $h(\mathbf{x}) = f(\mathbf{x})e^{-i\mathbf{x}\cdot\mathbf{t}}$, *then*

$$SF_{g,\alpha,\beta}(h)(\mathbf{z},\boldsymbol{\nu}) = E(\mathbf{w})e^{2a\boldsymbol{\nu}\cdot\mathbf{w}}SF_{g,\alpha,\beta}(f)(\mathbf{z},\boldsymbol{\nu}-\mathbf{w}),$$

where $\mathbf{w} = \sin\gamma(t_1\cos\delta - t_2\sin\delta, t_1\sin\delta + t_2\cos\delta)$.

4. *Additive property: If*

$$\alpha_1+\beta_1, \alpha_2+\beta_2, (\alpha_2+\alpha_1)+(\beta_2+\beta_1) \notin 2\pi\mathbb{Z},$$

and $SF_{g,\alpha_1,\beta_1}(f)(\mathbf{z},\cdot) \in L^2(\mathbb{R}^2)$, $\forall \mathbf{z} \in \mathbb{R}^2$, *then*

$$\mathcal{F}_{\alpha_2,\beta_2}(SF_{g,\alpha_1,\beta_1}(f)(\mathbf{z},\cdot))(\mathbf{u}) = SF_{g,\alpha_2+\alpha_1,\beta_2+\beta_1}(f)(\mathbf{z},\mathbf{u}),$$

5. *If* $f, g \in L^1(\mathbb{R}^2) \cap L^2(\mathbb{R}^2)$, *then*

$$\int_{\mathbb{R}^2} SF_{g,\alpha,\beta}(f)(\mathbf{z},\boldsymbol{\nu})d\mathbf{z} = C\mathcal{F}_{\alpha,\beta}(f)(\boldsymbol{\nu}),$$

where the $C = \int_{\mathbb{R}^2} \overline{g(\mathbf{y})}d\mathbf{y}$.

Proof

1. *We easily observe that SCFrFT is linear.*

2. *Using Theorem 38 (1), we get*

$$\begin{aligned}
& SF_{g,\alpha,\beta}(T_{\mathbf{t}}f)(\mathbf{z},\boldsymbol{\nu}) = \mathcal{F}_{\alpha,\beta}(T_{\mathbf{t}}fT_{\mathbf{z}}\overline{g})(\boldsymbol{\nu}) \\
= \; & \mathcal{F}_{\alpha,\beta}((fT_{\mathbf{z}-\mathbf{t}}g))(\boldsymbol{\nu}) \\
= \; & E^{-1}(\boldsymbol{\nu})E^{-1}(\mathbf{t})E(\boldsymbol{\nu}-\mathbf{v})\exp\{(b\nu_1+c\nu_2)t_1+(b\nu_2-c\nu_1)t_2\} \\
& \times\mathcal{F}_{\alpha,\beta}(fT_{\mathbf{z}-\mathbf{t}}g)(\boldsymbol{\nu}-\mathbf{v}), \\
& (\text{where } \mathbf{v} = \cos\gamma(t_1\cos\delta - t_2\sin\delta, t_1\sin\delta + t_2\cos\delta)\;) \\
= \; & E^{-1}(\boldsymbol{\nu})E^{-1}(\mathbf{t})E(\boldsymbol{\nu}-\mathbf{v})\exp\{(b\nu_1+c\nu_2)t_1+(b\nu_2-c\nu_1)t_2\} \\
& \times SF_{g,\alpha,\beta}(f)(\mathbf{z}-\mathbf{t},\boldsymbol{\nu}-\mathbf{v}).
\end{aligned}$$

3. *Using Theorem 38 (2), we get*

$$\begin{aligned}
SF_{g,\alpha,\beta}(h)(\mathbf{z},\boldsymbol{\nu}) &= \mathcal{F}_{\alpha,\beta}(hT_{\mathbf{z}}\overline{g})(\boldsymbol{\nu}) \\
&= E(\mathbf{w})e^{-2a\boldsymbol{\nu}\cdot\mathbf{w}}\mathcal{F}_{\alpha,\beta}(fT_{\mathbf{z}g})(\boldsymbol{\nu}-\mathbf{w}) \\
&\quad (\text{where } \mathbf{w} = \sin\gamma(t_1\cos\delta - t_2\sin\delta, t_1\sin\delta + t_2\cos\delta)) \\
&= E(\mathbf{w})e^{-2a\boldsymbol{\nu}\cdot\mathbf{w}}SF_{g,\alpha,\beta}(f)(\mathbf{z},\boldsymbol{\nu}-\mathbf{w}).
\end{aligned}$$

4. *Let* $\alpha_1+\beta_1, \alpha_2+\beta_2, (\alpha_2+\alpha_1)+(\beta_2+\beta_1) \notin 2\pi\mathbb{Z}$. *If* $g \in L^2(\mathbb{R}^2) \cap L^\infty(\mathbb{R}^2)$ *is such that* $SF_{g,\alpha_1,\beta_1}(f)(\mathbf{z},\cdot) \in L^2(\mathbb{R}^2)$, *then using inversion formula of CFrFT, we get*

$$\begin{aligned}
\mathcal{F}_{\alpha_2,\beta_2}(SF_{g,\alpha_1,\beta_1}(f)(\mathbf{z},\cdot))(\boldsymbol{\nu}) &= \mathcal{F}_{\alpha_2,\beta_2}(\mathcal{F}_{\alpha_1,\beta_1}(fT_{\mathbf{z}}\overline{g}))(\boldsymbol{\nu}) \\
&= \mathcal{F}_{\alpha_2+\alpha_1,\beta_2+\beta_1}(fT_{\mathbf{z}}\overline{g})(\boldsymbol{\nu}) \\
&= SF_{g,\alpha_2+\alpha_1,\beta_2+\beta_1}(f)(\mathbf{z},\boldsymbol{\nu}).
\end{aligned}$$

Hence this result follows for the case $g \in L^2(\mathbb{R}^2) \cap L^\infty(\mathbb{R}^2)$. *Using the density principle ([114, p.329]), we get this result for the case* $g \in L^2(\mathbb{R}^2)$.

5. *By integrating $SF_{g,\alpha,\beta}(f)(\mathbf{z},\boldsymbol{\nu})$ with respect to $\mathbf{z}$, we get*

$$\begin{aligned}\int_{\mathbb{R}^2} SF_{g,\alpha,\beta}(f)(\mathbf{z},\boldsymbol{\nu})d\mathbf{z} &= \int_{\mathbb{R}^2}\left(\int_{\mathbb{R}^2} f(\mathbf{x})\overline{g(\mathbf{x}-\mathbf{z})}K_{\alpha,\beta}(\mathbf{x},\boldsymbol{\nu})\,d\mathbf{x}\right)d\mathbf{z}\\ &= \int_{\mathbb{R}^2} f(\mathbf{x})\left(\int_{\mathbb{R}^2}\overline{g(\mathbf{x}-\mathbf{z})}d\mathbf{z}\right)K_{\alpha,\beta}(\mathbf{x},\boldsymbol{\nu})\,d\mathbf{x}\\ &= C(\mathcal{F}_{\alpha,\beta}(f))(\boldsymbol{\nu}), \quad \text{where } C=\textstyle\int_{\mathbb{R}^2}\overline{g(\mathbf{y})}d\mathbf{y}.\end{aligned}$$

Hence, the theorem follows.

8.6 UNCERTAINTY PRINCIPLE

Uncertainty principles in the fractional Fourier transform domains were obtained in [216, 277]. In this section, we derive an uncertainty relation for the coupled fractional Fourier transform. These results generalize those for the Fourier transform when $\alpha=\beta=\pi/2$. First, let us introduce some notation and definitions that are required to discuss the uncertainty principle.

Definition 38. *Let $f \in L^2(\mathbb{R}^2)$ be such that $P_k f, P_k\hat{f} \in L^2(\mathbb{R}^2)$, for $k=1,2$; see theorem 39 . We define*

(i) The spread in the time domain

$$\Delta\mathbf{x}_f^2 = \frac{1}{2\pi}\int_{\mathbb{R}^2}\|\mathbf{x}-\mathbf{x}^0\|^2|f(\mathbf{x})|^2 d\mathbf{x},$$

where $\mathbf{x}_f^0=(x_1^0,x_2^0)$ and $x_k^0=\frac{1}{2\pi\|f\|_2^2}\int_{\mathbb{R}^2}x_k|f(\mathbf{x})|^2d\mathbf{x}$.

(ii) The spread in the frequency domain

$$\Delta\mathbf{w}_f^2 = \frac{1}{2\pi}\int_{\mathbb{R}^2}\|\mathbf{w}-\mathbf{w}^0\|^2|\hat{f}(\mathbf{w})|^2 d\mathbf{w},$$

where $\mathbf{w}_f^0=(w_1^0,w_2^0)$ and $w_k^0=\frac{1}{2\pi\|f\|_2^2}\int_{\mathbb{R}^2}w_k|\hat{f}(\mathbf{w})|^2d\mathbf{w}$.

(iii) The spread in the CFrFT domain

$$\Delta\mathbf{u}_{\alpha,\beta}^2 = \frac{1}{2\pi}\int_{\mathbb{R}^2}\|\mathbf{u}-\mathbf{u}^{\alpha,\beta,0}\|^2|\mathcal{F}_{\alpha,\beta}(f)(\mathbf{u})|^2 d\mathbf{u},$$

where $\mathbf{u}^{\alpha,\beta,0}=(u_1^{\alpha,\beta,0},u_2^{\alpha,\beta,0})$ and

$$u_k^{\alpha,\beta,0} = \frac{1}{2\pi\|f\|_2^2}\int_{\mathbb{R}^2}u_k|\mathcal{F}_{\alpha,\beta}(f)(\mathbf{u})|^2 d\mathbf{u}.$$

An uncertainty principle for the CFrFT is stated in the next theorem whose proof can be found in [155].

Theorem 47 (Uncertainty principle for CFrFT). *Assume that f is a real valued function and $f \in L^2(\mathbb{R}^2)$. If $P_k f, P_k \hat{f} \in L^2(\mathbb{R}^2)$, for $k = 1,2$, then*

$$\Delta \mathbf{u}^2_{\alpha_1,\beta_1} \Delta \mathbf{u}^2_{\alpha_2,\beta_2} \geq (\|f\|_2^2 + \|\mathbf{w}_f^0\|^2 \Delta \mathbf{x}_f^2)\|f\|_2^2 \sin^2(\gamma_1 - \gamma_2)$$
$$+[(\sin\gamma_1 \sin\gamma_2)(\Delta \mathbf{w}_f^2 + \|\mathbf{w}_f^0\|^2 \|f\|^2) + (\cos\gamma_1 \cos\gamma_2)\Delta \mathbf{x}_f^2]^2, \tag{8.18}$$

where $\gamma_k = \frac{\alpha_k+\beta_k}{2}$, for $k = 1,2$. In particular, we have

$$\Delta \mathbf{u}^2_{\alpha_1,\beta_1} \Delta \mathbf{u}^2_{\alpha_2,\beta_2} \geq \|f\|_2^4 \sin^2(\gamma_1 - \gamma_2). \tag{8.19}$$

For uncertainty principles for the short-time coupled fractional Fourier transform (SCFrFT), we refer the reader to [155].

CHAPTER 9

The Linear Canonical Transform (LCT)

9.1 INTRODUCTION AND HISTORICAL OVERVIEW

The linear canonical transform (LCT) , which is a generalization of several integral transforms, including the fractional Fourier transform, is an integral transform that depends on four parameters (a,b,c,d) with $ad-bc=1$. More generally, the set of real canonical transforms may be viewed as a group of unitary transformations acting on $L^2(\mathbb{R})$ and it is a representation of the special linear group $SL(2,\mathbb{R})$ of unimodular matrices

$$SL(2,\mathbb{R}) = \left\{ \begin{pmatrix} a & b \\ c & d \end{pmatrix} : \quad a,b,c,d \in \mathbb{R} \text{ such that } ad-bc=1 \right\}.$$

The transform has been the focus of many research articles in recent years because of its applications in optics, radar system analysis and signal processing.

To the best of our knowledge the first appearance of the linear canonical transform in the literature was in the work of S. Collins in 1970 [78] in which he studied the propagation of coherent light through lens system, as well as, the relationship between diffraction theory and ray optics. He derived an integral transform relating the field of the input plane to the field of the output plane and in which the kernel of the integral transform is a function of the elements of the ray tracing matrix.

It turned out that that matrix is the same as the matrix associated with the linear canonical transform. It should be noted that some special cases of the linear canonical transform appeared prior to the work of Collins, such as the transform introduced by V. Bargmann in 1961 [26] and which is now called the Bargmann transform .

But the transform was formally introduced in 1971 by Moshinsky and Quesne in the context of unitary representations in quantum mechanics [213, 212] and they were the ones who referred to it as the *Linear Canonical Transform.*

9.2 DEFINITIONS AND SPECIAL CASES OF THE LINEAR CANONICAL TRANSFORM

Below we give a formal definition of the linear canonical transform

DOI: 10.1201/9781003089353-9

Definition 39. *Let $f \in L^1(\mathbb{R})\bigcap L^2(\mathbb{R})$ and a,b,c,d be four real parameters such that $ad-bc=1$. The linear canonical transform of f associated with the matrix*

$$M=\begin{pmatrix} a & b \\ c & d \end{pmatrix},$$

where $\det M = |M| = 1$, is defined as

$$\mathcal{L}_M[f](t)=F_M(t)=\begin{cases} \frac{1}{\sqrt{2\pi ib}}\int_{\mathbb{R}} f(x)\exp\left\{\frac{i}{2b}\left[ax^2-2xt+dt^2\right]\right\}dx & b\neq 0 \\ (e^{ict^2/2a}/\sqrt{a})f(t/a) & b=0. \end{cases}$$

In particular, when $b=0$ and $a=1$, we obtain that the transformation corresponding to the identity matrix

$$M=\begin{pmatrix} 1 & 0 \\ 0 & 1 \end{pmatrix},$$

is the identity transformation $\mathcal{L}_M[f](t)=f$. This suggests that the inverse LCT is given by

$$f(x)=\frac{1}{\sqrt{-2\pi ib}}\int_{\mathbb{R}} F_M(t)\exp\left\{\frac{-i}{2b}\left[dt^2-2xt+ax^2\right]\right\}dt,$$

which corresponds to the matrix

$$M^{-1}=\begin{pmatrix} d & -b \\ -c & a \end{pmatrix}, \text{ which is the inverse of the matrix } \quad M=\begin{pmatrix} a & b \\ c & d \end{pmatrix}$$

Because the case where $b=0$ is not very obvious, we will provide a proof for it. Let us denote the transformation Kernel by $K_M(x,t)$. Hence, we have

$$\begin{aligned} K_M(x,t) &= \frac{1}{\sqrt{2\pi ib}}\exp\left\{\frac{i}{2b}\left[ax^2+dt^2-2xt\right]\right\} \\ &= \frac{1}{\sqrt{2\pi ib}}\exp\left\{\frac{ia}{2b}\left[(x-\frac{t}{a})^2+t^2(\frac{d}{a}-\frac{1}{a^2})\right]\right\} \\ &= \frac{1}{\sqrt{2\pi ib}}\exp\left\{\frac{ia}{2b}\left[(x-\frac{t}{a})^2\right]+\frac{it^2}{2b}(\frac{ad-1}{a})\right\} \\ &= \frac{1}{\sqrt{2\pi ib}}\exp\left\{\frac{ict^2}{2a}\right\}\exp\left\{\frac{ia}{2b}(x-\frac{t}{a})^2\right\} \end{aligned}$$

Finally, if we set $w=\sqrt{2ib/a}$ and use the fact that

$$\lim_{w\to 0}\frac{e^{-x^2/w^2}}{\sqrt{\pi}w}=\delta(x), \quad \text{see Eq. } (1.26),$$

we obtain

$$\lim_{b\to 0} K_M(x,t)=\frac{e^{ict^2/2a}}{\sqrt{a}}\delta(x-t/a).$$

As special cases of the linear canonical transform, we obtain other integral transforms which we list together with their corresponding matrices:

1. The Fourier transform
$$M = \begin{pmatrix} 0 & -1 \\ 1 & 0 \end{pmatrix}, \text{ or } \qquad M = \begin{pmatrix} 0 & 1 \\ -1 & 0 \end{pmatrix},$$
depending on whether the kernel of the Fourier transform is taken as e^{ixt} or e^{-ixt}.

2. The fractional Fourier transform,
$$M = \begin{pmatrix} \cos\theta & -\sin\theta \\ \sin\theta & \cos\theta \end{pmatrix},$$

3. The Bargmann transform ,
$$M = \begin{pmatrix} \frac{1}{\sqrt{2}} & \frac{-i}{\sqrt{2}} \\ \frac{-i}{\sqrt{2}} & \frac{1}{\sqrt{2}} \end{pmatrix},$$

4. The Weierstrass transform ,
$$M = \begin{pmatrix} 1 & -i\sigma \\ 0 & 1 \end{pmatrix},$$

5. The Fresnel transform ,
$$M = \begin{pmatrix} 1 & \frac{\lambda z}{2\pi} \\ 0 & 1 \end{pmatrix},$$

It should be noted that the following three sets of matrices
$$M_1 = \begin{pmatrix} a & 0 \\ 0 & a^{-1} \end{pmatrix}, M_2 = \begin{pmatrix} 1 & b \\ 0 & 1 \end{pmatrix}, M_3 = \begin{pmatrix} 1 & 0 \\ c & 1 \end{pmatrix},$$

each constitutes a one-parameter subgroup of $SL(2,\mathbb{R})$. It follows that each element of $SL(2,\mathbb{R})$ can be decomposed as a product of elements of these three subgroups as shown below
$$M = \begin{pmatrix} a & b \\ c & d \end{pmatrix} = \begin{pmatrix} 1 & 0 \\ c/a & 1 \end{pmatrix} \begin{pmatrix} a & 0 \\ 0 & a^{-1} \end{pmatrix} \begin{pmatrix} 1 & b/a \\ 0 & 1 \end{pmatrix}.$$

Moreover, we have
$$M = \begin{pmatrix} a & b \\ c & d \end{pmatrix} = \begin{pmatrix} b & 0 \\ d & b^{-1} \end{pmatrix} \begin{pmatrix} 0 & 1 \\ -1 & 0 \end{pmatrix} \begin{pmatrix} 1 & 0 \\ a/b & 1 \end{pmatrix}.$$
in which the middle matrix
$$\begin{pmatrix} 0 & 1 \\ -1 & 0 \end{pmatrix}$$
corresponds to the Fourier transformation if the kernel is e^{-ixt}, and

$$M = \begin{pmatrix} a & b \\ c & d \end{pmatrix} = \begin{pmatrix} 1 & a/c \\ 0 & 1 \end{pmatrix} \begin{pmatrix} 0 & -1 \\ 1 & 0 \end{pmatrix} \begin{pmatrix} c & d \\ 0 & 1/c \end{pmatrix}.$$
if the kernel is e^{ixt},

9.3 PROPERTIES OF THE LINEAR CANONICAL TRANSFORM

9.3.1 Basic Properties

In this section, we list some properties of the LCT, some of which are derived from the properties of the kernel of the transform

1. **Linearity**: The LCT is linear, i.e.

$$\mathcal{L}_M[af+bg] = a\mathcal{L}_M[f]+b\mathcal{L}_M[g], \text{ where } a,b \in \mathbb{C}.$$

2. **Differentiation**:

$$\mathcal{L}_M[f'](t) = a\frac{\partial}{\partial t}\mathcal{L}_M[f](t) - ict\mathcal{L}_M[f](t)$$

3. **Multiplication by the input variable**:

$$\mathcal{L}_M[xf(x)](t) = ib\frac{\partial}{\partial t}\mathcal{L}_M[f](t) + dt\mathcal{L}_M[f](t).$$

4. **Frequency Shift**:

$$\mathcal{L}_M\left[e^{i\lambda x}f(x)\right] = e^{id\lambda(t-b\lambda/2)}\mathcal{L}_M[f](t-b\lambda)$$

5. **Translation**:

$$\mathcal{L}_M[f(x-\lambda)](t) = e^{ic\lambda(t-a\lambda/2)}\mathcal{L}_M[f](t-a\lambda).$$

In addition, we have

$$\mathcal{L}_M\left[f(x-\lambda)e^{\frac{-ia}{b}\lambda(x-\lambda/2)}\right](t) = F_M(t)e^{-i\lambda t/b}.$$

6. **Poisson Summation Formula**:

$$\frac{1}{\sqrt{2\pi ib}}\sum_{k\in\mathbb{Z}} f(t+2\sigma k)e^{\frac{i}{2b}(t+2\sigma k)^2} = \frac{1}{2\sigma}\sum_{k\in\mathbb{Z}} F_M(t_k)e^{-\frac{id}{2b}t_k^2}e^{ik\pi t/\sigma},$$

where $t_k = k\pi b/\sigma$.

7. **Convolution**: For an appropriate convolution operation $\star$ of f and g, given below, we have the following relation: if

$$h(x) = (f\star g), \quad \text{then } H_M(x) = e^{-idx^2/2b}F_M(x)G_M(x).$$

8. **Parseval's Relation**:

$$\langle F_M, G_M\rangle = \langle f,g\rangle$$

9. **Composition (Additive Formula)** :

$$\mathcal{L}_{M_2}\mathcal{L}_{M_1}[f](t) = \mathcal{L}_{M_{21}}[f](t),$$

where $M_{12} = M_2 M_1$.

10. **Sampling Formula:** If f is bandlimited to $[-\sigma, \sigma]$ in the LCT domain, then

$$f(x) = e^{-iax^2/2b} \sum_{n\in\mathbb{Z}} \frac{\sin \frac{\sigma}{b}(x-t_n)}{\frac{\sigma}{b}(x-t_n)} f(t_n) e^{iat_n^2/2b},$$

where $t_n = nb\pi/\sigma$.

11. **Eigenfunctions**: The eigenfunctions of the LCT are

$$\psi_n(x) = (\sqrt{\pi}2^n n!\lambda)^{-1/2} H_n(x/\lambda) e^{-x^2(1+i\alpha)/2\lambda^2},$$

where α is a real number, $\lambda > 0$ and $H_n(x)$ is the Hermite polynomial of degree n. The constants α and λ depend on the transform parameters.

The proof of these properties will follow.
Linearity is obvious. We will prove properties (2) and (3) together. Let us recall that

$$K_M(x,t) = \frac{1}{\sqrt{2\pi ib}} \exp\left\{\frac{i}{2b}\left[ax^2 + dt^2 - 2xt\right]\right\}.$$

If we denote the transform of f by $\mathcal{L}_M[f](t)$, then under the assumption that

$$f(x)K_M(x,t)|_{x=-\infty}^{\infty} = 0,$$

we have

$$\begin{aligned}
\mathcal{L}_M[f'](t) &= \int_{\mathbb{R}} f'(x)K(x,t)dx = -\int_{\mathbb{R}} f(x)\frac{\partial}{\partial x}K_M(x,t)dx \\
&= -\frac{i}{b}\int_{\mathbb{R}} f(x)\left[ax-t\right]K(x,t)dx \\
&= -\frac{i}{b}\left\{a\mathcal{L}_M[xf(x)](t) - t\mathcal{L}_M[f](t)\right\},
\end{aligned}$$

or

$$ib\mathcal{L}_M[f'](t) = a\mathcal{L}_M[xf(x)](t) - t\mathcal{L}_M[f](t). \tag{9.1}$$

On the other hand, by differentiating the transform with respect to t, we have

$$ib\frac{\partial}{\partial t}\mathcal{L}_M[f](t) + dt\mathcal{L}_M[f](t) = \mathcal{L}_M[xf(x)](t). \tag{9.2}$$

by combining Eqs. (9.1) and (9.2), we have

$$ib\mathcal{L}_M[f'](t) = t\mathcal{L}_M[f](t)(ad-1) + iab\frac{\partial}{\partial t}\mathcal{L}_M[f](t),$$

hence,

$$i\mathcal{L}_M[f'](t) = ct\mathcal{L}_M[f](t) + ia\frac{\partial}{\partial t}\mathcal{L}_M[f](t).$$

Therefore, we have

$$\mathcal{L}_M[f'](t) = a\frac{\partial}{\partial t}\mathcal{L}_M[f](t) - ict\mathcal{L}_M[f](t) \tag{9.3}$$

and

$$\mathcal{L}_M[xf(x)](t) = ib\frac{\partial}{\partial t}\mathcal{L}_M[f](t) + dt\mathcal{L}_M[f](t). \tag{9.4}$$

We note that Equations (9.3) and (9.4) are generalization of the following relations of the Fourier transform. If we denote the Fourier transform of f by

$$\mathcal{F}[f](t) = \frac{1}{\sqrt{2\pi}}\int_{\mathbb{R}} f(x)e^{-itx}dx,$$

we have

$$\mathcal{F}[f'](t) = it\mathcal{F}[f](t), \qquad \text{and } \mathcal{F}[xf(x)](t) = i\frac{\partial}{\partial t}\mathcal{F}[f](t),$$

which corresponds to the Fourier matrix

$$M = \begin{pmatrix} 0 & 1 \\ -1 & 0 \end{pmatrix}.$$

For properties (4) and (5), we have

$$\begin{aligned}
\mathcal{L}_M\left[e^{i\lambda x}f(x)\right](t) &= \frac{1}{\sqrt{2\pi ib}}\int_{\mathbb{R}} f(x)\exp\left\{\frac{i}{2b}\left[ax^2+dt^2-2x(t-b\lambda)\right]\right\}dx \\
&= \frac{1}{\sqrt{2\pi ib}}\int_{\mathbb{R}} f(x)\exp\left\{\frac{i}{2b}\left[ax^2+d(t-b\lambda)^2-2x(t-b\lambda)-db\lambda(b\lambda-2t)\right]\right\}dx \\
&= e^{id\lambda(t-b\lambda/2)}\mathcal{L}_M[f](t-b\lambda).
\end{aligned}$$

And

$$\begin{aligned}
\mathcal{L}_M\left[f(x-\lambda)\right](t) &= \frac{1}{\sqrt{2\pi ib}}\int_{\mathbb{R}} f(x-\lambda)\exp\left\{\frac{i}{2b}\left[ax^2+dt^2-2xt\right]\right\}dx \\
&= \frac{1}{\sqrt{2\pi ib}}\int_{\mathbb{R}} f(y)\exp\left\{\frac{i}{2b}\left[a(y+\lambda)^2+dt^2-2t(y+\lambda)\right]\right\}dy \\
&= \frac{1}{\sqrt{2\pi ib}}\int_{\mathbb{R}} f(y)\exp\left\{\frac{i}{2b}\left[ay^2+d(t-a\lambda)^2-2y(t-a\lambda)+\lambda(ad-1)\left[2t-a\lambda\right]\right]\right\}dy \\
&= e^{ic\lambda[t-a\lambda/2]}\mathcal{L}_M[f](t-a\lambda).
\end{aligned}$$

Property (6) follows from the classical Poisson Summation Formula for the Fourier transform, which may be stated as follows: if $F(t) = \frac{1}{\sqrt{2\pi}}\int_{\mathbb{R}} f(x)e^{-itx}dx$, then

$$\frac{1}{\sqrt{2\pi}}\sum_{k\in Z} f(t+2\sigma k) = \frac{1}{2\sigma}\sum_{k\in Z} F(k\pi/\sigma)e^{ik\pi t/\sigma}, \qquad \sigma > 0.$$

Since the LCT of f may be written as

$$\tilde{F}(tb) = \frac{1}{\sqrt{2\pi}} \int_{\mathbb{R}} \tilde{f}(x) e^{-itx} dx,$$

where

$$\tilde{f}(x) = f(x) e^{iax^2/2b}, \qquad \tilde{F}(t) = \sqrt{ib} e^{-idt^2/2b} F_M(t)$$

it immediately follows that

$$\frac{1}{\sqrt{2\pi}} \sum_{k \in Z} \tilde{f}(t+2\sigma k) = \frac{1}{2\sigma} \sum_{k \in Z} e^{ik\pi t/\sigma} \tilde{F}(k\pi b/\sigma),$$

or

$$\frac{1}{\sqrt{2\pi ib}} \sum_{k \in \mathbb{Z}} f(t+2\sigma k) e^{\frac{ia}{2b}(t+2\sigma k)^2} = \frac{1}{2\sigma} \sum_{k \in \mathbb{Z}} F_M(t_k) e^{-\frac{id}{2b} t_k^2} e^{ik\pi t/\sigma},$$

where $t_k = \frac{k\pi b}{\sigma}$.

9.3.2 Convolution Theorems

Next, we will present the convolution theorem, but first let us introduce the following definition.

Definition 40. *We define the convolution operation $\star$ of two functions f, g as follows:*

$$h(x) = (f \star g)(x) = \frac{e^{-iax^2/2b}}{\sqrt{bi}} (\tilde{f} * \tilde{g}),$$

where $\tilde{f}(t) = f(t) e^{iat^2/2b}$, the same for $\tilde{g}$, and $*$ stands for the convolution operation associated with the Fourier transform, namely

$$(f * g)(x) = \frac{1}{\sqrt{2\pi}} \int_{\mathbb{R}} f(x-t) g(t) dt.$$

This definition enables us to establish the next convolution theorem for the LCT.

Theorem 48. *Let H_M, F_M, G_M be the LCT of $, h, f, g$, respectively, where $h(x) = (f \star g)(x)$. Then*

$$H_M(x) = e^{-idx^2/2b} F_M(x) G_M(x).$$

Proof *First, to simplify the notation, let $C = \frac{1}{\sqrt{2\pi ib}}$. We have*

$$\begin{aligned} H_M(x) &= \int_{\mathbb{R}} h(u) K_M(u,x) du = C \int_{\mathbb{R}} e^{-iau^2/2b} K_M(u,x) du \int_{\mathbb{R}} \tilde{f}(u-t) \tilde{g}(t) dt \\ &= C \int_{\mathbb{R}} g(t) e^{iat^2/2b} dt \int_{\mathbb{R}} e^{-iau^2/2b} K_M(u,x) f(u-t) e^{ia(u-t)^2/2b} du \\ &= C \int_{\mathbb{R}} g(t) e^{iat^2/2b} dt \int_{\mathbb{R}} f(y) e^{iay^2/2b} K_M(y+t,x) e^{-ia(y+t)^2/2b} dy. \qquad (9.5) \end{aligned}$$

In view of the fact that

$$K_M(y+t,x) = C\exp\left\{\frac{i}{2b}\left[a(y+t)^2+dx^2-2x(y+t)\right]\right\}$$

the exponential factor inside (9.5) can be simplified to

$$\exp\left\{-\frac{i}{2b}(dx^2)\right\}\exp\left\{\frac{i}{2b}\left[ay^2+dx^2-2yx\right]\right\}\exp\left\{\frac{i}{2b}\left[at^2+dx^2-2tx\right]\right\},$$

which in turn reduces (9.5) to the desired form

$$H_M(x) = \exp\left\{-idx^2/2b\right\}F_M(x)G_M(x).$$

In the next theorem we prove Parseval's relation.

Theorem 49. *Let $F_M(t)$ and $G_M(t)$ be the LCT of $f(x)$ and $g(x)$, respectively. Then*

$$\langle F_M, G_M\rangle = \langle f, g\rangle,$$

where

$$\langle F_M, G_M\rangle = \int_{\mathbb{R}} F_M(t)\overline{G}_M(t)dt, \quad \langle f, g\rangle = \int_{\mathbb{R}} f(x)\overline{g}(x)dx.$$

Proof

$$\begin{aligned}\langle F_M, G_M\rangle &= \int_{\mathbb{R}} F_M(t)\overline{G}_M(t)dt = \int_{\mathbb{R}} dt\int_{\mathbb{R}} f(x)K_M(x,t)dx\int_{\mathbb{R}}\overline{g}(u)\overline{K}_M(u,t)du\\ &= \int_{\mathbb{R}} f(x)dx\int_{\mathbb{R}}\overline{g}(u)du\int_{\mathbb{R}} K_M(x,t)\overline{K}_M(u,t)dt, \qquad (9.6)\end{aligned}$$

but we have

$$\begin{aligned}\int_{\mathbb{R}} K_M(x,t)\overline{K}_M(u,t)dt &= \frac{1}{2\pi b}\int_{\mathbb{R}}\exp\left\{\frac{i}{2b}\left[a(x^2-u^2)+2t(u-x)\right]\right\}dt\\ &= \frac{1}{2\pi b}\exp\left\{\frac{i}{2b}\left[a(x^2-u^2)\right]\right\}\int_{\mathbb{R}}\exp\left\{\frac{i}{b}t(u-x)\right\}dt\\ &= \exp\left\{\frac{i}{2b}\left[a(x^2-u^2)\right]\right\}\delta(u-x).\end{aligned}$$

Therefore, by substituting this last equation into (9.6), we obtain the result.

For related work, see [273, 300].

9.3.3 Additive Property of the Linear Canonical Transform

The proof of the additive property, which is somewhat technical, is given in the next theorem

Theorem 50. *Let*

$$M_1 = \begin{pmatrix} a_1 & b_1\\ c_1 & d_1\end{pmatrix}, \quad M_2 = \begin{pmatrix} a_2 & b_2\\ c_2 & d_2\end{pmatrix},$$

where $|M_1| = |M_2| = 1$. *The following additive relation holds*

$$\mathcal{L}_{M_2}\mathcal{L}_{M_1}[f](t) = \mathcal{L}_{M_{21}}[f](t),$$

where

$$M_{21} = \begin{pmatrix} a_2 & b_2 \\ c_2 & d_2 \end{pmatrix}\begin{pmatrix} a_1 & b_1 \\ c_1 & d_1 \end{pmatrix} = \begin{pmatrix} a_2a_1+b_2c_1 & a_2b_1+b_2d_1 \\ c_2a_1+d_2c_1 & c_2b_1+d_2d_1 \end{pmatrix} = \begin{pmatrix} a_{12} & b_{12} \\ c_{12} & d_{12} \end{pmatrix}$$

Proof *To simplify the notation, let us denote the kernel* K_{M_i} *by* K_i, $i = 1,2$. *Since*

$$\mathcal{L}_{M_1}[f](t) = \int_{\mathbb{R}} f(x)K_1(x,t)dx$$

we have

$$\mathcal{L}_{M_2}\left[\mathcal{L}_{M_1}[f]\right](w) = \int_{\mathbb{R}}\int_{\mathbb{R}} f(x)K_1(x,t)K_2(t,w)dxdt$$

$$= \frac{1}{2\pi i\sqrt{b_2b_1}}\int_{\mathbb{R}} f(x)\exp\left\{\frac{i}{2}\left[\frac{a_1x^2}{b_1}+\frac{d_2w^2}{b_2}\right]\right\}J(x,w)dx \tag{9.7}$$

where

$$J(x,w) = \int_{\mathbb{R}} \exp\left\{\frac{i}{2}\left[(\frac{d_1}{b_1}+\frac{a_2}{b_2})t^2\right] - i(\frac{x}{b_1}+\frac{w}{b_2})t\right\}dt.$$

In view of the relation (4.8),

$$\int_{\mathbb{R}} e^{-p^2x^2 \pm iqx}dx = \sqrt{\pi/p^2}e^{-q^2/4p^2},$$

we obtain

$$J(x,w) = \sqrt{\frac{2\pi i b_1b_2}{d_1b_2+a_2b_1}}\exp\left\{-\frac{ib_1b_2}{2(d_1b_2+a_2b_1)}\left(\frac{x^2}{b_1^2}+\frac{w^2}{b_2^2}+\frac{2xw}{b_1b_2}\right)\right\}.$$

By substituting $J(x,w)$ *back into (9.7), we have*

$$\begin{aligned} = & \frac{1}{\sqrt{2\pi i(d_1b_2+a_2b_1)}}\int_{\mathbb{R}} f(x)\exp\left\{\frac{-ixw}{(d_1b_2+a_2b_1)}\right\} \\ \times & \exp\left\{ix^2\left(\frac{a_1}{2b_1}-\frac{b_2}{2b_1(d_1b_2+a_2b_1)}\right)\right\} \\ \times & \exp\left\{iw^2\left(\frac{d_2}{2b_2}-\frac{b_1}{2b_2(d_1b_2+a_2b_1)}\right)\right\}dx \end{aligned} \tag{9.8}$$

Let us recall that if

$$\begin{pmatrix} a_2 & b_2 \\ c_2 & d_2 \end{pmatrix}\begin{pmatrix} a_1 & b_1 \\ c_1 & d_1 \end{pmatrix} = \begin{pmatrix} a_{12} & b_{12} \\ c_{12} & d_{12} \end{pmatrix},$$

we have

$$\begin{pmatrix} a_{12} = a_1a_2 + b_2c_1 & b_{12} = a_2b_1 + b_2d_1 \\ c_{12} = a_1c_2 + c_1d_2 & d_{12} = b_1c_2 + d_1d_2 \end{pmatrix}$$

Therefore, we may write (9.8) as

$$\begin{aligned} \mathcal{L}_{M_2}\left[\mathcal{L}_{M_1}[f]\right](w) &= \frac{1}{\sqrt{2\pi i b_{12}}} \int_{\mathbb{R}} f(x) \exp\left\{\frac{-ixw}{(b_{12})}\right\} \\ &\times \exp\left\{ix^2\left(\frac{a_1}{2b_1} - \frac{b_2}{2b_1b_{12}}\right)\right\} \\ &\times \exp\left\{iw^2\left(\frac{d_2}{2b_2} - \frac{b_1}{2b_2b_{12}}\right)\right\} dx. \end{aligned} \tag{9.9}$$

But in addition, we have

$$\begin{aligned} \left(\frac{a_1}{2b_1} - \frac{b_2}{2b_1b_{12}}\right) &= \frac{b_2(a_1d_1 - 1) + a_1a_2b_1}{2b_1(d_1b_2 + a_2b_1)} \\ &= \frac{b_2b_1c_1 + a_1a_2b_1}{2b_1(d_1b_2 + a_2b_1)} \\ &= \frac{b_1(b_2c_1 + a_1a_2)}{2b_1(d_1b_2 + a_2b_1)} = \frac{a_{12}}{2b_{12}}, \end{aligned}$$

and similarly

$$\begin{aligned} \left(\frac{d_2}{2b_2} - \frac{b_1}{2b_2b_{12}}\right) &= \frac{b_1(a_2d_2 - 1) + d_1d_2b_2}{2b_2(d_1b_2 + a_2b_1)} \\ &= \frac{b_2(b_1c_2 + d_1d_2)}{2b_2(d_1b_2 + a_2b_1)} \\ &= \frac{d_{12}}{2b_{12}}. \end{aligned}$$

Therefore, we can write (9.9) as

$$\mathcal{L}_{M_2}\left[\mathcal{L}_{M_1}[f]\right](w) = \frac{1}{\sqrt{2\pi i b_{12}}} \int_{\mathbb{R}} f(x) \exp\left\{\frac{i}{2b_{12}}\left[a_{12}x^2 + d_{12}w^2 - 2xw\right]\right\} dx, \tag{9.10}$$

which completes the proof.

9.3.4 Sampling Theorem

The next theorem establishes the sampling formula for functions that are bandlimited in the LCT domain; for related material, see [17, 18, 105, 124, 125, 136, 172, 179, 181, 184, 270, 271, 274, 288, 295, 365, 367, 368, 369]. But it should be pointed out that another sampling theorem for the LCT will be presented in the next section after we introduce the linear canonical transform in higher dimensions because that theorem deals with sampling in two dimensions for functions that are bandlimited to a disc centered at the origin.

Theorem 51. *Let f be bandlimited to $[-\sigma,\sigma]$ in the LCT domain, i.e., the support of its LCT, F_M is $[-\sigma,\sigma]$. Then*

$$f(x) = e^{-iax^2/2b} \sum_{n\in\mathbb{Z}} \frac{\sin\frac{\sigma}{b}(x-t_n)}{\frac{\sigma}{b}(x-t_n)} f(t_n) e^{iat_n^2/2b},$$

where $t_n = nb\pi/\sigma$.

Proof *From the inversion formula of the LCT, we have*

$$f(x) = \frac{1}{\sqrt{-2\pi ib}} \int_{-\sigma}^{\sigma} F_M(t) \exp\left\{\frac{-i}{2b}\left[dt^2 + ax^2 - 2xt\right]\right\} dt,$$

which can be rewritten as

$$\tilde{f}(x) = \int_{-\sigma}^{\sigma} \tilde{F}(t) e^{ixt/b} dt,$$

where

$$\tilde{f}(x) = f(x) e^{iax^2/2b}, \qquad \tilde{F}(t) = \frac{1}{\sqrt{-2\pi ib}} e^{-idt^2/2b} F_M(t).$$

By a simple change of variable, we have

$$\tilde{f}(x) = \int_{-\sigma/b}^{\sigma/b} \tilde{F}(bt) e^{ixt} dt. \tag{9.11}$$

Since $\left\{e^{-inb\pi t/\sigma}\right\}_{n\in\mathbb{Z}}$ is an orthogonal basis for $L^2[-\sigma/b,\sigma/b]$, it is easy to see that

$$e^{ixt} = \sum_{n\in\mathbb{Z}} \frac{\sin\frac{\sigma}{b}(x-nb\pi/\sigma)}{\frac{\sigma}{b}(x-nb\pi/\sigma)} e^{inb\pi t/\sigma}.$$

Let $t_n = nb\pi/\sigma$, hence

$$e^{ixt} = \sum_{n\in\mathbb{Z}} \frac{\sin\frac{\sigma}{b}(x-t_n)}{\frac{\sigma}{b}(x-t_n)} e^{it_nt}.$$

In view of the uniform convergence of the series, it follows that by substituting this last series in (9.11) and interchanging summation and integration signs, we obtain

$$\tilde{f}(x) = \sum_{n\in\mathbb{Z}} \frac{\sin\frac{\sigma}{b}(x-t_n)}{\frac{\sigma}{b}(x-t_n)} \tilde{f}(t_n),$$

or

$$f(x) = e^{-iax^2/2b} \sum_{n\in\mathbb{Z}} \frac{\sin\frac{\sigma}{b}(x-t_n)}{\frac{\sigma}{b}(x-t_n)} f(t_n) e^{iat_n^2/2b},$$

which is the desired result.

9.3.5 Eigenfunctions and Eigenvalues

The next theorem describes the eigenfunctions and eigenvalues of the LCT.

Theorem 52. *Let*

$$\psi_n(x) = (\sqrt{\pi}2^n n!\lambda)^{-1/2} H_n(x/\lambda) e^{-x^2(1+i\alpha)/2\lambda^2},$$

where α is a real number, $\lambda > 0$ and $H_n(x)$ is the Hermite polynomial of degree n. Then

$$\int_{\mathbb{R}} \psi_n(x) K_M(x,y)dx = e^{-i(n+1/2)\theta}\psi_n(y),$$

where

$$K_M(x,y) = \frac{1}{\sqrt{2\pi i B}} \exp\left\{\frac{i}{2B}\left[Ax^2 + Dy^2 - 2xy\right]\right\},$$

$$\alpha = \frac{A-D}{\sqrt{4-(A+D)^2}}, \quad \lambda^2 = \frac{2B}{\sqrt{4-(A+D)^2}},$$

and

$$\theta = \cos^{-1}[(A+D)/2].$$

In other words, the functions ψ_n are the eigenfunctions of the LCT with eigenvalues $e^{-i(n+1/2)\theta}$.

Proof *From the orthogonality relation of the Hermite functions it follows that $\{\psi_n\}_{n=0}^{\infty}$ are orthonormal since*

$$\int_{\mathbb{R}} \psi_n(x)\psi_m^*(x)dx = (\sqrt{\pi}2^n n!)^{-1} \int_{\mathbb{R}} H_n(x)H_m(x)e^{-x^2}dx = \delta_{m,n},$$

where $*$ denotes the conjugate of a complex number. Now we derive the generating function of the product of the ψ_n

$$\begin{aligned}\sum_{n=0}^{\infty} \psi_n(y)\psi_n^*(x)z^n &= \sum_{n=0}^{\infty} \frac{1}{\sqrt{\pi}2^n n!\lambda} H_n(y/\lambda)H_n(x/\lambda)e^{-x^2(1-i\alpha)/2\lambda^2} \\ &\times \; e^{-y^2(1+i\alpha)/2\lambda^2} z^n \\ &= \sum_{n=0}^{\infty} \frac{1}{\sqrt{\pi}2^n n!\lambda} H_n(Y)H_n(X)\exp\left\{-X^2(1-i\alpha)/2 - Y^2(1+i\alpha)/2\right\} z^n,\end{aligned}$$

which, in view of Mehler's formula

$$\sum_{n=0}^{\infty} \frac{H_n(x)H_n(y)}{2^n n!} z^n = \frac{1}{\sqrt{1-z^2}} \exp\left(\frac{2xyz - (x^2+y^2)z^2}{1-z^2}\right),$$

implies

$$\sum_{n=0}^{\infty} \psi_n(y)\psi_n^*(x)z^n = \frac{1}{\sqrt{\pi(1-z^2)\lambda^2}} \exp\left\{\frac{4xyz - (x^2+y^2)(1+z^2)}{2\lambda^2(1-z^2)}\right\} e^{i\alpha(x^2-y^2)/2\lambda^2}.$$

Setting $z = e^{-i\theta}$, $0 < \theta < 2\pi$, *we obtain after some simplifications*

$$\begin{aligned}\sum_{n=0}^{\infty} \psi_n(y)\psi_n^*(x)e^{-in\theta} &= \frac{e^{i\theta/2}}{\sqrt{2\pi i\lambda^2 \sin\theta}} \exp\left\{\frac{2xy-(x^2+y^2)\cos\theta}{2i\lambda^2\sin\theta}\right\} e^{\frac{i\alpha(x^2-y^2)}{2\lambda^2}} \\ &= \frac{e^{i\theta/2}}{\sqrt{2\pi i\lambda^2 \sin\theta}} \exp\left\{\frac{i\left[x^2(\cos\theta+\alpha\sin\theta)+y^2(\cos\theta-\alpha\sin\theta)-2xy\right]}{2\lambda^2\sin\theta}\right\} \\ &= \frac{e^{i\theta/2}}{\sqrt{2\pi iB}} \exp\left\{\frac{i}{2B}\left[Ax^2+Dy^2-2xy\right]\right\} = e^{i\theta/2}K_M(x,y), \end{aligned} \tag{9.12}$$

where

$$K_M(x,y) = \frac{1}{\sqrt{2\pi iB}} \exp\left\{\frac{i}{2B}\left[Ax^2+Dy^2-2xy\right]\right\},$$

and

$$M = \begin{pmatrix} A & B \\ C & D \end{pmatrix},$$

with

$$\begin{aligned} A &= \cos\theta+\alpha\sin\theta, \\ D &= \cos\theta-\alpha\sin\theta, \\ B &= \lambda^2\sin\theta, \end{aligned}$$

and hence

$$C = -\frac{\alpha^2+1}{\lambda^2}\sin\theta.$$

The constants α, λ *and* θ *can be given in terms of the transform parameters provided that* $|A+D| < 2$. *In fact,*

$$\begin{aligned} \alpha &= \frac{A-D}{\sqrt{4-(A+D)^2}}, \\ \lambda^2 &= \frac{2B}{\sqrt{4-(A+D)^2}}, \\ \theta &= \cos^{-1}\left[(A+D)/2\right]. \end{aligned}$$

By multiplying Eq. (9.12) by $\psi_n(x)$ *and integrating, we obtain*

$$\int_{\mathbb{R}} \psi_n(x)K_M(x,y)dx = e^{-i(n+1/2)\theta}\psi_n(y).$$

As a special case, if $\alpha = 0$, and $\lambda = 1$, we obtain the eigenfunctions and eigenvalues of the fractional Fourier transform but in which the factor $e^{-i\theta}$ is part of the kernel.

For the case where $|A+D| \geq 2$, we refer the reader to [237, 234]. For related work, see [156, 178].

9.4 THE METAPLECTIC REPRESENTATION AND CONVOLUTION

The extension of the linear canonical transform to n dimensions can be accomplished in different ways. One way to do that is through metaplectic representations to which we alluded in Chapter 5, see also [39, 261].

Throughout this section we consider $2n \times 2n$ matrices $\mathbf{M}$ of the form

$$\mathbf{M} = \begin{pmatrix} A & B \\ C & D \end{pmatrix},$$

where A, B, C, D are $n \times n$ matrices and denote by $\mathcal{J}$ the matrix

$$\mathcal{J} = \begin{pmatrix} 0 & I_n \\ -I_n & 0 \end{pmatrix},$$

where I_n denotes the n-dimensional identity matrix. It is easy to see that

$$\mathcal{J}^T = \mathcal{J}^{-1} = -\mathcal{J},$$

where T stands for the transpose. We recall that the matrix $\mathbf{M}$ is said to be a free symplectic matrix if

$$\mathbf{M}^T \mathcal{J} \mathbf{M} = \mathcal{J}, \quad \text{and} \quad |\det B| \neq 0.$$

The set of all $2n \times 2n$ real symplectic matrices forms the symplectic group $Sp(n, \mathbb{R})$. The symplectic group may also be defined as follows. Let

$$[\mathbf{x}, \mathbf{y}] = \mathbf{x}^T \mathcal{J} \mathbf{y}$$

denote the symplectic form on $\mathbb{R}^{2n}$, where $\mathbf{x}, \mathbf{y} \in \mathbb{R}^{2n}$. A matrix $\mathbf{M} \in Sp(n, \mathbb{R})$ if and only if

$$[\mathbf{Mx}, \mathbf{My}] = [\mathbf{x}, \mathbf{y}] \qquad \forall \mathbf{x}, \mathbf{y} \in \mathbb{R}^{2n}.$$

It is easy to see that $Sp(n, \mathbb{R})$ is a group since if

$$\mathbf{A}^T \mathcal{J} \mathbf{A} = \mathcal{J} = \mathbf{B}^T \mathcal{J} \mathbf{B},$$

then

$$(\mathbf{AB})^T \mathcal{J} \mathbf{AB} = \mathbf{B}^T \mathbf{A}^T \mathcal{J} \mathbf{AB} = \mathbf{B}^T \mathcal{J} \mathbf{B} = \mathcal{J},$$

and

$$\mathbf{A}^T \mathcal{J} \mathbf{A} = \mathcal{J} \Rightarrow \quad \mathcal{J} = (\mathbf{A}^\mathbf{T})^{-1} \mathcal{J} \mathbf{A}^{-1} = (\mathbf{A}^{-\mathbf{1}})^T \mathcal{J} \mathbf{A}^{-1}.$$

The transpose matrix corresponding to the symplectic matrix $\mathbf{M}$ is given by

$$\mathbf{M}^T = \begin{pmatrix} A^T & C^T \\ B^T & D^T \end{pmatrix}.$$

The inverse of $\mathbf{M}^{-1}$ is given by

$$\mathbf{M}^{-1} = \begin{pmatrix} D^T & -B^T \\ -C^T & A^T \end{pmatrix}.$$

It follows from the fact that $\mathbf{M}^{-1}\mathbf{M} = \mathbf{M}\mathbf{M}^{-1}$ that

$$AD^T - BC^T = I_n, \quad BA^T = AB^T, \quad CD^T = DC^T,$$

$$D^T A - B^T C = I_n, A^T C = C^T A, \quad D^T B = B^T D.$$

To simplify the notation, we may write the $2n \times 2n$ matrices in the form $\mathbf{M} = (A, B; C, D)$, and hence, we have $\mathcal{J} = (0, I_n; -I_n, \mathbf{0})$. The transpose and inverse matrices corresponding to the symplectic matrix $\mathbf{M} = (A, B; C, D)$ are given by $\mathbf{M}^T = \left(A^T, C^T; B^T, D^T\right)$ and $\mathbf{M}^{-1} = \left(D^T, -B^T; -C^T, A^T\right)$, respectively.

We recall the definition of the free metaplectic transformation and some of its fundamental properties [101].

Definition 41. *Given a free symplectic matrix $\mathbf{M} = (A, B; C, D)$, the free metaplectic transformation of $f \in L^2(\mathbb{R}^n)$, denoted by $L_{\mathbf{M}}[f]$ is defined as*

$$F_M(\mathbf{w}) = L_{\mathbf{M}}[f](\mathbf{w}) = \int_{\mathbb{R}^n} f(\mathbf{x}) \mathcal{K}_M(\mathbf{x}, \mathbf{w}) \, d\mathbf{x}, \tag{9.13}$$

where the kernel $\mathcal{K}_{\mathbf{M}}(\mathbf{x}, \mathbf{w})$ is given by

$$\begin{aligned} \mathcal{K}_{\mathbf{M}}(\mathbf{x}, \mathbf{w}) &= \frac{1}{(2\pi i)^{n/2} |\det B|^{1/2}} \\ &\times \exp\left\{\frac{i}{2}\left(\mathbf{w}^T D B^{-1} \mathbf{w} - 2\mathbf{w}^T B^{-T} \mathbf{x} + \mathbf{x}^T B^{-1} A \mathbf{x}\right)\right\}. \end{aligned} \tag{9.14}$$

We have the following special cases of Definition 41:

1. For $A = D = 0, B = C = I_n$, we obtain the n-dimensional Fourier transform,
2. For

$$\begin{aligned} A &= \operatorname{diag}(a_{11}, a_{22}, \ldots, a_{nn}), \quad B = \operatorname{diag}(b_{11}, b_{22}, \ldots, b_{nn}) \\ C &= \operatorname{diag}(c_{11}, c_{22}, \ldots, c_{nn}), \quad D = \operatorname{diag}(d_{11}, d_{22}, \ldots, d_{nn}), \end{aligned}$$

   we obtain the separable linear canonical transform.
3. For

$$A = D = \operatorname{diag}(\cos\theta_1, \ldots, \cos\theta_n), B = -C = \operatorname{diag}(\sin\theta_1, \ldots, \sin\theta_n)$$

   we obtain the separable fractional Fourier transform.

9.5 ELEMENTARY PROPERTIES OF THE METAPLECTIC TRANSFORMATIONS

1. **Linearity:** The transformation is linear,

$$L_{\mathbf{M}}[af+bg](\mathbf{w}) = aL_{\mathbf{M}}[f](\mathbf{w}) + L_{\mathbf{M}}[g](\mathbf{w}), \quad \text{where } a,b \in \mathbb{C}.$$

2. **Inversion Formula:** The inverse transformation is given by

$$L_{\mathbf{M}}^{-1}[F](\mathbf{x}) \int_{\mathbb{R}^n} F_M(\mathbf{w}) \overline{\mathcal{K}}_M(\mathbf{x},\mathbf{w})\, d\mathbf{w},$$

that is

$$f(\mathbf{x}) = \int_{\mathbb{R}^N} L_{\mathbf{M}}[f](\mathbf{w}) \overline{\mathcal{K}}_{\mathbf{M}}(\mathbf{x},\mathbf{w})\, d\mathbf{w}, \tag{9.15}$$

where

$$\overline{\mathcal{K}_{\mathbf{M}}}(\mathbf{x},\mathbf{w}) = \frac{1}{(-2\pi i)^{n/2}|\det B|^{1/2}} \times \exp\left\{\frac{-i}{2}(\mathbf{w}^T DB^{-1}\mathbf{w} - 2\mathbf{w}^T B^{-T}\mathbf{x} + \mathbf{x}^T B^{-1}A\mathbf{x})\right\}. \tag{9.16}$$

3. **Parseval's relation :** The following relation holds

$$\langle f, g \rangle = \Big\langle L_{\mathbf{M}}[f], L_{\mathbf{M}}[g] \Big\rangle, \quad \forall f,g \in L^2(\mathbb{R}^n). \tag{9.17}$$

4. **Convolution:** For an appropriate convolution operation $\circledast_{\mathbf{M}}$, defined below (see Definition 42) the following relation holds

$$L_{\mathbf{M}}\Big[(f \circledast_{\mathbf{M}} g)(\mathbf{t})\Big](\mathbf{w}) = \exp\left\{(\frac{-i}{2})(\mathbf{w}^T DB^{-1}\mathbf{w})\right\} \times L_{\mathbf{M}}\Big[f\Big](\mathbf{w})\, L_{\mathbf{M}}\Big[g\Big](\mathbf{w}). \tag{9.18}$$

To derive the inversion formula, let us first observe that

$$\int_{\mathbb{R}^n} \overline{\mathcal{K}}_M(\mathbf{x},\mathbf{w}) \mathcal{K}_M(\mathbf{y},\mathbf{w})\, d\mathbf{w} = \frac{1}{(2\pi)^n |B|} \exp\left\{\frac{i}{2}\left[\mathbf{y}^T B^{-1}A\mathbf{y} - \mathbf{x}^T B^{-1}A\mathbf{x}\right]\right\} \times \int_{\mathbb{R}^n} \exp\left\{i\left[\mathbf{w}^T B^{-T}(\mathbf{x}-\mathbf{y})\right]\right\} d\mathbf{w}. \tag{9.19}$$

Setting $\omega^T = \mathbf{w}^T B^{-T}$, we obtain after some calculations that $d\mathbf{w} = |B| d\omega$. Hence,

$$\frac{1}{(2\pi)^n |B|} \int_{\mathbb{R}^n} \exp\left\{i\left[\mathbf{w}^T B^{-T}(\mathbf{x}-\mathbf{y})\right]\right\} d\mathbf{w} = \frac{1}{(2\pi)^n} \int_{\mathbb{R}^n} \exp\left\{i\left[\omega^T(\mathbf{x}-\mathbf{y})\right]\right\} d\omega = \delta(\mathbf{x}-\mathbf{y}). \tag{9.20}$$

Therefore, from (9.20), we have

$$\begin{aligned} L_{\mathbf{M}}^{-1}[F](\mathbf{x}) &= \int_{\mathbb{R}^n} F_M(\mathbf{w})\overline{\mathcal{K}}_M(\mathbf{x},\mathbf{w})\,d\mathbf{w} \\ &= \int_{\mathbb{R}^n} \overline{\mathcal{K}}_M(\mathbf{x},\mathbf{w})\,d\mathbf{w}\int_{\mathbb{R}^n} f(\mathbf{y})\mathcal{K}_M(\mathbf{y},\mathbf{w})\,d\mathbf{y} \\ &= \int_{\mathbb{R}^n} f(\mathbf{y})\,d\mathbf{y}\int_{\mathbb{R}^n} \overline{\mathcal{K}}_M(\mathbf{x},\mathbf{w})\mathcal{K}_M(\mathbf{y},\mathbf{w})\,d\mathbf{w} \\ &= \int_{\mathbb{R}^n} f(\mathbf{y})\delta(\mathbf{x}-\mathbf{y})d\mathbf{y} = f(\mathbf{x}). \end{aligned}$$

To prove Parseval's relation, we use (9.20) again to obtain

$$\begin{aligned} \Big\langle L_{\mathbf{M}}[f], L_{\mathbf{M}}[g]\Big\rangle &= \int_{\mathbb{R}^n} L_{\mathbf{M}}[f](\mathbf{w})\,\overline{L_{\mathbf{M}}}[g](\mathbf{w})d\mathbf{w} \\ &= \int_{\mathbb{R}^n} f(\mathbf{x})d\mathbf{x}\int_{\mathbb{R}^n} \overline{g}(\mathbf{y})d\mathbf{y}\int_{\mathbb{R}^n} \mathcal{K}_M(\mathbf{x},\mathbf{w})\overline{\mathcal{K}}_M(\mathbf{y},\mathbf{w})d\mathbf{w} \\ &= \int_{\mathbb{R}^n} f(\mathbf{x})d\mathbf{x}\int_{\mathbb{R}^n} \overline{g}(\mathbf{y})d\mathbf{y}\delta(\mathbf{x}-\mathbf{y}) = \langle f,g\rangle. \end{aligned}$$

Next we introduce the notion of metaplectic convolution that leads to the convolution theorem [261].

Definition 42. *Let $f,g \in L^2(\mathbb{R}^n)$. We define the free metaplectic convolution , denote by $\circledast_{\mathbf{M}}$, as*

$$\begin{aligned} (f \circledast_{\mathbf{M}} g)(\mathbf{t}) = {} & \frac{1}{(2\pi i)^{n/2}|\det B|^{1/2}}\int_{\mathbb{R}^N} f(\mathbf{x})\,g(\mathbf{t}-\mathbf{x}) \\ & \times \exp\left\{\frac{i}{2}(\mathbf{x}^T B^{-1}A(\mathbf{x}-\mathbf{t}) + (\mathbf{x}-\mathbf{t})^T B^{-1}A\mathbf{x})\right\}d\mathbf{x}. \end{aligned} \tag{9.21}$$

Theorem 53. (Convolution Theorem): *For any pair of functions $f,g \in L^2(\mathbb{R}^n)$, we have*

$$\begin{aligned} L_{\mathbf{M}}\Big[(f \circledast_{\mathbf{M}} g)(\mathbf{t})\Big](\mathbf{w}) = {} & \exp\left\{(\frac{-i}{2})(\mathbf{w}^T DB^{-1}\mathbf{w})\right\} \\ & \times L_{\mathbf{M}}\Big[f\Big](\mathbf{w})\,L_{\mathbf{M}}\Big[g\Big](\mathbf{w}). \end{aligned} \tag{9.22}$$

Proof. *Using Definition 41, we have*

$$\begin{aligned} L_{\mathbf{M}}\Big[(f \circledast_{\mathbf{M}} g)(\mathbf{t})\Big](\mathbf{w}) = {} & \frac{1}{(2\pi i)^n|\det B|}\int_{\mathbb{R}^n} f(\mathbf{x})\bigg\{\int_{\mathbb{R}^n} g(\mathbf{t}-\mathbf{x}) \\ & \times \exp\left\{\frac{i}{2}(\mathbf{x}^T B^{-1}A(\mathbf{x}-\mathbf{t}) + (\mathbf{x}-\mathbf{t})^T B^{-1}A\mathbf{x})\right\} \\ & \times \exp\left\{\frac{i}{2}(\mathbf{w}^T DB^{-1}\mathbf{w} - 2\mathbf{w}^T B^{-T}\mathbf{t} + \mathbf{t}^T B^{-1}A\mathbf{t})\right\}d\mathbf{t}\bigg\}d\mathbf{x} \\ = {} & \frac{1}{(2\pi i)^n|\det B|}\int_{\mathbb{R}^n} f(\mathbf{x})\bigg\{\int_{\mathbb{R}^n} g(\mathbf{z}) \end{aligned}$$

$$\times \exp\left\{(\frac{-i}{2})(\mathbf{x}^T B^{-1} A\mathbf{z} + \mathbf{z}^T B^{-1} A\mathbf{x})\right\}$$
$$\times \exp\left\{\frac{i}{2}(\mathbf{w}^T DB^{-1}\mathbf{w} - 2\mathbf{w}^T B^{-T}(\mathbf{x}+\mathbf{z})\right.$$
$$\left.\left. + (\mathbf{x}+\mathbf{z})^T B^{-1} A(\mathbf{x}+\mathbf{z}))\right\} d\mathbf{z}\right\} d\mathbf{x}$$
$$= \exp\left\{(\frac{-i}{2})(\mathbf{w}^T DB^{-1}\mathbf{w})\right\}$$
$$\times \int_{\mathbb{R}^n} f(\mathbf{x}) \left\{\int_{\mathbb{R}^n} g(\mathbf{z}) \mathcal{K}_{\mathbf{M}}(\mathbf{z},\mathbf{w})\, d\mathbf{z}\right\} \mathcal{K}_{\mathbf{M}}(\mathbf{x},\mathbf{w})\, d\mathbf{x}$$
$$= \exp\left\{(\frac{-i}{2})(\mathbf{w}^T DB^{-1}\mathbf{w})\right\} L_{\mathbf{M}}\left[f\right](\mathbf{w})\, L_{\mathbf{M}}\left[g\right](\mathbf{w}).$$

This completes the proof.

Remark 1. *A Convolution formula and theorem for an n-dimensional integral transform that is a special case of the Metaplectic transformation can be obtained from Theorem 53 by choosing the appropriate matrix $\mathbf{M} = (A, B; C, D)$.*

9.6 TWO-DIMENSIONAL SAMPLING THEOREM FOR THE LINEAR CANONICAL TRANSFORM

9.6.1 Two-Dimensional LCT in Polar Coordinates

In this section, we discuss a special case of the extension of the linear canonical transform given above. We focus our attention on the two-dimensional case and derive some elementary results that will be needed to prove the sampling theorem for signals bandlimited to a disc in the LCT domain [338].

Let $\mathbf{t} = (t_1, \ldots, t_n), \mathbf{x} = (x_1, \ldots, x_n)$, $\mathbf{x} \cdot \mathbf{t} = x_1 t_1 + \cdots + x_n t_n$ and $|x|^2 = x_1^2 + \cdots + x_n^2$. The n-dimensional linear canonical transform is defined as

$$F(\mathbf{t}) = \frac{1}{(2\pi b)^{n/2}} \int_{\mathbb{R}^n} f(\mathbf{x}) e^{\frac{i}{2b}\left(a|x|^2 + d|t|^2 - 2\mathbf{x}\cdot\mathbf{t}\right)} d\mathbf{x},$$

where $d\mathbf{x} = dx_1 \cdots dx_n$.

In two dimensions the transform takes the form:

$$F(\mathbf{t}) = \frac{1}{(2\pi b)} \int_{\mathbb{R}^2} f(\mathbf{x}) e^{\frac{i}{2b}\left(a|x|^2 + d|t|^2 - 2(x_1 t_1 + x_2 t_2)\right)} dx_1 dx_2.$$

Upon using the polar coordinates

$$x_1 = r\cos\theta,\ x_2 = r\sin\theta,\ t_1 = \rho\cos\phi,\ t_2 = \rho\sin\phi,$$

we obtain

$$F(\rho,\phi) = \frac{1}{(2\pi b)} \int_{\mathbb{R}^2} f(r,\theta) e^{\frac{i}{2b}\left(ar^2 + d\rho^2 - 2r\rho\cos(\theta-\phi)\right)} r dr d\theta.$$

Hence,

$$\tilde{F}(\rho,\phi) = \frac{1}{2\pi b} \int_0^{2\pi} \int_0^{\infty} \tilde{f}(r,\theta) e^{-\frac{i}{b} r\rho\cos(\theta-\phi)} r dr d\theta,$$

where

$$\tilde{F}(\rho,\phi) = e^{-id\rho^2/2b}F(\rho,\phi), \text{ and } \tilde{f}(r,\theta) = e^{iar^2/2b}f(r,\theta).$$

In view of the relation [117, p. 973],

$$e^{-it\sin\psi} = \sum_{n=-\infty}^{\infty} J_n(t)e^{-in\psi} \tag{9.23}$$

or equivalently,

$$e^{-it\cos(\theta-\phi)} = \sum_{n=-\infty}^{\infty} (-i)^n J_n(t)e^{-in(\theta-\phi)}, \tag{9.24}$$

we obtain

$$\tilde{F}(\rho,\phi) = \frac{1}{2\pi b}\int_{\mathbb{R}^2} \tilde{f}(r,\theta) \sum_{n=-\infty}^{\infty} (-i)^n J_n(r\rho/b)e^{-in(\theta-\phi)}rdrd\theta, \tag{9.25}$$

where $J_\nu(z)$ is the Bessel function of the first kind and order $\nu \geq -1/2$. Let us denote the positive zeros of $J_\nu(z)$ by

$$0 < z_{\nu,1} < z_{\nu,2} < \cdots < z_{\nu,n} < \cdots.$$

From the relation [175, p. 128]

$$\int_0^a rJ_\nu(\alpha r)J_\nu(\beta r)dr = \frac{a\beta J_\nu(\alpha a)J'_\nu(\beta a) - a\alpha J_\nu(\beta a)J'_\nu(\alpha a)}{\alpha^2-\beta^2}, \tag{9.26}$$

we obtain by setting $\alpha = \alpha_{\nu,n} = z_{\nu,n}/a$ and $\beta = \ell/a$

$$\int_0^a J_\nu(\alpha_{\nu,n}r)J_\nu(\ell r/a)rdr = \frac{a^2 z_{\nu,n}J_{\nu+1}(z_{\nu,n})J_\nu(\ell)}{z_{\nu,n}^2-\ell^2}. \tag{9.27}$$

In the last equation, we used the relation

$$\frac{d}{dz}\left[z^{-\nu}J_\nu(z)\right] = -z^{-\nu}J_{\nu+1}(z),$$

which implies that $J'_\nu(z_{\nu,n}) = -J_{\nu+1}(z_{\nu,n})$. Again from Eq. (9.27), we obtain by setting $\beta = z_{\nu,m}/a$, i.e. $\ell = z_{\nu,m}$, that

$$\int_0^a J_\nu(\alpha_{\nu,n}r)J_\nu(\alpha_{\nu,m}r)rdr = 0 \quad \text{if } m \neq n. \tag{9.28}$$

Taking the limit of Eq. (9.27) as $\ell \to z_{\nu,n}$, we obtain

$$\int_0^a J_\nu^2(\alpha_{\nu,n}r)rdr = \frac{a^2}{2}\left[J'_\nu(z_{\nu,n})\right]^2 = \frac{a^2}{2}J_{\nu+1}^2(z_{\nu,n}), \tag{9.29}$$

that is

$$\int_0^a J_\nu(\alpha_{\nu,n}x)J_\nu(\alpha_{\nu,m}x)xdx = \begin{cases} 0, & m \neq n \\ \frac{a^2}{2}J_{\nu+1}^2(z_{\nu,n}), & m = n \end{cases} \tag{9.30}$$

9.6.2 Sampling Theorem for LCT

To derive the main result of this section, which is the sampling theorem for functions that are bandlimited to a disc in the linear canonical transform domain, we need the following lemmas for which an abridged proof will be given.

Lemma 6. *Consider $J_\nu(\rho x)$ where $0 \leq x \leq a$ and $\rho \geq 0$, and let $\alpha_{\nu,n}$, $z_{\nu,n}$ be defined as before. Then*

$$J_\nu(\rho x) = \sum_{n=1}^{\infty} \frac{2z_{\nu,n} J_\nu(a\rho) J_\nu(\alpha_{\nu,n} x)}{\left(z_{\nu,n}^2 - a^2\rho^2\right) J_{\nu+1}(z_{\nu,n})}.$$

Proof *Expand $J_\nu(\rho x)$ in terms of the orthogonal system given by Eq. (9.30)*

$$J_\nu(\rho x) = \sum_{n=1}^{\infty} b_n(\rho) J_\nu(\alpha_{\nu,n} x).$$

Therefore, from Eqs. (9.27) and (9.30), we have

$$\begin{aligned} \int_0^a J_\nu(\rho x) J_\nu(\alpha_{\nu,n} x) x dx &= b_n(\rho) \frac{a^2}{2} J_{\nu+1}^2(z_{\nu,n}) \\ &= \frac{a^2 z_{\nu,n} J_{\nu+1}(z_{\nu,n}) J_\nu(a\rho)}{z_{\nu,n}^2 - a^2\rho^2}. \end{aligned} \tag{9.31}$$

By solving for $b_n(\rho)$ we obtain the result.

Lemma 7. *Let*

$$F(\rho) = \int_0^a f(r) J_\nu(\rho r) r dr.$$

Then F can be reconstructed from its samples via the formula

$$F(\rho) = \sum_{j=1}^{\infty} F(\alpha_{\nu,j}) \frac{2z_{\nu,j} J_\nu(a\rho)}{\left(z_{\nu,j}^2 - a^2\rho^2\right) J_{\nu+1}(z_{\nu,j})}. \tag{9.32}$$

Proof *With the aid of Lemma 6, we have*

$$\begin{aligned} F(\rho) &= \int_0^a f(r) J_\nu(\rho r) r dr \\ &= \int_0^a f(r) \left(\sum_{j=1}^{\infty} \frac{2z_{\nu,j} J_\nu(a\rho) J_\nu(\alpha_{\nu,j} r)}{\left(z_{\nu,j}^2 - a^2\rho^2\right) J_{\nu+1}(z_{\nu,j})} \right) r dr \\ &= \sum_{j=1}^{\infty} \frac{2z_{\nu,j} J_\nu(a\rho)}{\left(z_{\nu,j}^2 - a^2\rho^2\right) J_{\nu+1}(z_{\nu,j})} \int_0^a f(r) J_\nu(\alpha_{\nu,j} r) r dr \\ &= \sum_{j=1}^{\infty} F(\alpha_{\nu,j}) \frac{2z_{\nu,j} J_\nu(a\rho)}{\left(z_{\nu,j}^2 - a^2\rho^2\right) J_{\nu+1}(z_{\nu,j})} \end{aligned}$$

Lemma 8. *Let $f(r,t)$ be a signal periodic with period T and highest frequency N/T, that is*

$$f(r,t) = \sum_{n=-N}^{N} c_n(r) e^{2\pi int/T}.$$

Then f can be reconstructed from $2N+1$ samples via the formula

$$f(r,t) = \sum_{k=-N}^{N} f\left(r, \frac{kT}{2N+1}\right) \sigma_k(t), \tag{9.33}$$

where

$$\sigma_k(t) = \frac{\sin\left[(2N+1)\frac{\pi}{T}\left(t - \frac{kT}{2N+1}\right)\right]}{(2N+1)\sin\left[\frac{\pi}{T}(t - \frac{kT}{2N+1})\right]} \tag{9.34}$$

Proof *We have*

$$f(r,t) = \sum_{n=-N}^{N} c_n(r) e^{2\pi int/T}, \tag{9.35}$$

and if we put $\eta = T/(2N+1)$, it will follow that

$$\sum_{k=-N}^{N} f(r,k\eta) e^{-2\pi imk/(2N+1)} = \sum_{k,n=-N}^{N} c_n(r) e^{ik\tau}, \tag{9.36}$$

where $\tau = \frac{2\pi l}{2N+1}$ with $l = n - m$. The second summation on the right-hand side, i.e., the summation over k can be written in the form

$$\begin{aligned} \sum_{k=-N}^{N} e^{ik\tau} &= \frac{e^{-iN\tau}\left(1 - e^{(2N+1)i\tau}\right)}{1 - e^{i\tau}} \\ &= \frac{\sin \pi l}{\sin(\pi l/(2N+1))} = 0, \quad \textit{if } l \neq 0, \end{aligned}$$

and when $l = 0$, i.e., $n = m$, we have $\tau = 0$ and $\sum_{k=-N}^{N} e^{i\tau k} = 2N+1$. By substituting this result into (9.36), we obtain

$$\sum_{k=-N}^{N} f(r,k\eta) e^{-2\pi imk/(2N+1)} = (2N+1)c_m.$$

Solving for c_n and substituting it into Eq. (9.35), we obtain

$$\begin{aligned} f(r,t) &= \frac{1}{2N+1} \sum_{k,n=-N}^{N} e^{2\pi int/T} f(r,k\eta) e^{-2\pi ink/(2N+1)} \\ &= \frac{1}{2N+1} \sum_{k=-N}^{N} f(r,k\eta) \sum_{n=-N}^{N} e^{inx} \end{aligned} \tag{9.37}$$

where $x = \frac{2\pi}{T}\left(t - \frac{kT}{2N+1}\right)$. *The second summation is easily seen to be*

$$\sum_{n=-N}^{N} e^{inx} = \frac{\sin(N+1/2)x}{\sin(x/2)}.$$

Thus, Eq. (9.37) takes the desired form given by (9.33).

Lemma 9. *Let*

$$\sigma_k(t) = \frac{\sin\left[(2N+1)\frac{\pi}{T}\left(t - \frac{Tk}{2N+1}\right)\right]}{(2N+1)\sin\left[\frac{\pi}{T}(t - \frac{Tk}{2N+1})\right]}; \quad -N \le k \le N.$$

Then

$$\int_0^T \sigma_k(t)e^{-i2\pi nt/T}dt = \begin{cases} \frac{T}{2N+1}e^{-2\pi ikn/(2N+1)} & -N \le n \le N, \\ 0 & \textit{otherwise}. \end{cases}$$

Proof *Let* $x_k = \frac{2\pi}{T}\left(t - \frac{kT}{2N+1}\right)$. *Hence, by (9.34)*

$$(2N+1)\sigma_k(t) = e^{-iNx_k} + \cdots + e^{iNx_k}.$$

Taking into account that for $-N \le l \le N$, *we have*

$$\begin{aligned} \int_0^T e^{ilx_k}e^{-i2\pi nt/T}dt &= \int_0^T e^{il\frac{2\pi}{T}(t-\frac{kT}{2N+1})}e^{-2\pi int/T}dt \\ = \quad e^{-i2\pi kl/(2N+1)}\int_0^T e^{i\frac{2\pi}{T}(l-n)t}dt &= Te^{-2i\pi kl/(2N+1)}\delta_{l,n}, \end{aligned}$$

it follows that

$$(2N+1)\int_0^T \sigma_k(t)e^{-i2\pi nt/T}dt = Te^{-2\pi ikn/(2N+1)}.$$

Now we are able to prove the main theorem

Theorem 54. *Let* f *be bandlimited to a disc centered at the origin with radius* R *and with highest frequency* $N/(2\pi)$, *that is*

$$f(r,\theta) = \sum_{n=-N}^{N} c_n(r)e^{in\theta}, \quad 0 \le r \le R.$$

Let $F(\rho,\phi)$ *be its linear canonical transform. Then* F *can be reconstructed from its samples according to the following formula*

$$\tilde{F}(\rho,\phi) = \frac{1}{(2N+1)b}\sum_{k,n=-N}^{N} e^{in\phi}\sum_{j=1}^{\infty}\Phi_{n,j}(\rho/b)\tilde{F}(b\alpha_{n,j},\tau k)e^{-ikn\tau},$$

where

$$\tilde{F}(\rho,\phi) = e^{-id\rho^2/2b}F(\rho,\phi),$$

$$\Phi_{n,j}(\rho/b) = \frac{2z_{n,j}J_n(R\rho/b)}{(z_{n,j}^2 - R^2\rho^2/b^2)J_{n+1}(z_{n,j})}, \tag{9.38}$$

and $\tau = 2\pi/(2N+1)$.

Proof *The linear canonical transform of $f(r,\theta)$ is given by*

$$F(\rho,\phi)=\frac{1}{2\pi b}\int_{\mathbb{R}^2} f(r,\theta)K(r,\rho,\theta,\phi)rdrd\theta,$$

where $K(r,\rho,\theta,\phi)=\exp\left[\frac{i}{2b}\left(ar^2+d\rho^2-2r\rho\cos(\theta-\phi)\right)\right]$. By setting

$$\tilde{F}(\rho,\phi)=e^{-id\rho^2/2b}F(\rho,\phi),\text{ and } \tilde{f}(r,\theta)=e^{iar^2/2b}f(r,\theta),$$

and using Eq. (9.25), we have

$$\begin{aligned}\tilde{F}(\rho,\phi) &= \frac{1}{2\pi b}\int_0^{2\pi}\int_0^R \tilde{f}(r,\theta) \\ &\times \sum_{n=-\infty}^{\infty}(-i)^n J_n(r\rho/b)e^{-in(\theta-\phi)}rdrd\theta \\ &= \frac{1}{2\pi b}\sum_{n=-\infty}^{\infty}(-i)^n e^{in\phi}\sum_{m=-N}^{N}\int_0^{2\pi} e^{i\theta(m-n)}d\theta \\ &\times \int_0^R c_m(r)e^{iar^2/2b}J_n(r\rho/b)rdr \\ &= \frac{1}{b}\sum_{n=-N}^{N} e^{in\phi}\hat{c}_n(\rho), \end{aligned} \tag{9.39}$$

where

$$\hat{c}_n(\rho)=\int_0^R C_n(r)J_n(r\rho/b)rdr,$$

which is the Hankel transform of $C_n(r)=(-i)^n c_n(r)e^{iar^2/2b}$ scaled by $1/b$. Hence, by a change of scale, we have

$$\hat{c}_n(b\rho)=\int_0^R C_n(r)J_n(r\rho)rdr.$$

Therefore, from the sampling formula for the Hankel transform Eq. (9.32), we have

$$\hat{c}_n(b\rho)=\sum_{j=1}^{\infty}\hat{c}_n(b\alpha_{n,j})\Phi_{n,j}(\rho), \tag{9.40}$$

where $\alpha_{n,j}=z_{n,j}/R$ and $z_{n,j}$ is the j-th zero of the Bessel function $J_n(x)$ and $\Phi_{n,j}(\rho)$ is given by

$$\Phi_{n,j}(\rho)=\frac{2z_{n,j}J_n(R\rho)}{(z_{n,j}^2-R^2\rho^2)J_{n+1}(z_{n,j})}. \tag{9.41}$$

Or equivalently

$$\hat{c}_n(\rho)=\sum_{j=1}^{\infty}\hat{c}_n(b\alpha_{n,j})\Phi_{n,j}(\rho/b), \tag{9.42}$$

with

$$\Phi_{n,j}(\rho/b) = \frac{2z_{n,j}J_n(R\rho/b)}{(z_{n,j}^2 - R^2\rho^2/b^2)J_{n+1}(z_{n,j})}. \tag{9.43}$$

Since

$$\tilde{F}(\rho,\phi) = \frac{1}{b}\sum_{n=-N}^{N} e^{in\phi}\hat{c}_n(\rho),$$

it follows from Lemma 8 , that

$$\tilde{F}(\rho,\phi) = \frac{1}{b}\sum_{k=-N}^{N} \tilde{F}(\rho,k\tau)\sigma_k(\phi), \tag{9.44}$$

where $\sigma_k(\phi)$ is given by (9.34) and $\tau = 2\pi/(2N+1)$. From (9.39), we also have

$$\tilde{F}(\rho,k\tau) = \frac{1}{b}\sum_{k=-N}^{N} e^{ink\tau}\hat{c}_n(\rho).$$

But again from (9.39), we obtain

$$\hat{c}_n(\rho) = \frac{b}{2\pi}\int_0^{2\pi} \tilde{F}(\rho,\phi)e^{-in\phi}d\phi.$$

Hence, from Eq. (9.44), we have

$$\begin{aligned}
\hat{c}_n(b\alpha_{n,j}) &= \frac{b}{2\pi}\int_0^{2\pi} \tilde{F}(b\alpha_{n,j},\phi)e^{-in\phi}d\phi \\
&= \frac{1}{2\pi}\int_0^{2\pi} e^{-in\phi}\left(\sum_{k=-N}^{N} \tilde{F}(b\alpha_{n,j},\tau k)\sigma_k(\phi)\right)d\phi \\
&= \frac{1}{2\pi}\sum_{k=-N}^{N} \tilde{F}(b\alpha_{n,j},\tau k)\int_0^{2\pi} \sigma_k(\phi)e^{-in\phi}d\phi \\
&= \frac{1}{2N+1}\sum_{k=-N}^{N} \tilde{F}(b\alpha_{n,j},\tau k)e^{-ikn\tau},
\end{aligned}$$

where the last equation follows from Lemma 9 with $T = 2\pi$. Thus, from Eq. (9.42), we have

$$\hat{c}_n(\rho) = \frac{1}{2N+1}\sum_{j=1}^{\infty}\Phi_{n,j}(\rho/b)\sum_{k=-N}^{N} \tilde{F}(b\alpha_{n,j},\tau k)e^{-ikn\tau}. \tag{9.45}$$

Finally by substituting Eq. (9.45) into (9.39), we obtain

$$\begin{aligned}
\tilde{F}(\rho,\phi) &= \frac{1}{b}\sum_{n=-N}^{N} e^{in\phi}\hat{c}_n(\rho) \\
&= \frac{1}{(2N+1)b}\sum_{k,n=-N}^{N} e^{in\phi}e^{-ikn\tau}\sum_{j=1}^{\infty}\Phi_{n,j}(\rho/b)\tilde{F}(b\alpha_{n,j},\tau k),
\end{aligned}$$

which completes the proof.

Remark. By putting $a = \cos\theta = d, b = \sin\theta = -c$, in the above theorem, we obtain a sampling formula for signals that are bandlimited to a disc in the fractional Fourier transform domain.

For related refrences, see [55, 124, 125, 135, 219, 269, 282, 286, 310, 312, 327, 355, 364].

CHAPTER 10

The Special Affine Fourier Transform (SAFT)

10.1 INTRODUCTION AND HISTORICAL REMARKS

The Special Affine Fourier Transformations or the (SAFT) generalize a number of well known unitary transformations, as well as, signal processing and optical operations [59] . Among those special cases of the SAFT are the Fourier, fractional Fourier (FrFT), Fresnel and the linear canonical transformations (LCT). The special affine Fourier transformations are becoming increasingly popular in physics, signal processing and communication engineering because they provide good models for *phase-space* and *time-frequency* representations .

The SAFT was first introduced by Abe and Sheridan [1] in their study to develop an operator formalism to show how the fractional Fourier transformation of a wave function can be derived from the rotation of the corresponding Wigner distribution function in phase-space. In their formalism, an arbitrary area-preserving linear transformation in phase space was introduced and called *the Special Affine Fourier Transformation,* which they showed that it was the most general inhomogeneous, lossless linear mapping in phase-space.

It turned out that such a transformation can also model a number of optical operations. This was shown in a subsequent paper [2] by Abe and Sheridan in which they extended their work and represented well-known optical operations on light wave functions, such as lens transformation, free space propagation, rotation and magnification, in a unified way from the viewpoint of one-parameter Abelian subgroups of SAFT.

10.2 DEFINITIONS

The SAFT depends on 6 parameters $(a,b,c,d;p,q)$ with $ad-bc=1$. In phase space if we denote the position and wave-number by x and k respectively, the transformation action maybe described by the equation

$$\begin{pmatrix} x' \\ k' \end{pmatrix} = \begin{pmatrix} a & b \\ c & d \end{pmatrix} \begin{pmatrix} x \\ k \end{pmatrix} + \begin{pmatrix} p \\ q \end{pmatrix}.$$

DOI: 10.1201/9781003089353-10

The transformation maps any convex body into another convex body and the condition $ad-bc=1$ guarantees that the area of the body is preserved by the transformation. The integral representation of SAFT is given in the next definition.

Definition 43. *The Special Affine Fourier transform of* $f\in L^1(\mathbb{R})$*, denoted by* $\mathcal{L}_M[f](\omega)$ *or* $\widehat{f}_{\mathbf{\Lambda}_\mathsf{S}}(\omega)$ *is given by*

$$\mathcal{L}_M[f](\omega)=\widehat{f}_{\mathbf{\Lambda}_\mathsf{S}}(\omega)=\frac{1}{\sqrt{2\pi|b|}}\int_{\mathbb{R}}f(t)\exp\left\{\frac{i}{2b}\left[at^2+d\omega^2+2t(p-\omega)-2\omega\lambda\right]\right\}dt, \tag{10.1}$$

where $b\neq 0$, $\lambda=dp-bq$ *and* Λ_S *stands for the six parameters* $(a,b,c,d;p,q)$.

The definition can be easily extended to all $f\in L^2(\mathbb{R})$ as done for the FrFT.

The kernel $\kappa_{\mathbf{\Lambda}_\mathsf{S}}(t,\omega)$ in (10.1) is parameterized by SAFT matrix $\mathbf{\Lambda}_\mathsf{S}=\mathbf{\Lambda}_\mathsf{S}^{(2\times 3)}$ which may be written in the form

$$\mathbf{\Lambda}_\mathsf{S}=\left[\begin{array}{c|c}\mathbf{\Lambda} & \underline{\lambda}\end{array}\right]$$

where $\mathbf{\Lambda}$ is the LCT matrix and $\underline{\lambda}$ is an offset vector, where

$$\mathbf{\Lambda}=\begin{bmatrix}a & b\\ c & d\end{bmatrix}\ \text{with } ad-bc=1 \quad\text{and}\quad \underline{\lambda}=\begin{bmatrix}p\\ q\end{bmatrix}.$$

If we let f^* denote the complex–conjugate of f and

$$\langle f,g\rangle=\int f(t)\,g^*(t)\,dt,$$

be the standard L^2 inner–product, we may write the SAFT operation,

$$\mathscr{T}_{\mathsf{SAFT}}:f\to\widehat{f}_{\mathbf{\Lambda}_\mathsf{S}}$$

defined as,

$$\widehat{f}_{\mathbf{\Lambda}_\mathsf{S}}(\omega)=\langle f,\kappa_{\mathbf{\Lambda}_\mathsf{S}}(\cdot,\omega)\rangle \quad b\neq 0.$$

where,

$$\begin{aligned}&\kappa_{\mathbf{\Lambda}_\mathsf{S}}(t,\omega)=\\ &K_b^*\exp\left(-\frac{i}{2b}\left(at^2+d\omega^2+2t(p-\omega)-2\omega(dp-bq)\right)\right),\end{aligned} \tag{10.2}$$

where $K_b=\frac{1}{\sqrt{2\pi|b|}}$.

10.3 THE OFFSET LINEAR CANONICAL TRANSFORM

It should be noted that the definition of the special affine Fourier transform given above is slightly different from the *Offset linear canonical transform* or the OLCT defined in [236] which is given in the following definition.

Definition 44. *The Offset Linear Canonical Transform of $f \in L^1(\mathbb{R})$ is given by*

$$\widehat{f}_{\mathbf{\Lambda}_\mathsf{S}}(\omega) = \begin{cases} \frac{e^{iq\omega}}{\sqrt{2i\pi b}} \int_\mathbb{R} f(t) \exp\left\{\frac{i}{2b}\left[at^2 + d(\omega-p)^2 - 2t(\omega-p)\right]\right\} dt, & \text{if } b \neq 0 \\ \sqrt{d} e^{i\frac{cd}{2}(\omega-p)^2 + i\omega q} f\left(d\left(\omega-p\right)\right), & \text{if } b = 0, \end{cases} \tag{10.3}$$

One minor advantage of using (10.3) over (10.1) is that the former is defined for $b = 0$. On the other hand, as we will see, the transform given by Eq. (10.1) has a simpler inversion formula than the one given by (10.3).

First let us prove the existence of (10.3) for $b = 0$. Let

$$K(t,\omega) = \frac{e^{iq\omega}}{\sqrt{2i\pi b}} \exp\left\{\frac{i}{2b}\left[at^2 + d(\omega-p)^2 - 2t(\omega-p)\right]\right\}.$$

If we set $\eta = (\omega - p)/a$, we have

$$\begin{aligned} K(t,\omega) &= \frac{e^{iq\omega}}{\sqrt{2i\pi b}} \exp\left\{\frac{ia}{2b}\left[(t-\eta)^2 + \eta^2(ad-1)\right]\right\} \\ &= \frac{e^{iq\omega}}{\sqrt{2i\pi b}} \exp\left\{\frac{ia}{2b}\left[(t-\eta)^2 + \eta^2 bc\right]\right\} \\ &= \frac{e^{iq\omega}}{\sqrt{2i\pi b}} \exp\left\{\frac{ia}{2b}\left[(t-\eta)^2\right]\right\} \exp\left\{\frac{iac}{2}\eta^2\right\} \\ &= e^{iq\omega} \exp\left\{\frac{ic}{2a}(\omega-p)^2\right\} \frac{1}{\sqrt{2i\pi b}} \exp\left\{\frac{ia}{2b}\left[(t-\eta)^2\right]\right\}. \end{aligned} \tag{10.4}$$

In view of the relation given by (1.26), that

$$\lim_{w\to 0} \frac{e^{-x^2/w^2}}{\sqrt{\pi} w} = \delta(x),$$

we obtain after taking the limit of (10.4) as $b \to 0$ that

$$\lim_{b\to 0} K(t,\omega) = e^{iq\omega} \exp\left\{\frac{ic}{2a}(\omega-p)^2\right\} \delta\left(\sqrt{a}(t-\eta)\right).$$

Therefore, because when $b = 0, d = 1/a$, we have

$$\begin{aligned} \widehat{f}_{\mathbf{\Lambda}_\mathsf{S}}(\omega) &= \lim_{b\to 0} \int_\mathbb{R} f(t) K(t,\omega) dt \\ &= \int_\mathbb{R} f(t) e^{iq\omega} \exp\left\{\frac{icd}{2}(\omega-p)^2\right\} \delta\left(\sqrt{a}(t-\eta)\right) dt \\ &= \sqrt{d} e^{iq\omega} \exp\left\{\frac{icd}{2}(\omega-p)^2\right\} f\left(d(\omega-p)\right). \end{aligned}$$

To get the inverse of the SAFT (or the iSAFT), we will need the matrix $\mathbf{\Lambda}_\mathsf{S}^{\text{inv}}$ with parameters,

$$\mathbf{\Lambda}_\mathsf{S}^{\text{inv}} \stackrel{\text{def}}{=} \left[\begin{array}{cc|c} +d & -b & bq-dp \\ -c & +a & cp-aq \end{array}\right] = \left[\begin{array}{cc|c} +d & -b & p_0 \\ -c & +a & q_0 \end{array}\right]. \tag{10.5}$$

We will show that the inverse transform (iSAFT) is given by

$$f(t) = \left\langle \widehat{f}_{\mathbf{\Lambda}_\mathsf{S}}, \kappa_{\mathbf{\Lambda}_\mathsf{S}^{\text{inv}}}(\cdot, t) \right\rangle \tag{10.6}$$

In addition, we shall develop a convolution structure for the SAFT denoted by $*_{\mathbf{\Lambda}_\mathsf{S}}$ so that we can obtain a representation of the form,

$$\mathscr{T}_{\text{SAFT}}\left[f *_{\mathbf{\Lambda}_\mathsf{S}} g\right] \propto \mathscr{T}_{\text{SAFT}}[f]\, \mathscr{T}_{\text{SAFT}}[g]$$

which is consistent with the result of Fourier convolution theorem.

In the next theorem we derive the inverse of the SAFT.

Theorem 55. *Let us denote the kernel associated with $\mathbf{\Lambda}_\mathsf{S}^{inv}$ by $\kappa_{\mathbf{\Lambda}_\mathsf{S}}^{-1}$; that is*

$$\kappa_{\mathbf{\Lambda}_\mathsf{S}}^{-1}(\omega, t) = \frac{1}{\sqrt{2\pi|b|}} \exp\left\{\frac{-i}{2b}\left[d\omega^2 + at^2 + 2\omega(p_0 - t) - 2t\lambda_0\right]\right\},$$

where

$$\lambda_0 = ap_0 + bq_0, \quad p_0 = bq - dp, \quad q_0 = cp - aq.$$

Therefore, the inverse of SAFT is given by

$$f(t) = \int_{\mathbb{R}} \widehat{f}(\omega)\kappa_{\mathbf{\Lambda}_\mathsf{S}}^{-1}(\omega, t)d\omega.$$

Proof *Let*

$$\begin{aligned} I(x) = \int_{\mathbb{R}} \widehat{f}(\omega)\kappa_{\mathbf{\Lambda}_\mathsf{S}}^{-1}(\omega, x)d\omega &= \int_{\mathbb{R}} \kappa_{\mathbf{\Lambda}_\mathsf{S}}^{-1}(\omega, x)d\omega \int_{\mathbb{R}} f(t)\kappa_{\mathbf{\Lambda}_\mathsf{S}}(t, \omega)dt \\ &= \int_{\mathbb{R}} f(t)dt \int_{\mathbb{R}} \kappa_{\mathbf{\Lambda}_\mathsf{S}}^{-1}(\omega, x)\kappa_{\mathbf{\Lambda}_\mathsf{S}}(t, \omega)d\omega \end{aligned} \tag{10.7}$$

But in view of the fact that

$$\lambda + p_0 = dp - bq + p_0 = 0, \quad \textit{and } \lambda_0 = ap_0 + bq_0 = -p$$

it is easy to see that

$$\begin{aligned} \kappa_{\mathbf{\Lambda}_\mathsf{S}}^{-1}(\omega, x)\kappa_{\mathbf{\Lambda}_\mathsf{S}}(t, \omega) &= \frac{1}{2\pi|b|} \exp\left\{\frac{i}{2b}\left[at^2 + 2tp\right]\right\} \\ &\times \exp\left\{\frac{-i}{2b}\left[ax^2 - 2x\lambda_0\right]\right\} \exp\left\{\frac{-i}{b}\left[\omega(t - x)\right]\right\}. \end{aligned} \tag{10.8}$$

Therefore, we have

$$\begin{aligned} I(x) &= \frac{1}{2\pi|b|} \exp\left\{\frac{-i}{2b}\left[ax^2 + 2xp\right]\right\} \int_{\mathbb{R}} f(t) \exp\left\{\frac{i}{2b}\left[at^2 + 2tp\right]\right\} dt \\ &\times \int_{\mathbb{R}} \exp\left\{\frac{-i}{b}\left[\omega(t - x)\right]\right\} d\omega \\ &= \exp\left\{\frac{-i}{2b}\left[ax^2 + 2xp\right]\right\} \int_{\mathbb{R}} f(t) \exp\left\{\frac{i}{2b}\left[at^2 + 2tp\right]\right\} \delta(t - x)dt \\ &= f(x). \end{aligned} \tag{10.9}$$

10.4 ELEMENTARY PROPERTIES OF THE SPECIAL AFFINE FOURIER TRANSFORM

In this section, we present some properties of the SAFT with their proofs.

1. **Linearity:** The SAFT is linear, i.e.,

$$\mathcal{L}_M[af+bg] = a\mathcal{L}_M[f]+b\mathcal{L}_M[g]; \quad a,b\in\mathbb{C}.$$

2. **Translation:**

$$\mathcal{L}_M[f(t-\beta)] = e^{i\beta[c\omega-ac\beta/2-(pc-aq)]}\mathcal{L}_M[f](\omega-a\beta)$$

3. **Frequency Shift:**

$$\mathcal{L}_M\left[e^{i\beta t}f(t)\right] = e^{i\beta[d\omega-db\beta/2-\lambda]}\mathcal{L}_M[f](\omega-b\beta).$$

4. **Differentiation:** Assuming that

$$f(t)\kappa_{\mathbf{\Lambda}_\mathsf{S}}(t,\omega)=0, \qquad \text{at } t=\pm\infty,$$

we have

$$\mathcal{L}_M[f'](\omega) = a\frac{\partial}{\partial\omega}\mathcal{L}_M[f](\omega)+i[c(p-\omega)-aq]\mathcal{L}_M[f](\omega)$$

and

$$\mathcal{L}_M[tf(t)](\omega) = ib\frac{\partial}{\partial\omega}\mathcal{L}_M[f](\omega)+(d\omega-\lambda)\mathcal{L}_M[f](\omega).$$

5. **Poisson Summation Formula:** The Poisson summation formula for the SAFT takes the form

$$\sum_{k=-\infty}^{\infty} f(t+2\sigma k)\exp\frac{i}{2b}[a(t+2\sigma k)^2+2p(t+2\sigma k)]$$
$$= \frac{\sqrt{|b|}}{2\sigma}\sum_{k=-\infty}^{\infty}\widehat{f}_{\mathbf{\Lambda}_\mathsf{S}}(\omega)(k\pi b/\sigma)\exp\frac{-i}{2b}[d(k\pi b/\sigma)^2-2\lambda(k\pi b/\sigma)]e^{ik\pi t/\sigma}$$

provided that both summations exist.

6. **Convolution:** Let f and g be two given functions for which a convolution $*_{\mathbf{\Lambda}_\mathsf{S}}$, written as

$$h(t) = (f *_{\mathbf{\Lambda}_\mathsf{S}} g)(t).$$

exists, then we have,

$$h(t) = (f *_{\mathbf{\Lambda}_\mathsf{S}} g)(t) \xrightarrow{\mathsf{SAFT}} \widehat{h}_{\mathbf{\Lambda}_\mathsf{S}}(\omega) = \eta^*(\omega)\widehat{f}_{\mathbf{\Lambda}_\mathsf{S}}(\omega)\widehat{g}_{\mathbf{\Lambda}_\mathsf{S}}(\omega),$$

where

$$\eta(\omega) = \exp\left\{\frac{i}{2b}\left[d\omega^2+\Omega\omega\right]\right\}, \quad \Omega = 2(bq-dp).$$

Moreover, let,

$$h(t) = \eta_{\mathbf{\Lambda}_S^{\text{inv}}}(t) f(t) g(t)$$

where

$$\eta_{\mathbf{\Lambda}_S^{\text{inv}}}(t) = \exp\left\{\frac{i}{2b}\left[at^2 - 2t\lambda_0\right]\right\},$$

and $\lambda_0 = ap_0 + bq_0$. We then have,

$$\widehat{h}_{\mathbf{\Lambda}_S}(\omega) = \left(\widehat{f}_{\mathbf{\Lambda}_S^{\text{inv}}} *_{\mathbf{\Lambda}_S^{\text{inv}}} \widehat{g}_{\mathbf{\Lambda}_S^{\text{inv}}}\right)(\omega).$$

Now we provide proof for those elementary properties. Linearity is trivial.
Translation:

Proof

$$\mathcal{L}_M[f(t-\beta)] = K_b \int_{\mathbb{R}} f(t-\beta) \exp\left\{\frac{i}{2b}\left[at^2 + d\omega^2 + 2t(p-\omega) - 2\lambda\omega\right]\right\} dt$$

$$= K_b \int_{\mathbb{R}} f(x) \exp\left\{\frac{i}{2b}\left[a(x+\beta)^2 + d\omega^2 + 2(x+\beta)(p-\omega) - 2\lambda\omega\right]\right\} dx$$

$$= C(\omega) K_b \int_{\mathbb{R}} f(x) \exp\left\{\frac{i}{2b}\left[ax^2 + d(\omega - a\beta)^2 + 2x(p-\omega+a\beta) - 2\lambda(\omega - a\beta)\right]\right\} dx$$

$$= C(\omega)\mathcal{L}_M[f](\omega - a\beta),$$

where

$$C(\omega) = e^{i\beta[c\omega - ac\beta/2 - (pc - aq)]}, \qquad \textit{and } \lambda = (dp - bq)$$

Frequency Shift:

Proof

$$\mathcal{L}_M\left[e^{i\beta t} f(t)\right] = K_b \int_{\mathbb{R}} e^{i\beta t} f(t) \exp\left\{\frac{i}{2b}\left[at^2 + d\omega^2 + 2t(p-\omega) - 2\lambda\omega\right]\right\} dt$$

$$= K_b \int_{\mathbb{R}} f(t) \exp\left\{\frac{i}{2b}\left[at^2 + d\omega^2 + 2t(p-\omega+b\beta) - 2\lambda\omega\right]\right\} dt$$

$$= \tilde{C}(\omega) K_b \int_{\mathbb{R}} f(t) \exp\left\{\frac{i}{2b}\left[at^2 + d(\omega - b\beta)^2 + 2t(p-\omega+b\beta) - 2\lambda(\omega - b\beta)\right]\right\} dt$$

$$= \tilde{C}(\omega)\mathcal{L}_M[f](\omega - b\beta),$$

where

$$\tilde{C}(\omega) = e^{i\beta[d\omega - db\beta/2 - \lambda]}.$$

Theorem 56 (Differentiation). *Assuming that*

$$f(t)\kappa_{\mathbf{\Lambda}_S}(t,\omega) = 0, \qquad \textit{at } t = \pm\infty,$$

we have

$$\mathcal{L}_M\left[f'\right](\omega) = a\frac{\partial}{\partial\omega}\mathcal{L}_M[f](\omega) + i\left[c(p-\omega) - aq\right]\mathcal{L}_M[f](\omega)$$

and

$$\mathcal{L}_M[tf(t)](\omega) = ib\frac{\partial}{\partial\omega}\mathcal{L}_M[f](\omega) + (d\omega - \lambda)\mathcal{L}_M[f](\omega)$$

Proof *From the definition of the SAFT, we have*

$$\begin{aligned}\mathcal{L}_M[f'](\omega) &= \int_{\mathbb{R}} f'(t)\kappa_{\mathbf{\Lambda_S}}(t,\omega)dt = -\int_{\mathbb{R}} f(t)\frac{\partial}{\partial t}\kappa_{\mathbf{\Lambda_S}}(t,\omega)dt \\ &= -\frac{i}{b}\int_{\mathbb{R}} f(t)\kappa_{\mathbf{\Lambda_S}}(t,\omega)(at+p-\omega)\,dt,\end{aligned}$$

hence

$$ib\mathcal{L}_M[f'](\omega) = a\mathcal{L}_M[tf](\omega) + (p-\omega)\mathcal{L}_M[f](\omega). \tag{10.10}$$

Moreover,

$$\frac{\partial}{\partial\omega}\mathcal{L}_M[f](\omega) = \frac{i}{b}\int_{\mathbb{R}} f(t)\kappa_{\mathbf{\Lambda_S}}(t,\omega)(d\omega - t - \lambda)dt$$

or

$$ib\frac{\partial}{\partial\omega}\mathcal{L}_M[f](\omega) = \mathcal{L}_M[tf(t)](\omega) + (\lambda - d\omega)\mathcal{L}_M[f(t)](\omega).$$

Therefore,

$$\mathcal{L}_M[tf](\omega) = ib\frac{\partial}{\partial\omega}\mathcal{L}_M[f](\omega) + (d\omega - \lambda)\mathcal{L}_M[f](\omega). \tag{10.11}$$

By solving Eqs. (10.10) and (10.11), we obtain after some calculations

$$\mathcal{L}_M[f'](\omega) = a\frac{\partial}{\partial\omega}\mathcal{L}_M[f](\omega) + i[c(p-\omega) - aq]\mathcal{L}_M[f](\omega)$$

10.5 POISSON SUMMATION FORMULA FOR SAFT

To derive the Poisson summation formula for the SAFT, we note that the SAFT may be written as

$$\tilde{F}(\omega) = \frac{1}{\sqrt{2\pi}}\int_{\mathbb{R}} \tilde{f}(t)e^{-i\omega t}dt,$$

where

$$\tilde{F}(\omega) = \sqrt{|b|}\widehat{f}_{\mathbf{\Lambda_S}}(b\omega)\exp\frac{-i}{2b}[d(b\omega)^2 - 2\lambda b\omega] \quad \text{and } \tilde{f}(t) = f(t)\exp\frac{i}{2b}[at^2 + 2pt].$$

By using the Poisson summation formula for the Fourier transform

$$\frac{1}{\sqrt{2\pi}}\sum_k h(t+2k\sigma) = \frac{1}{2\sigma}\sum_k \widehat{h}(k\pi/\sigma)e^{ik\pi t/\sigma},$$

we obtain

$$\begin{aligned}&\frac{1}{\sqrt{2\pi}}\sum_{k=-\infty}^{\infty} f(t+2\sigma k)\exp\frac{i}{2b}[a(t+2\sigma k)^2 + 2p(t+2\sigma k)] \\ =\ &\frac{\sqrt{|b|}}{2\sigma}\sum_{k=-\infty}^{\infty} \widehat{f}_{\mathbf{\Lambda_S}}(k\pi b/\sigma)\exp\frac{-i}{2b}[d(k\pi b/\sigma)^2 - 2\lambda(k\pi b/\sigma)]e^{ik\pi t/\sigma}.\end{aligned}$$

We present another version of the Poisson summation formula for SAFT using a more direct approach and then relate it to sampling of bandlimited functions. For some related work, see [371].

Theorem 57. *The Poisson summation formula for the SAFT is given by*

$$\sqrt{2\pi b}\sum_{k=-\infty}^{+\infty} f(t+k\Delta)\exp\left\{\frac{i}{2b}\left[a(t+k\Delta)^2+2p(t+k\Delta)\right]\right\}$$
$$=\sum_{n=-\infty}^{+\infty}\exp\left\{\frac{-i}{2b}\left[dn^2+\Omega n-2nt\right]\right\}F_{\mathbf{\Lambda}_S}(n),$$

where $\Delta=2\pi b$.

Proof *Let* $f(t)$ *be an integrable function and define*

$$\tilde{f}(t)=\sqrt{2\pi b}\sum_{k=-\infty}^{+\infty} f(t+k\Delta)\exp\left\{\frac{i}{2b}\left[a(t+k\Delta)^2+2p(t+k\Delta)\right]\right\}.$$

It is easy to see that $\tilde{f}(t)$ *is a* Δ*–periodic function. Therefore, it can be expanded as a Fourier series in terms of the orthogonal family* $\left\{e^{int/b}\right\}$. *Hence,*

$$\begin{aligned}
\tilde{f}(t) &= \frac{1}{2\pi b}\sum_{n=-\infty}^{+\infty} e^{int/b}\int_0^{\Delta} e^{-iny/b}\tilde{f}(y)dy \\
&= \frac{1}{\sqrt{2\pi b}}\sum_{n=-\infty}^{+\infty} e^{int/b}\sum_{k=-\infty}^{\infty}\int_0^{\Delta} e^{-iny/b}f(y+k\Delta) \\
&\times \exp\left\{\frac{i}{2b}\left[a(y+k\Delta)^2+2p(y+k\Delta)\right]\right\}dy \\
&= \frac{1}{\sqrt{2\pi b}}\sum_{n=-\infty}^{+\infty} e^{int/b}\sum_{k=-\infty}^{\infty}\int_0^{\Delta} f(y+k\Delta) \\
&\times \exp\left\{\frac{i}{2b}\left[a(y+k\Delta)^2+2p(y+k\Delta)\right]-2ny\right\}dy \\
&= \frac{1}{\sqrt{2\pi b}}\sum_{n=-\infty}^{\infty} e^{int/b}\sum_{k=-\infty}^{\infty}\int_{k\Delta}^{(k+1)\Delta} f(x) \\
&\times \exp\left\{\frac{i}{2b}\left[ax^2+2px-2n(x-k\Delta)\right]\right\}dx \\
&= \frac{1}{\sqrt{2\pi b}}\sum_{n=-\infty}^{\infty}\exp\left\{\frac{-i}{2b}\left[dn^2+\Omega n-2nt\right]\right\} \\
&\times \int_{-\infty}^{\infty} f(x)\exp\left\{\frac{i}{2b}\left[ax^2+dn^2-2nx+\Omega n+2px\right]\right\}dx \\
&= \sum_{n=-\infty}^{+\infty}\exp\left\{\frac{-i}{2b}\left[dn^2+\Omega n-2nt\right]\right\}F_{\mathbf{\Lambda}_S}(n).
\end{aligned}$$

That is,

$$\begin{aligned}&\sqrt{2\pi b}\sum_{k=-\infty}^{+\infty} f(t+k\Delta)\exp\left\{\frac{i}{2b}\left[a(t+k\Delta)^2+2p(t+k\Delta)\right]\right\}\\&=\sum_{n=-\infty}^{+\infty}\exp\left\{\frac{-i}{2b}\left[dn^2+\Omega n-2nt\right]\right\}F_{\mathbf{\Lambda}_S}(n).\end{aligned}$$

In particular, if $F_{\mathbf{\Lambda}_S}$ is an interpolating function, i.e., $F_{\mathbf{\Lambda}_S}(n)=\delta(n)$, then

$$\sum_{k=-\infty}^{+\infty} f(t+k\Delta)\exp\left\{\frac{i}{2b}\left[a(t+k\Delta)^2+2p(t+k\Delta)\right]\right\}=\frac{1}{\sqrt{2\pi b}}.$$

Lemma 10. *If g is bandlimited to $(-1,1)$ in the SAFT domain, then*

$$\sum_k |g(t_k)|^2=\frac{1}{2\pi b}\int_{\mathbb{R}}|G_{\mathbf{\Lambda}_S}(\omega)|^2\,d\omega.$$

Proof *Let f be bandlimited to $(-1,1)$ in the SAFT domain. The Poisson summation formula for $t=0$ and $t_k=k\Delta$ becomes*

$$\sqrt{2\pi b}\sum_k f(t_k)\zeta(t_k)=F_{\mathbf{\Lambda}_S}(0), \tag{10.12}$$

where $\zeta(t)=\exp\left\{\frac{i}{2b}\left(at^2+2pt\right)\right\}$. Setting $f(t)=\overline{\zeta}(t)g(t)\overline{h}(t)$ in the last equation yields

$$\sum_k g(t_k)\overline{h}(t_k)=\frac{1}{\sqrt{2\pi b}}F_{\mathbf{\Lambda}_S}(0),$$

but

$$\begin{aligned}F_{\mathbf{\Lambda}_S}(\omega)&=\frac{1}{\sqrt{2\pi b}}\int_{\mathbb{R}}\overline{\zeta}(t)g(t)\overline{h}(t)\exp\left\{\frac{i}{2b}\left[at^2+d\omega^2+\Omega\omega+2t(p-\omega)\right]\right\}dt\\&=\frac{1}{\sqrt{2\pi b}}\int_{\mathbb{R}}g(t)\overline{h}(t)\exp\left\{\frac{i}{2b}\left[d\omega^2+\Omega\omega-2\omega t\right]\right\}dt;\end{aligned}$$

hence

$$F_{\mathbf{\Lambda}_S}(0)=\frac{1}{\sqrt{2\pi b}}\int_{\mathbb{R}}g(t)\overline{h}(t)dt.$$

Therefore, we have

$$\sum_k g(t_k)\overline{h}(t_k)=\frac{1}{2\pi b}\int_{\mathbb{R}}g(t)\overline{h}(t)dt,$$

which leads to

$$\sum_k |g(t_k)|^2=\frac{1}{2\pi b}\int_{\mathbb{R}}|g(t)|^2=\frac{1}{2\pi b}\int_{\mathbb{R}}|G_{\mathbf{\Lambda}_S}(\omega)|^2.$$

10.6 CONVOLUTION AND PRODUCT THEOREMS FOR SPECIAL AFFINE FOURIER TRANSFORM

10.6.1 Modulation and Convolution Operations

The focus of this section is the convolution property of the SAFT, [35, 261, 370]. But before we define the convolution operation in the SAFT domain, let us introduce the chirp modulation operation.

Definition 45 (Chirp Modulation). *Let* $\mathbf{A} = [a_{j,k}]$ *be a* 2×2 *matrix. We define the modulation function associated with the matrix* $\mathbf{A}$

$$m_{\mathbf{A}}(t) \stackrel{\text{def}}{=} \exp\left(i\frac{a_{11}t^2}{2a_{12}}\right).$$

Furthermore, for a given function f*, we define its chirp modulated functions associated with the matrix* $\mathbf{A}$ *as,*

$$\vec{f}(t) \stackrel{\text{def}}{=} m_{\mathbf{A}}(t)\, f(t) \quad \textit{and} \quad \overleftarrow{f}(t) \stackrel{\text{def}}{=} m^*_{\mathbf{A}}(t)\, f(t). \tag{10.13}$$

For example, for $\mathbf{A} = \mathbf{\Lambda}_{\mathsf{S}}$, we have,

$$\vec{f}(t) = m_{\mathbf{\Lambda}_{\mathsf{S}}}(t) f(t) = e^{i\frac{at^2}{2b}} f(t),$$

and for the case when $\mathbf{A} = \mathbf{\Lambda}_{\mathsf{S}}^{\text{inv}}$, we obtain,

$$\vec{f}(t) = m_{\mathbf{\Lambda}_{\mathsf{S}}^{\text{inv}}}(t)\, f(t) = e^{-i\frac{dt^2}{2b}} f(t)$$

.

Next we define the SAFT convolution operation.

Definition 46 (SAFT Convolution). *Let* f *and* g *be two given functions and* $*$ *denote the usual convolution operation for the Fourier transorm. The SAFT convolution is defined as,*

$$h(t) = (f *_{\mathbf{\Lambda}_{\mathsf{S}}} g)(t) \stackrel{\text{def}}{=} \frac{m^*_{\mathbf{\Lambda}_{\mathsf{S}}}}{\sqrt{|b|}}(t)\left(\vec{f}(t) * \vec{g}(t)\right). \tag{10.14}$$

10.6.2 Convolution Theorem

In this section, we introduce a convolution operation associated with the SAFT for two functions, but in a later section we will introduce another convolution operation for functions and sequences of numbers. The latter will be useful for the discrete SAFT, as well as, for deriving the Poisson summation formula and the Zak transform associated with SAFT .

Figure 10.1 illustrates the block diagram for SAFT convolution defined above. We now state the convolution and product theorem for the SAFT domain.

Theorem 58. *[35] [SAFT Convolution Theorem] Let f and g be two given functions for which the convolution $*_{\mathbf{\Lambda}_\mathsf{S}}$ exists and set,*

$$h(t) = (f *_{\mathbf{\Lambda}_\mathsf{S}} g)(t).$$

Furthermore, let $\widehat{f}_{\mathbf{\Lambda}_\mathsf{S}}(\omega), \widehat{g}_{\mathbf{\Lambda}_\mathsf{S}}(\omega)$ and $\widehat{h}_{\mathbf{\Lambda}_\mathsf{S}}(\omega)$ be the SAFT of f, g and h, respectively. Then we have,

$$h(t) = (f *_{\mathbf{\Lambda}_\mathsf{S}} g)(t) \xrightarrow{\mathsf{SAFT}} \widehat{h}_{\mathbf{\Lambda}_\mathsf{S}}(\omega) = \eta^*(\omega)\,\widehat{f}_{\mathbf{\Lambda}_\mathsf{S}}(\omega)\,\widehat{g}_{\mathbf{\Lambda}_\mathsf{S}}(\omega),$$

where

$$\eta(\omega) = \exp\left\{\frac{i}{2b}[d\omega^2 + \Omega\omega]\right\} \quad \Omega = 2(bq - dp).$$

Proof *We begin by computing the SAFT of h,*

$$\begin{aligned}
\widehat{h}_{\mathbf{\Lambda}_\mathsf{S}}(\omega) &\overset{(10.1)}{=} \mathscr{T}_{SAFT}[h](\omega) \\
&= \langle h(t), \kappa_{\mathbf{\Lambda}_\mathsf{S}}(t,\omega)\rangle \\
&= \int_{\mathbb{R}} h(t)\,\kappa^*{}_{\mathbf{\Lambda}_\mathsf{S}}(t,\omega)\,dt \\
&\overset{(10.14)}{=} \int_{\mathbb{R}} \left(K_b m^*_{\mathbf{\Lambda}_\mathsf{S}}(t) \int_{\mathbb{R}} \vec{f}(z)\,\vec{g}(t-z)\,dz\right) \kappa^*{}_{\mathbf{\Lambda}_\mathsf{S}}(t,\omega)\,dt \\
&= \underbrace{K_b^2 e^{i\frac{d\omega^2}{2b}} e^{-i\frac{\omega}{b}(dp-bq)}}_{C(\omega)} \int_{\mathbb{R}} e^{i\frac{t}{b}(p-\omega)} \boxed{m_{\mathbf{\Lambda}_\mathsf{S}}(t)} \\
&\quad \times \left(\boxed{m^*_{\mathbf{\Lambda}_\mathsf{S}}(t)} \int_{\mathbb{R}} \vec{f}(z)\,\vec{g}(t-z)\,dz\right) dt.
\end{aligned}$$

In the above development, note that the items in the box cancel one another because $m^*_{\mathbf{\Lambda}_\mathsf{S}} m_{\mathbf{\Lambda}_\mathsf{S}} = 1$ (see Definition 45). Setting $t - z = v$ and using (10.13), we obtain an integral of separable form, that is, $\widehat{h}_{\mathbf{\Lambda}_\mathsf{S}}(\omega) = I_f(\omega)\,I_g(\omega)$ because,

$$\begin{aligned}
\widehat{h}_{\mathbf{\Lambda}_\mathsf{S}}(\omega) &= \underbrace{K_b^2 \eta(\omega) \int_{\mathbb{R}}\int_{\mathbb{R}} f(z)\,m_{\mathbf{\Lambda}_\mathsf{S}}(z)\,g(v)\,m_{\mathbf{\Lambda}_\mathsf{S}}(v)\,e^{i\frac{v+z}{b}(p-\omega)}\,dt\,dv}_{\widehat{h}_{\mathbf{\Lambda}_\mathsf{S}}(\omega)} \\
&= I_f(\omega)\,I_g(\omega),
\end{aligned} \tag{10.15}$$

where, for a given function f, we define,

$$I_f(\omega) \overset{\text{def}}{=} K_b \sqrt{\eta(\omega)} \int_{\mathbb{R}} f(z)\,m_{\mathbf{\Lambda}_\mathsf{S}}(z)\,e^{i\frac{z}{b}(p-\omega)}\,dz, \tag{10.16}$$

and similar expression for I_g. Indeed, using (10.16) and (10.1), it is easy to see that,

$$I_f(\omega) = \sqrt{\eta(\omega)}\eta^*(\omega)\,\widehat{f}_{\mathbf{\Lambda}_\mathsf{S}}(\omega) \tag{10.17}$$

and this result extends to $I_g(\omega)$ by symmetry.

We conclude that,

$$\begin{aligned}
\widehat{h}_{\mathbf{\Lambda}_{\mathrm{S}}}(\omega) &\overset{(10.15)}{=} I_f(\omega)\, I_g(\omega) \\
&\overset{(10.17)}{=} \sqrt{\eta(\omega)\eta^*(\omega)}\, \widehat{f}_{\mathbf{\Lambda}_{\mathrm{S}}}(\omega) \cdot \sqrt{\eta(\omega)\eta^*(\omega)}\, \widehat{g}_{\mathbf{\Lambda}_{\mathrm{S}}}(\omega) \\
&= \eta^*(\omega)\, \widehat{f}_{\mathbf{\Lambda}_{\mathrm{S}}}(\omega)\, \widehat{g}_{\mathbf{\Lambda}_{\mathrm{S}}}(\omega)
\end{aligned}$$

which completes the proof.

10.6.3 Product Theorem

Now we establish the product theorem for the SAFT:

Theorem 59. *Let*

$$h(t) = \eta_{\mathbf{\Lambda}_{\mathrm{S}}^{inv}}(t)\, f(t)\, g(t),$$

where

$$\eta_{\mathbf{\Lambda}_{\mathrm{S}}^{inv}} = \exp\left\{ \frac{i}{2b}[at^2 - 2t\lambda_0 \right\}, \qquad \lambda_0 = ap_0 + bq_0.$$

Then

$$\widehat{h}(\omega) = \left(\widehat{f}_{\mathbf{\Lambda}_{\mathrm{S}}} *_{\mathbf{\Lambda}_{\mathrm{S}}^{inv}} \widehat{g}_{\mathbf{\Lambda}_{\mathrm{S}}} \right)(\omega),$$

Proof *From the definition of SAFT, we have*

$$\begin{aligned}
\widehat{h}(\omega) &= \int_{\mathbb{R}} \exp\left\{ \frac{i}{2b}\left[at^2 - 2t\lambda_0\right] \right\} f(t) g(t) \kappa_{\mathbf{\Lambda}_{\mathrm{S}}}(t,\omega) dt \\
&= \int_{\mathbb{R}} \exp\left\{ \frac{i}{2b}\left[at^2 - 2t\lambda_0\right] \right\} \kappa_{\mathbf{\Lambda}_{\mathrm{S}}}(t,\omega) dt \\
&\times \int_{\mathbb{R}} \widehat{f}_{\mathbf{\Lambda}_{\mathrm{S}}}(x) \kappa_{\mathbf{\Lambda}_{\mathrm{S}}}^{-1}(x,t) dx \int_{\mathbb{R}} \widehat{g}_{\mathbf{\Lambda}_{\mathrm{S}}}(y) \kappa_{\mathbf{\Lambda}_{\mathrm{S}}}^{-1}(y,t) dy \\
&= \int_{\mathbb{R}} \widehat{f}_{\mathbf{\Lambda}_{\mathrm{S}}}(x) dx \int_{\mathbb{R}} \widehat{g}_{\mathbf{\Lambda}_{\mathrm{S}}}(y) dy\, I(x,y,\omega),
\end{aligned} \tag{10.18}$$

where

$$I(x,y,\omega) = \int_{\mathbb{R}} \exp\left\{ \frac{i}{2b}\left[at^2 - 2t\lambda_0\right] \right\} \kappa_{\mathbf{\Lambda}_{\mathrm{S}}}(t,\omega) \kappa_{\mathbf{\Lambda}_{\mathrm{S}}}^{-1}(x,t) \kappa_{\mathbf{\Lambda}_{\mathrm{S}}}^{-1}(y,t) dt.$$

After some calculations, we obtain

$$\begin{aligned}
I(x,y,\omega) &= K_b^3 \exp\left\{ \frac{-i}{2b}\left[dx^2 + 2xp_0\right] \right\} \exp\left\{ \frac{-i}{2b}\left[dy^2 + 2yp_0\right] \right\} \\
&\times \exp\left\{ \frac{i}{2b}\left[d\omega^2 - 2\omega\lambda\right] \right\} \int_{\mathbb{R}} \exp\left\{ \frac{it}{b}(p - \omega + x + y + \lambda_0) \right\} dt
\end{aligned} \tag{10.19}$$

But since $p + \lambda_0 = 0$, the last integral is reduced to

$$\int_{\mathbb{R}} \exp\left\{ \frac{it}{b}(x + y - \omega) \right\} dt = 2\pi b \delta(x + y - \omega).$$

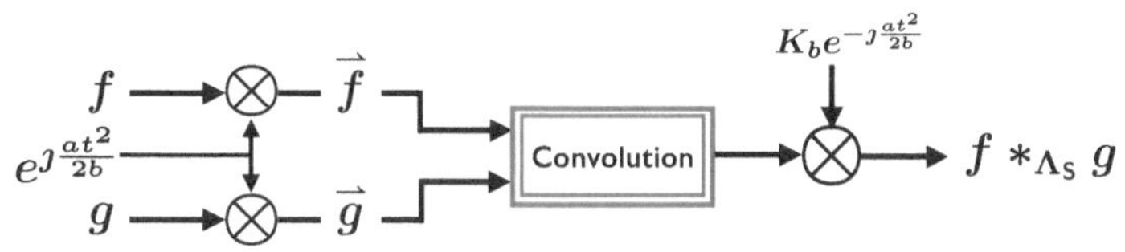

Figure 10.1 SAFT convolution

By substituting I back into (10.18), we have in view of the fact that $\lambda + p_0 = 0$

$$
\begin{aligned}
\widehat{h}(\omega) &= K_b \exp\left\{\frac{i}{2b}\left[d\omega^2 - 2\omega\lambda\right]\right\} \int_{\mathbb{R}} \widehat{f}_{\Lambda_S}(x) \exp\left\{\frac{-i}{2b}\left[dx^2 + 2xp_0\right]\right\} dx \\
&\times \int_{\mathbb{R}} \widehat{g}_{\Lambda_S}(\omega - x) \exp\left\{\frac{-i}{2b}\left[d(\omega-x)^2 + 2p_0(\omega - x)\right]\right\} dx \\
&= K_b \exp\left\{\frac{i}{2b}\left[d\omega^2 - 2\omega\lambda - 2p_0\omega\right]\right\} \int_{\mathbb{R}} \left(\widehat{f}_{\Lambda_S}(x) e^{-idx^2/2b}\right) \\
&\times \left(\widehat{g}_{\Lambda_S}(\omega - x) e^{-id(\omega-x)^2/2b}\right) dx \\
&= K_b \exp\left\{\frac{i}{2b}\left[d\omega^2\right]\right\} \int_{\mathbb{R}} \left(\widehat{f}_{\Lambda_S}(x) e^{-idx^2/2b}\right)\left(\widehat{g}_{\Lambda_S}(\omega - x) e^{-id(\omega-x)^2/2b}\right) dx \\
&= \frac{1}{\sqrt{|b|}} m^*_{\Lambda_S^{inv}}(\omega) \left(\vec{f}_{\Lambda_S}(x) * \vec{g}_{\Lambda_S}(x)\right)(\omega) = \left(f_{\Lambda_S} *_{\Lambda_S} g_{\Lambda_S}\right)(\omega).
\end{aligned}
$$

Remark. Because the SAFT generalizes a number of known transformations, such as the Fourier, fractional Fourier, Fresnel and the linear canonical transforms, the aforementioned convolution and product theorems are applicable to all those transformations, see also [370].

10.7 SHIFT-INVARIANT SPACES FOR THE SPECIAL AFFINE FOURIER TRANSFORM

10.7.1 Preliminaries

Shift-invariant spaces were discussed in Chapters 1 and 5. In this section, we study shift-invariant spaces for the special affine Fourier transform .

We consider spaces of the form

$$
\mathcal{V}(\varphi) = \left\{ f(t) = \sum_{n=-\infty}^{+\infty} c[n]\varphi(t-n), \varphi \in L^2(\mathbb{R}), \{c[n]\} \in \ell_2 \right\}. \tag{10.20}
$$

The closure of $\mathcal{V}(\varphi)$ is a closed subspace of L^2. Furthermore, it is shift-invariant in the sense that for all $f \in \mathcal{V}(\varphi)$, its shifted version, $f(\cdot - k) \in \mathcal{V}(\varphi)$, $k \in \mathbb{Z}$, where $\mathbb{Z}$ denotes the set of integers.

Recall that for the basis functions to be stable, it is required that the family of functions $\{\varphi(t-n)\}_{n=-\infty}^{\infty}$ forms a Riesz basis or equivalently, there exists two positive constants $0 < \eta_1, \eta_2 < +\infty$, such that

$$
\forall c \in \ell_2, \quad \eta_1 \|c\|^2_{\ell_2} \leqslant \left\| \sum_{n=-\infty}^{\infty} c[k]\varphi(t-k) \right\|^2_{L^2} \leqslant \eta_2 \|c\|^2_{\ell_2} \tag{10.21}
$$

where ℓ_2 is the space of all square-summable sequences and $\|c\|_{\ell_2}^2$ is the squared ℓ_2-norm of the sequence. The Fourier domain equivalence of (10.21) is

$$\eta_1 \leqslant \sum_{n=-\infty}^{+\infty} |\widehat{\varphi}(\omega+2\pi n)|^2 \leqslant \eta_2. \tag{10.22}$$

The basis is shift-orthonormal which means that $\langle \varphi, \varphi(\cdot - k)\rangle = \delta_k$ where $\langle x,y\rangle = \int_{-\infty}^{+\infty} x(t)y^*(t)dt$ is the L^2-inner product and $\delta_k = \begin{cases} 1, & \text{if } k=0 \\ 0, & \text{if } k\neq 0 \end{cases}$, denotes the Kronecker delta.

10.7.2 Discrete Special Affine Fourier Transform

Definition 47 (Discrete Special Affine Fourier Transform (DT–SAFT)). *Let $P=\{p(k)\}$ be a sequence in ℓ_2, that is, $\sum_k |p(k)|^2 < \infty$. We define the discrete time SAFT of P as*

$$\widehat{P}_{\mathbf{\Lambda}_S}(\omega) = \frac{1}{\sqrt{2\pi|b|}} \sum_k p(k) \exp\left\{\frac{i}{2b}\left(ak^2+d\omega^2+2k(p-\omega)+\Omega\omega\right)\right\}, \tag{10.23}$$

and define the convolution of a sequence P and a function $\phi \in L^2(\mathbb{R})$ as

$$h(t) = (P *_{\mathbf{\Lambda}_S} \phi)(t) = \frac{1}{\sqrt{2\pi|b|}} e^{-iat^2/2b} \sum_k e^{iak^2/2b} p(k) e^{ia(t-k)^2/2b} \phi(t-k). \tag{10.24}$$

Lemma 11. *Let P and ϕ be defined as above and $h(t) = (P *_{\mathbf{\Lambda}_S} \phi)(t)$. Then*

$$H(\omega) = \widehat{h}_{\mathbf{\Lambda}_S} = \overline{\eta}_{\mathbf{\Lambda}_S}(\omega)\widehat{P}_{\mathbf{\Lambda}_S}\Phi_{\mathbf{\Lambda}_S}(\omega),$$

where $\Phi_{\mathbf{\Lambda}_S}$ is the SAFT of ϕ, $\eta_{\mathbf{\Lambda}_S}(\omega) = \exp\left(\frac{i}{2b}\left(d\omega^2+\Omega\omega\right)\right)$ and $\Omega = 2(bq-dp)$. Moreover, $\left|\widehat{P}_{\mathbf{\Lambda}_S}\right|$ is periodic with period $\Delta = 2\pi b$.

Proof *From the definition of SAFT, we have*

$$H(\omega) = \int_{\mathbb{R}} h(t) k_{\mathbf{\Lambda}_S}(t,\omega)dt = \frac{1}{2\pi|b|}\sum_k e^{iak^2/2b}p(k)\int_{\mathbb{R}} e^{-iat^2/2b}\phi(t-k)e^{ia(t-k)^2/2b}$$
$$\times \exp\left\{\frac{i}{2b}\left(at^2+d\omega^2+2t(p-\omega)+\Omega\omega\right)\right\}dt$$
$$= \frac{1}{2\pi|b|}\sum_k e^{iak^2/2b}p(k)\int_{\mathbb{R}}\phi(u)\exp\left\{\frac{i}{2b}\left(au^2+d\omega^2-2\omega(u+k)+\Omega\omega+2p(u+k)\right)\right\}du$$
$$= \frac{1}{2\pi|b|}\sum_k \exp\left\{\frac{i}{2b}\left(ak^2-2k\omega+2pk\right)\right\}p(k)$$
$$\times \int_{\mathbb{R}}\phi(u)\exp\left\{\frac{i}{2b}\left(au^2+d\omega^2-2\omega u+\Omega\omega+2pu\right)\right\}du$$
$$= \frac{\overline{\eta}_{\mathbf{\Lambda}_S}(\omega)}{\sqrt{2\pi|b|}}\left[\sum_k \exp\left\{\frac{i}{2b}\left(ak^2+d\omega^2-2k\omega+\Omega\omega+2pk\right)\right\}p(k)\right]\Phi_{\mathbf{\Lambda}_S}(\omega)$$
$$= \overline{\eta}_{\mathbf{\Lambda}_S}(\omega)\widehat{P}_{\mathbf{\Lambda}_S}(\omega)\Phi_{\mathbf{\Lambda}_S}(\omega).$$

Furthermore, since $e^{-ik\Delta/b} = e^{-2ik\pi} = 1$, *we have*

$$\begin{aligned} \hat{P}_{\mathbf{\Lambda}_\mathrm{S}}(\omega+\Delta) &= \frac{1}{\sqrt{2\pi|b|}} \sum_k p(k) \\ \times \quad & \exp\left\{\frac{i}{2b}\left(ak^2 + d(\omega+\Delta)^2 - 2k(\omega+\Delta) + \Omega(\omega+\Delta) + 2pk\right)\right\} \\ = \quad & \frac{1}{\sqrt{2\pi|b|}} \sum_k p(k) \exp\left\{\frac{i}{2b}\left(ak^2 + d\omega^2 - 2k\omega + \Omega\omega + 2pk\right)\right\} \\ \times \quad & \exp\left\{\frac{i}{2b}\left(d\Delta^2 + 2d\omega\Delta - 2k\Delta + \Omega\Delta\right)\right\} \\ = \quad & \hat{P}_{\mathbf{\Lambda}_\mathrm{S}}(\omega) \exp\left\{\frac{i}{2b}\left(d\Delta^2 + 2d\omega\Delta + \Omega\Delta\right)\right\} \end{aligned}$$

Thus,

$$\left|\hat{P}_{\mathbf{\Lambda}_\mathrm{S}}(\omega+\Delta)\right| = \left|\hat{P}_{\mathbf{\Lambda}_\mathrm{S}}(\omega)\right|.$$

10.7.3 Riesz Basis for Shift-Invariant Spaces in the SAFT Domain

In the next theorem we give a necessary and sufficient condition for a function $\phi(t)$ to be a generator for a shift-invariant space in terms of its SAFT.

Theorem 60. *Let* $P = \{p(n)\} \in \ell_2$, $\phi \in L^2(\mathbb{R})$ *and consider the chirp-modulated shift-invariant subspaces of* $L^2(\mathbb{R})$

$$\mathcal{V}(\phi) = \text{closure}\left\{ f \in L^2 : f(t) = (P *_{\mathbf{\Lambda}_\mathrm{S}} \phi)(t) \right\}.$$

Then $\left\{e^{i(t-k)^2/2b}\phi(t-k)\right\}$ *is a Riesz basis for* $\mathcal{V}(\phi)$ *if and only if there exist two positive constants* $\beta_1, \beta_2 > 0$ *such that*

$$\beta_1 \le \sum_{k=-\infty}^{+\infty} |\Phi_{\mathbf{\Lambda}_\mathrm{S}}(w+k)|^2 \le \beta_2 \tag{10.25}$$

for all $w \in [0, \Delta]$, *and* $\Phi_{\mathbf{\Lambda}_\mathrm{S}}$ *is SAFT of* ϕ.

Proof *Since* $f(t) = (P *_{\mathbf{\Lambda}_\mathrm{S}} \phi)(t)$, *we have by the previous lemma,*

$$F_{\mathbf{\Lambda}_\mathrm{S}}(w) = \overline{\eta}_{\mathbf{\Lambda}_\mathrm{S}}(w)\hat{P}_{\mathbf{\Lambda}_\mathrm{S}}(w)\Phi_{\mathbf{\Lambda}_\mathrm{S}}(w);$$

and hence

$$|F_{\mathbf{\Lambda}_\mathrm{S}}(w)|^2 = \left|\hat{P}_{\mathbf{\Lambda}_\mathrm{S}}(w)\right|^2 |\Phi_{\mathbf{\Lambda}_\mathrm{S}}(w)|^2,$$

where $\hat{P}_{\mathbf{\Lambda}_S}(w)$ *is given by Definition (47). Thus, because* $\left|\hat{P}_{\mathbf{\Lambda}_S}(\omega)\right|$ *is periodic with period* Δ

$$\begin{aligned}
\|F_{\mathbf{\Lambda}_S}(\omega)\|^2_{L^2(R)} &= \int_{-\infty}^{\infty} \left|\hat{P}_{\mathbf{\Lambda}_S}(\omega)\right|^2 |\Phi_{\mathbf{\Lambda}_S}(\omega)|^2 d\omega \\
&= \sum_{k=-\infty}^{\infty} \int_{k\Delta}^{(k+1)\Delta} \left|\hat{P}_{\mathbf{\Lambda}_S}(\omega)\right|^2 |\Phi_{\mathbf{\Lambda}_S}(\omega)|^2 d\omega \\
&= \sum_{k=-\infty}^{\infty} \int_0^{\Delta} \left|\hat{P}_{\mathbf{\Lambda}_S}(\omega+k\Delta)\right|^2 |\Phi_{\mathbf{\Lambda}_S}(\omega+k\Delta)|^2 d\omega. \\
&= \int_0^{\Delta} \left|\hat{P}_{\mathbf{\Lambda}_S}(\omega)\right|^2 \sum_{k=-\infty}^{\infty} |\Phi_{\mathbf{\Lambda}_S}(\omega+k\Delta)|^2 d\omega \\
&= \int_0^{\Delta} \left|\hat{P}_{\mathbf{\Lambda}_S}(\omega)\right|^2 G_{\phi,\mathbf{\Lambda}_S}(\omega) d\omega,
\end{aligned}$$

where $G_{\phi,\mathbf{\Lambda}_S}(\omega) = \sum_k |\Phi_{\mathbf{\Lambda}_S}(\omega+k\Delta)|^2$ *is the Grammian of* ϕ. *But*

$$\int_0^{\Delta} \left|\hat{P}_{\mathbf{\Lambda}_S}(\omega)\right|^2 d\omega = \frac{1}{2\pi b} \int_0^{\Delta} \sum_{k,l} p(k)\overline{p}(l) \exp\left\{\frac{i}{2b}\left[a(k^2-l^2) - 2\omega(k-l) + 2p(k-l)\right]\right\} d\omega$$

and since

$$\int_0^{\Delta} e^{i(-2\omega)(k-l)/2b} d\omega = b \int_0^{2\pi} e^{iu(l-k)} du = 2\pi b \delta_{k,l},$$

it follows that

$$\left\|\hat{P}_{\mathbf{\Lambda}_S}\right\|^2_{L^2[0,\Delta]} = \int_0^{\Delta} \left|\hat{P}_{\mathbf{\Lambda}_S}(\omega)\right|^2 d\omega = \sum_k |p(k)|^2 = \|p(k)\|^2_{\ell^2}.$$

Since

$$0 < \beta_1 \le G_{\phi,\mathbf{\Lambda}_S}(\omega) \le \beta_2 < \infty,$$

and

$$\|p(k)\|^2_{\ell^2} = \left\|e^{j(ak^2)/2b} p(k)\right\|^2_{\ell^2},$$

we have

$$\beta_1 \left\|\hat{P}_{\mathbf{\Lambda}_S}\right\|^2 = \beta_1 \|p(k)\|^2 \le \|F_{\mathbf{\Lambda}_S}\|^2 \le \beta_2 \|p(k)\|^2 \le \beta_2 \left\|\hat{P}_{\mathbf{\Lambda}_S}\right\|^2$$

which completes the proof.

To get orthonormal basis for $\mathcal{V}(\phi)$, we use the standard trick of putting

$$H_{\mathbf{\Lambda}_S}(\omega) = \frac{\Phi_{\mathbf{\Lambda}_S}(\omega)}{\sqrt{G_{\phi,\mathbf{\Lambda}_S}(\omega)}}$$

so that

$$\sum_k |H_{\mathbf{\Lambda}_S}(\omega+k\Delta)|^2 = \frac{1}{G_{\phi,\mathbf{\Lambda}_S}(\omega)} \sum_k |\Phi_{\mathbf{\Lambda}_S}(\omega+k\Delta)|^2 = 1.$$

10.8 ZAK TRANSFORM ASSOCIATED WITH THE SAFT

We now turn our attention to the Zak transform associated with SAFT . We will introduce the Zak transform and derive some of its properties that generalize those of the standard Zak transform. To proceed, we recall the following properties of the Zak transform:

$$Z_f(t,\omega+1) = Z_f(t,\omega) \text{ and } Z_f(t+1,\omega) = e^{2\pi i\omega} Z_f(t,\omega)$$

and that the Zak transform is a unitary transformation from $L^2(\mathbb{R})$ onto $L^2(Q)$, with

$$\|Z_f\|_{L^2(Q)} = \|f\|_{L^2(\mathbb{R})},$$

where Q is the unit square $Q = [0,1] \times [0,1]$.

If f belongs to a sampling space with sampling function $\psi \in L^2(\mathbb{R}) \cap L^1(\mathbb{R})$, i.e., $f(t) = \sum_{k=-\infty}^{+\infty} f(k)\psi(t-k)$, then

$$\hat{f}(w) = \hat{F}(w)\hat{\psi}(w), \quad \text{where} \quad \hat{F}(w) = \sum_{k=-\infty}^{+\infty} f(k)e^{-2\pi ikw}.$$

where $\hat{F}(w)$ is periodic with period one. Therefore,

$$\left|\hat{f}(w+k)\right| = \left|\hat{F}(w)\right| \left|\hat{\psi}(w+k)\right|,$$

which implies that

$$G_f(w) = |Z_f(0,w)|^2 G_\psi(w),$$

where G_g denotes the Grammian of $g \in L^2(\mathbb{R})$, defined by

$$G_g(w) = \sum_{k\in\mathbb{Z}} |\hat{g}(w+k)|^2, \qquad \text{where } 0 < A \leq G_g(w) \leq B < \infty.$$

Therefore, for such a function f, we have

$$A|Z_f(0,w)|^2 \leq G_f(w) \leq B|Z_f(0,w)|^2, \tag{10.26}$$

for some $A, B > 0$. From this relation, we obtain

$$A\sum_{k=-\infty}^{+\infty} |f(k)|^2 \leq \int_0^1 G_f(w)dw \leq B\sum_{k=-\infty}^{+\infty} |f(k)|^2, \tag{10.27}$$

and

$$\int_0^1 \frac{\sum_{k\in\mathbb{Z}} \left|\hat{f}(w+k)\right|}{|Z_f(0,w)|} dw = \int_{\mathbb{R}} \left|\hat{\psi}(w)\right| dw < \infty, \tag{10.28}$$

whenever $\hat{\psi} \in L^1(\mathbb{R})$.

Definition 48. *We define the Zak transform associated with the SAFT of a signal f as*

$$Z_{f,\mathbf{\Lambda_S}}(t,\omega) = \frac{1}{\sqrt{2\pi b}} \sum_k f(t+k) \exp\left\{\frac{i}{2b}\left[d\omega^2 + ak^2 + 2k(p-\omega) + \Omega\omega\right]\right\}$$

In light of the above definition, we have the following theorem

Theorem 61. *[34] The Zak transform given by definition 48 is an isometry between $L^2(\mathbb{R})$ and $L^2(B)$, where $B = [0,1] \times [0,\Delta]$.*

Proof *First, let us observe that since*

$$e^{-ik\Delta/b} = e^{-2ik\pi} = 1,$$

it follows that

$$\begin{aligned} Z_{\mathbf{\Lambda_S}}(t,\omega+\Delta) &= \frac{1}{\sqrt{2\pi b}}\sum_k f(t+k) \\ &\times \exp\left\{\frac{i}{2b}\left[d(\omega+\Delta)^2 + ak^2 - 2k(\omega+\Delta) + \Omega(\omega+\Delta) + 2pk\right]\right\} \\ &= \exp\left\{\frac{i\Delta}{2b}\left[d\Delta + 2d\omega + +\Omega\right]\right\} Z_{\mathbf{\Lambda_S}}(t,\omega) \end{aligned}$$

Thus $|Z_{\mathbf{\Lambda_S}}(t,\omega+\Delta)|^2 = |Z_{\mathbf{\Lambda_S}}(t,\omega)|^2$, and we have

$$\begin{aligned} \int_0^\Delta |Z_{\mathbf{\Lambda_S}}(t,\omega)|^2 d\omega &= \frac{1}{2\pi b}\sum_{k,l} f(t+k)\overline{f}(t+l) \\ &\times \int_0^\Delta \exp\left\{\frac{i}{2b}\left[a(k^2-l^2) - 2\omega(k-l) + 2p(k-l)\right]\right\} d\omega \\ &= \frac{1}{2\pi b}\sum_{k,l} f(t+k)\overline{f}(t+l)\exp\left\{\frac{i}{2b}\left[a(k-l)^2 + 2p(k-l)\right]\right\} \\ &\times \int_0^\Delta e^{-i\omega(k-l)/b} d\omega \\ &= \frac{1}{2\pi b}\sum_{k,l} f(t+k)\overline{f}(t+l)\exp\left\{\frac{i}{2b}\left[a(k^2-l^2) + 2p(k-l)\right]\right\} \\ &\times \int_0^{2\pi} e^{-ix(k-l)} b\,dx \\ &= \sum_k |f(t+k)|^2. \end{aligned}$$

Therefore

$$\begin{aligned} \|Z_{\mathbf{\Lambda_S}}\|^2_{L^2(B)} &= \int_0^1\int_0^\Delta |Z_{\mathbf{\Lambda_S}}(t,\omega)|^2 d\omega dt = \int_0^1 \sum_k |f(t+k)|^2 dt \\ &= \int_{-\infty}^{\infty} |f(t)|^2 dt = \|f\|^2_{L^2(\mathbb{R})}. \end{aligned}$$

Now we derive the analogue of formulae (10.28) . It is easy to see that

$$Z_{\mathbf{\Lambda_S},f}(0,\omega) = \hat{F}_{\mathbf{\Lambda_S}}(\omega,\{k\}),$$

where $\hat{F}_{\mathbf{\Lambda_S}}(\omega,\{k\})$ is the SAFT of the sequence of samples $\{f(k)\}$ of f. Let f be in the sampling space generated by $\phi(t)$, i.e., from Theorem 60 and Eq. (10.24), we have

$$f(t) = \left(\hat{F}_{\mathbf{\Lambda_S}} *_{\mathbf{\Lambda_S}} \phi\right) \frac{1}{\sqrt{2\pi b}} e^{-iat^2/2b} \sum_k e^{iak^2/2b} f(k) e^{ia(t-k)^2/2b} \phi(t-k).$$

By Lemma 11 we have

$$\hat{F}_{\mathbf{\Lambda_S}}(\omega) = \overline{\eta}_{\mathbf{\Lambda_S}}(\omega) \hat{F}_{\mathbf{\Lambda_S}}(\omega,\{k\}) \Phi_{\mathbf{\Lambda_S}}(\omega) = \overline{\eta}_{\mathbf{\Lambda_S}}(\omega) Z_{\mathbf{\Lambda_S},f}(0,\omega) \Phi_{\mathbf{\Lambda_S}}(\omega)$$

Hence,

$$\left|\hat{F}_{\mathbf{\Lambda_S}}(\omega)\right|^2 = \left|Z_{\mathbf{\Lambda_S},f}(0,\omega)\right|^2 \left|\Phi_{\mathbf{\Lambda_S}}(\omega)\right|^2$$

and because the modulus of the Zak transform is periodic with period Δ, we have

$$\left|\hat{F}_{\mathbf{\Lambda_S}}(\omega+k\Delta)\right|^2 = \left|Z_{\mathbf{\Lambda_S},f}(0,\omega)\right|^2 \left|\Phi_{\mathbf{\Lambda_S}}(\omega+k\Delta)\right|^2$$

which implies that

$$G_{f,\mathbf{\Lambda_S}}(\omega) = \left|Z_{A,f}(0,\omega)\right|^2 G_{\mathbf{\Lambda_S},\phi}(\omega),$$

where $G_{f,\mathbf{\Lambda_S}}$ is the Grammian of f. Since

$$\int_0^{\Delta} G_{\mathbf{\Lambda_S},\phi}(\omega) d\omega = \sum_{k=-\infty}^{+\infty} \int_0^{\Delta} \left|\hat{\Phi}_{\mathbf{\Lambda_S}}(\omega+k\Delta)\right|^2 = \left\|\Phi_{\mathbf{\Lambda_S}}\right\|^2 < \infty,$$

it follows that $G_{\phi,\mathbf{\Lambda_S}}(\omega) < \infty$ almost everywhere. Using (10.25), and the fact that $\beta_1 \leq G_{\mathbf{\Lambda_S},\phi}(\omega) \leq \beta_2$, we obtain

$$\beta_1 \leq \frac{G_{\mathbf{\Lambda_S},f}(\omega)}{\left|Z_{\mathbf{\Lambda_S},f}(0,\omega)\right|^2} \leq \beta_2.$$

Thus,

$$\beta_1 \left|Z_{\mathbf{\Lambda_S},f}(0,\omega)\right|^2 \leq G_{\mathbf{\Lambda_S},f}(\omega) \leq \beta_2 \left|Z_{\mathbf{\Lambda_S},f}(0,\omega)\right|^2.$$

On the other hand, we have

$$\begin{aligned} \left|Z_{\mathbf{\Lambda_S},f}(0,\omega)\right|^2 &= Z_{\mathbf{\Lambda_S},f}(0,\omega) \overline{Z}_{\mathbf{\Lambda_S},f}(0,\omega) \\ &= \sum_{k,l} f(k)\overline{f}(l) \exp\left\{\frac{i}{2b}\left[a(k^2-l^2) - 2\omega(k-l) + 2p(k-l)\right]\right\}. \end{aligned}$$

Therefore, since $\int_0^{\Delta} \exp(-i\omega(k-l)/b)\, d\omega = 2\pi b \delta_{k,l}$, we have

$$\int_0^{\Delta} \left|Z_{\mathbf{\Lambda_S},f}(0,\omega)\right|^2 d\omega = \sum_k |f(k)|^2,$$

and it follows that

$$\beta_1 \sum_k |f(k)|^2 \leq \int_0^{\Delta} G_{\mathbf{\Lambda_S},f}(\omega) d\omega \leq \beta_2 \sum_k |f(k)|^2. \tag{10.29}$$

Moreover,

$$\int_0^{\Delta} \frac{\sum_k \left|F_{\mathbf{\Lambda_S}}(\omega+k\Delta)\right|}{\left|Z_{A,f}(0,\omega)\right|} d\omega = \int_0^{\Delta} \sum_k \left|\Phi_{\mathbf{\Lambda_S}}(\omega+k\Delta)\right| d\omega = \int_{\mathbb{R}} \left|\Phi_{\mathbf{\Lambda_S}}(\omega)\right| d\omega < \infty. \tag{10.30}$$

Relations (10.29) and (10.30) are the analogue of (10.27) and (10.28).

10.9 SHANNON'S SAMPLING THEOREM AND THE SAFT: REINTERPRETATION, EXTENSION AND APPLICATIONS

In this section, we revisit Shannon's sampling theorem for SAFT. We will provide a formal link between Shannon's sampling theorem for SAFT and its convolution. The key idea is that modulated and shifted versions of the sinc–function form a SAFT–bandlimited subspace. We then discuss strategies for reconstruction/interpolation of functions using arbitrary basis functions—not necessarily sinc functions. This serves as an extension of the sampling theorem to generic basis functions.

A form of Shannon's sampling theorem for the SAFT was presented in [246, 287, 323, 324, 326]. Here we present a different proof of the sampling theorem that uses the theory of reproducing-kernels.

Recall that if f is bandlimited to $[-\sigma, \sigma]$, in the Fourier transform domain, then

$$f(t) = \sum_n f(t_n) \frac{\sin[\sigma(t-t_n)]}{\sigma(t-t_n)}, \quad t_n = n\pi/\sigma$$

if and only if

$$\hat{f}(\omega) = \sqrt{\pi/(2\sigma^2)} \left(\sum_n f(t_n) e^{it_n \omega} \right) \chi_{[-\sigma,\sigma]},$$

where χ_A is the characteristic function of A.

Likewise, one can prove that if f is bandlimited to $[-\sigma, \sigma]$, in the SAFT domain, then

$$\tilde{f}(t) = \sum_n \tilde{f}(t_n) \frac{\sin[\sigma(t-t_n)/b]}{\sigma(t-t_n)/b}, \quad t_n = n\pi b/\sigma$$

if and only if

$$F_{\mathbf{\Lambda_S}}(\omega) = \frac{\pi b}{\sigma} F_{\mathbf{\Lambda_S}}(\{f(t_n)\}) \chi_{[-\sigma,\sigma]},$$

where $F_{\mathbf{\Lambda_S}}(\{f(t_n)\}$ is the SAFT of the sequence $\{f(t_n)\}$ as given by Definition 47 and $\tilde{f}(t) = \zeta(t) f(t)$. Although the sampling theorem can be derived either directly or from the Poisson summation formula as in [57], we will derive it using the theory of reproducing-kernels.

Theorem 62. *If f is bandlimited to $[-\sigma, \sigma]$ in the SAFT domain, then*

$$\tilde{f}(t) = \sum_k \tilde{f}(t_n) \frac{\sin\left[\frac{\sigma}{b}(t-t_n)\right]}{\frac{\sigma}{b}(t-t_n)}, \tag{10.31}$$

where

$$\tilde{f} = \overline{\zeta}(t) f(t), \quad \textit{and } t_n = n\pi b/\sigma.$$

Proof *It is easy to verify that the functions*

$$\phi_n(t) = \sqrt{\sigma/\pi b} \exp\left\{i\left[at^2 + 2pt\right]/2b\right\} \frac{\sin[\sigma(t-t_n)/b]}{\sigma(t-t_n)/b},$$

form a complete orthonormal family in $L^2(\mathbb{R})$ *with respect to the inner product*

$$\langle \phi_m, \phi_n \rangle = \int_{\mathbb{R}} \phi_m(t)\overline{\phi}_n(t)dt = \delta_{m,n},$$

and consequently we have from the theory of reproducing kernels that the reproducing kernel of the space of functions bandlimited to $[-\sigma, \sigma]$ *in SAFT domain is*

$$k(x,t) = \sum\nolimits_n \phi_n(t)\overline{\phi}_n(x).$$

Hence, with some calculations one can verify that

$$k(x,t) = e^{\frac{i}{2b}\left[a(t^2-x^2)+2p(t-x)\right]}\frac{\sin\left[\frac{\sigma}{b}(t-x)\right]}{\pi(t-x)},$$

so that

$$f(t) = \int_{\mathbb{R}} f(x)k(x,t)dx.$$

Thus,

$$\begin{aligned} f(t) &= \int_{\mathbb{R}} f(x)\sum \phi_n(t)\overline{\phi}_n(x)dx \\ &= \sum \phi_n(t)\int_{\mathbb{R}} f(x)\overline{\phi}_n(x)dx. \end{aligned}$$

But because $\phi_n(t_k) = \sqrt{\sigma/\pi b}\,\zeta(t_k)\delta_{n,k}$, *we have*

$$f(t_k) = \sum \phi_n(t_k)\int_{\mathbb{R}} f(x)\overline{\phi}_n(x)dx = \sqrt{\sigma/\pi b}\,\zeta(t_k)\int_{\mathbb{R}} f(x)\overline{\phi}_k(x)dx,$$

and it follows that

$$f(t) = \sqrt{\pi b/\sigma}\sum f(t_n)\overline{\zeta}(t_n)\phi_n(t),$$

which is Eq. (10.31).

The sampling theorem can be put in the following form: Let $f(t)$ be a continuous time signal such that $|\widehat{f}(\omega)| = 0, |\omega| > \sigma$. Then, $f(t)$ is completely determined by its equidistant samples spaced $T = \pi b/\sigma$ seconds apart.

The reconstruction formula is then specified by,

$$\begin{aligned} f(t) &= e^{iat^2/2b}\sum_{k\in\mathbb{Z}} f(kT)\,e^{-iat_n^2/2b+ip(t-t_n)/b}\,\mathrm{sinc}(t/T-k) \\ &= e^{iat^2/2b}\sum_{k\in\mathbb{Z}} f(kT)\,e^{-iat_n^2/2b+ip(t-t_n)/b}\frac{\sin\frac{\sigma}{\pi}(t-t_k)}{\frac{\sigma}{\pi}(t-t_k)} \end{aligned} \tag{10.32}$$

where $kT = t_k$.

Remark 2 (Generalization of Shannon's Sampling Theory)**.** *An immediate consequence of Theorem 62 is that it applies to all transforms that are special cases of the SAFT.*

For example, the sampling theorems for the Fresnel transform [113], fractional Fourier transform [350, 325, 296, 37, 36] and the linear canonical transform [368, 295, 274] are all now a straight–forward consequence of Theorem 62.

Unlike previous approaches in [324, 323, 287], we will show that sampling of SAFT–bandlimited signals amounts to orthogonal projection of the signal onto a subspace of SAFT bandlimited functions. Even more so, the projection amounts to filtering/SAFT–convolution of the signal with a low–pass filter followed by the sampling step. The reconstruction process in (10.32) is simply a semi–discrete SAFT convolution.

Let $\varphi_0 = \varphi$ and consider basis functions of form,

$$\varphi_n(t) = \frac{1}{\sqrt{T}} \exp\left\{\frac{-i}{2b}\left[at^2 - at_n^2 + 2p(t-t_n)\right]\right\} \frac{\sin\frac{\sigma}{b}(t-t_n)}{\frac{\sigma}{b}(t-t_n)}. \tag{10.33}$$

The family $\{\varphi_n\}_{n\in\mathbb{Z}}$ has two interesting properties.

(P1) Orthonormality: By construction, the basis functions are orthonormal. This is easy to check. We have, $\langle\varphi_n, \varphi_k\rangle = \delta_{n-k}$ [299],

$$\begin{aligned}\langle\varphi_n, \varphi_k\rangle &= \frac{1}{T}\exp\left\{\frac{i}{2b}\left[a(t_k^2 - t_n^2) + 2p(t_k - t_n)\right]\right\} \int_{\mathbb{R}} \frac{\sin\frac{\sigma}{b}(t-t_n)}{\frac{\sigma}{b}(t-t_n)} \frac{\sin\frac{\sigma}{b}(t-t_k)}{\frac{\sigma}{b}(t-t_k)} dt \\ &= \exp\left\{\frac{i}{2b}\left[a(t_k^2 - t_n^2) + 2p(t_k - t_n)\right]\right\} \int_{\mathbb{R}} \frac{\sin\pi(t-n)}{\pi(t-n)} \frac{\sin\pi(t-k)}{\pi(t-k)} dt = \delta_{k,n}.\end{aligned}$$

(P2) Bandlimitedness: Let $T\sigma = \pi b$. From the definition of SAFT it follows that

$$\widehat{\varphi}_{\mathbf{\Lambda}_\mathsf{S}}(\omega) = \frac{\eta(\omega)}{\sqrt{2\sigma}} \chi_{[-\sigma,\sigma]}(\omega),$$

where $\chi_\mathrm{D}(\cdot)$ is the characteristic function of the set D. Since,

$$\left|\widehat{\varphi}_{\mathbf{\Lambda}_\mathsf{S}}(\omega)\right| = \left|\frac{\chi_{[-\sigma,\sigma]}(\omega)}{\sqrt{2\sigma}}\right|,$$

we observe that $\widehat{\varphi}(\omega)$ is a bandlimited function.

Let us define the space of SAFT–bandlimited functions ,

$$\mathcal{V}_{\mathbf{\Lambda}}(\varphi) = \left\{ f(t) = \sum_{k=-\infty}^{\infty} f(t_k)\varphi_n(t) : \{f(t_k)\} \in \ell_2 \right\}.$$

Thanks to **(P1)**, for any $f \in L_2$, the orthogonal projection operator $P_{\mathcal{V}_{\mathbf{\Lambda}}(\varphi)} : L_2 \to \mathcal{V}_{\mathbf{\Lambda}}(\varphi)$ prescribed by,

$$P_{\mathcal{V}_{\mathbf{\Lambda}}(\varphi)} f = \sum_{k=-\infty}^{\infty} c[k]\varphi_n(t) \quad \text{with} \quad c = \langle f, \varphi_n\rangle$$

results in Shannon's Sampling theorem associated with the SAFT domain. The inner–product of the function f with the basis—functions φ_n is equivalent to pre-filtering (in sense of SAFT convolution) with the ideal low—pass filter, followed by uniform sampling.

$$\begin{aligned}
\langle f,\varphi_n\rangle &= \frac{1}{\sqrt{T}}\exp\left\{\frac{i}{2b}(at_n^2+2pt_n)\right\}\int_{\mathbb{R}} f(\tau)\exp\left\{\frac{-i}{2b}(at^2+2pt)\right\}\frac{\sin\frac{\sigma}{b}(t-t_n)}{\frac{\sigma}{b}(t-t_n)}dt \\
&= \exp\left\{\frac{i}{2b}(at_n^2+2pt_n)\right\}\int_{\mathbb{R}} f(\tau)\exp\left\{\frac{-i}{2b}(at^2+2pt)\right\}\frac{\sin\frac{\sigma}{b}(t-t_n)}{\pi(t-t_n)}dt \\
&= \sqrt{T}f(t_n)
\end{aligned}$$

Therefore, we have

$$\begin{aligned}
f(t) &= \sqrt{T}\sum_k f(t_k)\varphi_k(t) \\
&= \exp\left\{\frac{-i}{2b}(at^2+2pt)\right\}\sum_k f(t_k)\exp\left\{\frac{i}{2b}(at_n^2+2pt_n)\right\}\frac{\sin\frac{\sigma}{b}(t-t_n)}{\frac{\sigma}{b}(t-t_n)}.
\end{aligned}$$

Appendix

Useful Integrals

No.	*Integrals*
1)	$\int_{-\infty}^{\infty} e^{-\pi x^2} dx = 1$
2)	$\int_{-\infty}^{\infty} e^{-q^2 x^2} dx = \frac{\sqrt{\pi}}{q}$
3)	$\int_{-\infty}^{\infty} e^{-p^2 x^2 \pm qx} dx = \frac{\sqrt{\pi}}{p} e^{q^2/4p^2}$
4)	$\int_{-\infty}^{\infty} x^2 e^{-px} dx = \frac{4}{p^3}$
5)	$\int_{-\infty}^{\infty} x^2 e^{-px^2} dx = \frac{1}{2p} \sqrt{\frac{\pi}{p}}$
6)	$\sqrt{\frac{1}{2\pi}} \int_{-\infty}^{\infty} e^{i\omega x} e^{-px^2} dx = \frac{1}{\sqrt{2p}} e^{-\omega^2/4p}$
7)	$\int_{-\infty}^{\infty} x^2 e^{-qx^2+2px} dx = \frac{1}{2q} \sqrt{\frac{\pi}{q}} \left(1 + \frac{2p^2}{q}\right) e^{p^2/q}$

Bibliography

[1] S. Abe and J. Sheridan, Generalization of the fractional Fourier transformation to an arbitrary linear lossless transformation: an operator approach, *J. Phys. A: Math. Gen.*, **27**: 4179–4187, (1994).

[2] S. Abe and J. Sheridan, Optical operations on wave functions as the Abelian subgroups of the special affine Fourier transformation, *J. Optics Letters*, **19**(22): 1801–1803, (1994).

[3] G. S. Agarwal and R. Simon, A simple realization of fractional Fourier transform and relation to harmonic oscillator Green's function, *Opt. Commun.*, **110**: 23–26, (1994).

[4] O. Akay and G. Boudreaux-Bartels, Fractional convolution and correlation via operator methods and an application to detection of linear FM signals, *IEEE Trans. Signal Proc.*, **49**(5): 979–993, (2001).

[5] A. Aldroubi and K. Grochenig, Nonuniform sampling and reconstruction in shift-invariant spaces, *SIAM Review*, **43**:585-620, (2001)

[6] T. Alieva, M. Calvo, and M. Bastiaans, Power filtering of nth order in the fractional Fourier domain, *J. Phys. A*, **35**(36): 7779–7785, (2002)

[7] T. Alieva and M. Bastiaans, Wigner distribution and fractional Fourier transform for two-dimensional symmetric optical beams, *J. Opt. Soc. Amer. A*, **17**(12): 2319–2323, (2000).

[8] T. Alieva and M. Bastiaans, On fractional Fourier transform moments, *IEEE Signal Proc. Letts.*, **7**: 320–323, (2000).

[9] T. Alieva and M. Bastiaans, Mode analysis in optics through fractional transform, *Optics Letts.*, **24**: 1206–1208, (1999).

[10] T. Alieva, V. Lopez, F. Aguillo-Lopez, and L. B. Almeida, The fractional Fourier transform in optical propagation problems, *J. Mod. Opt.*, **41**: 1037–1040, (1994).

[11] T. Alieva, V. Lopez, F. Aguillo-Lopez, and L. B. Almeida, Reply to the comment on the fractional Fourier transform in optical propagation problems, *J. Mod. Opt.*, **42**: 2379–2383, (1995).

[12] J. Allen, Applications of the short time Fourier transform to speech processing and spectral analysis, *International Conference on Acoustics, Speech, and Signal Process*: 1012–1015, (1982).

[13] J. Allen, Short term spectral analysis, synthesis, and modification by discrete Fourier transform, *IEEE Transactions on Acoustics, Speech, and Signal Processing*, **23**(3): 235–238, (1997).

[14] L. B. Almeida, Product and convolution theorems for the fractional Fourier transform, *IEEE Trans. Signal Processing Letters*, **4**: 15–17, (1997).

[15] L. B. Almeida, The fractional Fourier transform and time-frequency representations, *IEEE Trans. Signal Processing*, **42**: 3084–3091, (1994).

[16] L. B. Almeida, An introduction to the angular Fourier transform, *Proc. IEEE Acoust., Speech, Signal Processing Conf.*, Minneapolis, Apr. (1993).

[17] M. H. Annaby, I. A. Al-Abdi, M. S. Abou-Dina, and A. F. Ghaleb, Regularized sampling reconstruction of signals in the linear canonical transform domain, *Signal Process*, **198**, 108569, (2022).

[18] M. H. Annaby and R. M. Asharabi, Derivative sampling expansions for the linear canonical transform: convergence and error analysis, *J. Comput. Math.*, **37**(3): 431–446, (2019).

[19] M. H. Annaby and A. I. Zayed, On the use of Green's function in sampling theory, *J. Integral Equations and Applications,* **10**: 117–139, (1998).

[20] J. P. Antoine, R. Murenzi, P. Vandergheynst, and S. T. Ali, *Two-Dimensional Wavelets and Their Relatives,* Cambridge University Press, Cambridge, United Kingdom (2004).

[21] N. Aronszajn, Theory of reproducing kernels, *Trans. Amer. Math. Soc.,* **68**: 337–404, (1950).

[22] L. Auslander and R. Tolimier, Radar amiguity functions and group theory, *SIAM J. Math. Anal.,* **16**: 577–601, (1985).

[23] M. Bahri, F. A. Shah and A. Y. Tantary, Uncertainty principles for the continuous shearlet transforms in arbitrary space dimensions, *Integ. Transf. Special Funct.,* (2020) https://doi.org/10.1080/10652469.2019.1707816.

[24] W. Bailey, The generating function of Jacobi polynomials, *J. London Math. Soc.,* **13**: 8–12, (1938).

[25] V. Bargmann, On a Hilbert space of analytic functions and an associated integral transform, Part II, A family of related function spaces application to distribution theory, *Comm. Pure Appl. Math.*, **20**: 1–101, (1967).

[26] V. Bargmann, On a Hilbert space of analytic functions and an associated integral transform, Part I, *Comm. Pure Appl. Math.*, **14**: 187–214, (1961).

[27] M. J. Bastiaans and A. J. van Leest, From the rectangular to the quincunx Gabor lattice via fractional Fourier transformation, *IEEE Signal Processing Lett.*, **5**: 203–205, (1998).

[28] M. J. Bastiaans, Gabor's expansion of a signal into Gaussian elementary signals, *Proceeding of the IEEE,* **68**: 538–539, (1980).

[29] W. Beckner, Pitt's inequality and the uncertainty principle, *Proc. Amer. Math. Soc.*, **123**: 1897–1905, (1995).

[30] J. Benedetto and A. Zayed, Sampling, Wavelets, and Tomography, *Birkhauser,* Boston, (2003).

[31] J. Benedetto and W. Heller, Irregular sampling and the theory of frames I, *Note di Math.,* **X**(1): 103–125, (1990)

[32] L. M. Bernardo, ABCD matrix formalism of fractional Fourier optics, *Opt. Eng.*, **35**: 732–740, (1996).

[33] A. Bhandari, O. Graf, F. Krahmer, and A. Zayed, One-bit sampling in fractional Fourier transform, *Proceedings of the 45th International Conference on Acoustics, Speech, and Signal Processing,* Barcelona, Spain, May 4–8, 2020, https://ieeexplore.ieee.org/document/9053505

[34] A. Bhandari and A. I. Zayed, Shift-invariant and sampling spaces associated with the special affine Fourier transform, *Appl. Comput. Harmon. Anal.,* **47**: 30–52, (2019).

[35] A. Bhandari and A. I. Zayed, Convolution and product theorems for the special affine Fourier transform, in *Frontiers in Orthogonal Polynomials and q-Series*, World Scientific, Eds: Nashed and Li, 119–137, (2018).

[36] A. Bhandari and A. Zayed, Shift-invariant and sampling spaces associated with the fractional Fourier transform domain, *IEEE Transactions on Signal Processing,* **60**(3): 1627–1637, (2012).

[37] A. Bhandari and P. Marziliano, Sampling and reconstruction of sparse signals in fractional Fourier domain, *IEEE Signal Proc. Letts.*, **17**(3): 221–224, (2010).

[38] A. Bhandari and P. Marziliano, Fractional delay filters based on generalized cardinal exponential splines, *IEEE Sig. Proc. Letters,* **17**(3): 225–228, (2010).

[39] Y. A. Bhat and N. A. Sheikh, Octonion offset linear canonical transform, *Anal. Math. Phys.* **12**: 95 (2022)

[40] R. Boas, Generalized Laplace integrals, *Bull. Amer. Math. Soc.*, **48**: 286–294, (1942).

[41] B. Boashash, editor, *Time-Frequency Signal Analysis-Method and Applications,* Halsted Press, New York, (1992).

[42] S. Bochner, A generalization of Poisson's summation formula, *Duke Math. J.,* **6**: 229–234, (1940).

[43] T. Boehme, The support of Mikusinski operators, *Trans. Amer. Math. Soc.,* **176**: 319–334, (1973).

[44] M. Born and E. Wolf, *Principles of Optics,* Pergamon Press, (1959).

[45] A. Boumenir and A. Zayed, The equivalence of Kramer and Shannon sampling theorems revisited, *J. of Sampling Theory in Image and Signal Proc.,* **4** (3): 251–269, (2005).

[46] B. Braaksma and B. Meulenbeld, Integral transforms with generalized Legendre functions as kernels, *Compositio Math.,* **18**: 235–287, (1967).

[47] R. Bracewell, *The Hartley Transform*, Oxford University Press, Oxford, U.K. (1986).

[48] R. Bracewell, H. Bartelt, A. Lohmann, and N. Streibl, Optical synthesis of Hartley transform, *Applied Optics,* **24**: 1401–1402, (1985).

[49] R. Bracewell, *The Fourier Transform and Its Applications,* McGraw-Hill, New York (1965).

[50] N. G. de Bruijn, A theory of generalized functions with applications to Wigner distribution and Weyl correspondence, *Nieuw Arch. Wisk.*, **21**: 205–280, (1973).

[51] Yu. A. Brychkov, H. Glaeske, A. Prudnikov, and Vu. K. Tuan, *Multidimensional Integral Transformations*, Gordon and Breach, New York (1992).

[52] Yu. A. Brychkov, and A. Prudnikov, *Integral Transforms of Generalized Functions*, Gordon and Breach, New York (1989).

[53] G. Buggle, *Die Kontorovich–Lebedev-Transformation und die Mehler–Fock Transformation verallgemeinerten Funcktionen*, Dissertation, TH. Darmstadt (1977).

[54] A. Bultan, A four-parameter atomic decomposition of chirps, *IEEE Trans. Signal Proc.,* **47**: 731–745, (1999).

[55] A. Bultheel, H. Martinez, et al., Recent developments in the theory of the fractional Fourier and linear canonical transforms, *Bull. Belgian Math. Soc.-Simon Stevin,* **13**(5): 971–1005, (2007).

[56] J. Bustoz and S. Suslov, Basic analog of Fourier series on a q-quadratic grid, *Methods and Applications of Analysis,* **5**: 1–38, (1998).

[57] P. Butzer, P. Ferreira, J. Higgins, G. Schmeisser, and R. Stens, The sampling theorem, Poisson's summation formula, general Parseval formula, reproducing kernel formula, and the Paley-Wiener theorem for bandlimited signals- their interconnections, *Appl. Anal.,* **90**(3-4): 431–461, (2011).

[58] P. Butzer, R. Stens, and M. Wehrens, The continuous Legendre transform, its inverse transform and applications, *Inter. J. Math & Math. Sci.*, **3**(1): 47–67, (1980).

[59] L. Z. Cai, Special affine Fourier transformation in frequency-domain, *Optics Communic.,* **185**: 271–276, (2000).

[60] L. L. Campbell, A comparison of the sampling theorems of Kramer and Whittaker. *J. Soc. Indust. Appl. Math.*, **12**: 117–130, (1964).

[61] C. Candan and H. M. Ozaktas, Sampling and series expansion theorems for fractional Fourier and other transforms, *Signal Process,* **83**: 1455–1457, (2003).

[62] C. Candan, A. Kutay, and M. Ozaktas, The discrete fractional Fourier transform, *IEEE Trans. Signal Proc.*, **48**: 1329–1337, (2000).

[63] E. Candés and D. Donoho, Ridgelets: A key to higher-dimensional intermittency, *Phil. Trans. Roy. Soc.Lond. A*, **357**: 2495–2509, (1999).

[64] C. Capus and K. Brown, Short-time fractional Fourier methods for the time-frequency representation of chirp signals, *J. Acoustical Soc. Amer.,* **113**(6): 6041–6047, (2012).

[65] G. Cariolaro, T. Erseghe, P. Kraniauskas, and N. Laurenti, Multiplicity of fractional Fourier transforms and their relationships, *IEEE Trans. Signal Proc.,* **48**: 227–241, (2000).

[66] G. Cariolaro, T. Erseghe, P. Kraniauskas, and N. Laurenti, A unified framework for the fractional Fourier transform, *IEEE Transactions on Signal Processing*, **46**: 3206–3219, (1998).

[67] P. Casazza, D. Han, and D. Larson, Frames for Banach spaces, *Contemp. Math., Amer. Math. Soc.*, Providence, RI, **247**: 149–182, (1999).

[68] L. Chen, K. Kou, and M. Liu, Pitt's inequality and the uncertainty principle associated with the quaternion Fourier transform, *J. Math. Anal. Appl.,* **423**: 681–700, (2015).

[69] W. Chena, Z. Fu, L. Grafakos, and Y. Wua, Fractional Fourier transforms on L^p and applications, *Appl. Comput. Harmonic Analys.*, **55**: 71–96, (2021).

[70] F. M. Cholewinski and D. T. Haimo, Inversion of the Hankel potential transform, *Pacific J. Math.*, **37**: 319–330, (1971).

[71] F. M. Cholewinski and D. T. Haimo, Weierstrass–Hankel convolution transform, *J. Anal. Math.*, **17**: 1–58, (1966).

[72] E. A. Coddington and N. Levinson, *Theory of Ordinary Differential Equations*, McGraw-Hill, New York, (1955).

[73] O. Christensen and D. Stoeva, p-frames in separable Banach spaces, *Adv. Comput. Math.,* **18**: 117–126, (2003).

[74] O. Christensen and C. Heil, Perturbations of Banach frames and atomic decomposition, *Math. Nachr,* **185**; 33–47, (1997).

[75] T. A. C. M. Classen and W. F. G. Mecklenbrauker, The Wigner distribution, part 2, *Philips Research Journal,* **35**: 276–300, (1980).

[76] T. A. C. M. Claasen and W. F. G. Mecklenbräuker, The Wigner distribution - a tool for time-frequency signal analysis. II: Discrete-time signals, part 2, *Philips J. Res.*, **35**: 276–300, (1980).

[77] L. Cohen, *Time-Frequency Analysis,* Prentice Hall, Endlewood Cliffes, NJ, (1995).

[78] S. A. Collins, Lens-system diffraction integral written in terms of matrix optics, *J. Optical Soc. Amer.*, **60**(9): 1168–1177, (1970)

[79] E. U. Condon, Immersion of the Fourier transform in a continouos group of functional transformations, *Proc. Natl. Acad. Sci, USA*, **23**(3): 158–164, (1937).

[80] R. G. Cooke, The inversion formula of Hardy and Titchmarsh, *Proc. London Math. Soc.,* **24**: 381–420, (1925).

[81] M. G. Cowling and J. F. Price, Bandwidth verses time concentration: the Heisenberg-Pauli-Weyl inequality, *SIAM J. Math. Anal.*, **15**: 151–65, (1994).

[82] A. Cusmariu, Fractional analytic signals, *Signal Process*, **82**: 267–272, (2002).

[83] H. Dai, Z. Zheng, and W. Wang, A new fractional wavelet transform, *Commun. Nonlinear Sci. Numer. Simulat.*, **44**: 19–36, (2017).

[84] I. Daubechies, *Ten Lectures on Wavelets*, SIAM Publications, Philadelphia, Pennsylvania, (1992).

[85] I. Daubechies, The wavelet transform, time-frequency localization and signal analysis, *IEEE Trans. Inform. Theory*, **36**: 961–1005, (1990).

[86] I. Daubechies, A. Grossmann, and Y. Meyer, Painless nonorthogonal expansions, *J. Math. Phys.*, **27**: 1271–1283, (1987).

[87] L. Debnath and F. A. Shah, *Lectuer Notes on Wavelet Transforms,* Birkhäuser, Boston, 2017.

[88] L. Debnath and F. A. Shah, *Wavelet Transforms and Their Applications,* Birkhäuser, New York, 2015.

[89] S. Deans, *The Radon Transform and Some of its Applications,* Wiley Interscience, New York (1983).

[90] E. Deeba and A. Zayed, Generalized Jacobi transforms, *Journal of Applicable Analysis,* **48**: 63–79, (1993).

[91] L. Dettori and A. Zayed, Texture Identification of Tissues Using Directional Wavelet, Ridgelet, and Curvelet Transforms, *Frames and Operator Theory in Analysis and Signal Processing,* Contemporary Math. Series, Amer. Math. Soc., **451**: 89–118, (2008).

[92] L. Dettori and A. Zayed, The Wavelet, Directional Wavelet, and Ridgelet Transforms With Applications in Texture Identification, *Modern Mathematical Models, Methods and Algorithms for Real World Systems,* edited by A.H. Siddiqi, I. Duff, and O. Christensen, Anamaya Publisher, New Delhi-London (2007).

[93] J. Du, M. Wong, and H. Zhu, Continuous and discrete inversion formulas for the Stockwell transform, *Integral. Transf. Spec. Funct.*, **18**(8): 537–543, (2007).

[94] R. Duffin and A. Schaeffer, A class of nonharmonic Fourier series, *Trans. Amer. Math. Soc.*, **72**: 341–366, (1952).

[95] Y. C. Elder and T. Michaeli, Beyond bandlimited sampling, *IEEE Signal Proc. Mag.*, **26**(3): 48–68, (2009).

[96] A. Erdelyi, W. Magnus, F. Oberhettinger, and F. Tricomi, *Tables of Integral Transforms,* Vol. I, Bateman Manuscript, McGraw-Hill, New York (1955).

[97] A. Erdelyi, W. Magnus, F. Oberhettinger, and F. Tricomi, *Higher Transcendental Functions,* Vol. II, Bateman Manuscript, McGraw-Hill, New York (1955).

[98] M. Erdogdu and M. Ozdemir, Generating four dimensional rotation matrices, https://www.researchgate.net/publication/283007638, May (2015).

[99] T. Erseghe, P. Kraniauskas, and G. Cariolaro, Unified fractional Fourier transform and sampling theorem, *IEEE Trans. Signal Proc.,* **47**: 3419–3423, (1999).

[100] T. Erseghe, P. Kraniauskas, and G. Carioraro, Unified fractional Fourier transform and sampling theorem, *IEEE Trans. Signal Proc.,* **47**(12): 3419–3423, (1999).

[101] G. Folland, *Harmonic Analysis in Phase Space,* Vol. 122, Annals of Mathematics Studies, Princeton University Press, (1989).

[102] G. B. Folland and A. Sitaram, The uncertainty principle: A mathematical survey, *J. Fourier Anal. Appl.,* **3**: 207–238, (1997).

[103] J. Fourier, *Théorie Analytique de la chaleur,* Paris (1822), *The Analytical Theory of Heat*, translated by A. Freeman, Cambridge, England (1878).

[104] D. Gabor, Theory of communications, *J. Inst. Elec. Eng.*, London, **93**: 429–457, (1946).

[105] W. B. Gao and B. Z. Li, The octonion linear canonical transform: definitionand properties, *Signal Proc.,* **188**: (2021), 108233, https://doi.org/10.1016/j.sigpro.2021.108233

[106] A. Garcia and A. Zayed, New sampling formulae for the fractional Fourier transform, *J. Signal Proc.,* **77**: 111–114, (1999).

[107] G. Gasper and M. Rahman, *Basic Hypergeometric Series,* Cambridge University Press, (1990).

[108] I. Gel'fand and G. Shilov, *Generalized Functions*, 1-4, Academic Press, New York (1964).

[109] H. Glaeske and A. Hess, On the convolution theorem of the Mehler–Fock transform for a class of generalized functions, *Math. Nachr.*, **136**: 119–129, (1988).

[110] H. Glaeske, Operational properties of a generalized Hermite transformation, *Aequationes Math.*, **32**: 155–170, (1987).

[111] H. Glaeske, A convolution connected with the Kontorovich–Lebedev transform of generalized functions, *Math. Z.*, **193**: 67–78, (1986).

[112] H. Glaeske, The laguerre transformof some elementary functions, *Z. Anal.*, **3**: 237–244, (1984).

[113] F. Gori, Fresnel transform and sampling theorem, *Optics Communications,* **39**(5): 293–297, (1981).

[114] K. Gröchenig, *Foundations of Time-Frequency Analysis,* (Applied and Numerical Harmonic Analysis), Birkh'auser, (2001).

[115] M. de Gosson and F. Luef, Metaplectic group, symplectic Cayley transform and fractional Fourier transfoms, *J. Math. Anal & Appl.,* **416**: 947–968, (2014).

[116] M. De Gosson, *The Wigner Transform,* Advanced books in Mathematics, World Scientific, (2017).

[117] I. S. Gradshteyn, *Tables of Integrals, Series, and Products.* Academic Press, New York (1965).

[118] A. Grossmann, and J. Morlet, Decomposition of Hardy functions into square integrable wavelets of constant shape, *SIAM J. Math. Anal.*, **15**(4): 723–736, (1984).

[119] K. Guanlei, W. Xiaotong, and X. Xiaogang, The logarithmic, Heisenberg's and short-time uncertainty principles associated with fractional Fourier transform, P(2014), *Sig. Process.*, **89**: 339–343, (2009).

[120] G. H. Hardy and J. F. Littlewood, Some properties of fractional integrals, *Math. Zeit.*, **27**: 265–606, (1928).

[121] G. H. Hardy, Some formulae in the theory of Bessel functions, *Proc. London Math. Soc., Ser. 2,* **23**: lxi–lxiii, (1925).

[122] G. Hardy, On Hilbert transforms, *Messenger of Math.,* **54**, (20-27), 81–88, (1924).

[123] R. V. Hartley, A more symmetrical Fourier analysis applied to tramsmission problems, *Proc. Inst. Radio Eng.,* **30**: 144–150, (1942).

[124] J. J. Healy, M. A. Kutay, H. M. Ozaktas, and J. T. Sheridan, *Linear Canonical Transforms: Theory and Applications*, Springer-Verlag, New York, NY, USA (2016).

[125] J. J. Healy and J. T. Sheridan, Cases where the linear canonical transform of a signal has compact support or is band-limited, *Opt. Lett.*, **33**: 228–230, (2008).

[126] F. Hlawatsch and G. F. Boudreaux-Bartels, Linear and quadratic time-frequency signal representations, *IEEE Signal Processing Magazine,* **9**(2): 21–67, (1992).

[127] C. E. Heil and D. F. Walnut, Continuous and discrete wavelet transforms, *SIAM Review,* **31**(4): 628–666, (1989).

[128] J. R. Higgins, Five short stories about the cardinal series, *Bul. Amer. Math. Soc.,* **12**(1): 45–89, (1985).

[129] J. R. Higgins, A sampling theorem for irregularly spaced sample points, *IEEE Trans. Inform. Theory,* **IT-22**: 621–622, (1976).

[130] J. R. Higgins, An interpolation series assocaited with the Bessel-Hankel transform, *J. London Math. Soc.*, **5**: 707–714, (1972).

[131] E. Hille and R. S. Philips, *Functional Analysis and semigroups*, Amer. Math Soc. Colloq. Publ., Vol. 31, revised ed., Providence, Rhode Island (1957).

[132] E. Hille, On semi-groups of transformations in Hilbert space, *Proc. Natl. Acad. Sci. USA*, **24**: 159–161, (1938).

[133] I. I. Hirschman and D. V. Widder, *The Convolution Transform*, Princeton University Press, Princeton, New Jersey (1955).

[134] I. I. Hirschman and D. V. Widder, General inversion formulas for convolution transforms, II, *Duke Math. J.*, **17**: 391–402, (1950).

[135] H. Huo, Uncertainty principles for the offset linear canonical transform, *Circ. Syst. Signal Process.,* **38**: 395–406, (2019).

[136] H. Huo and W. Sun, Sampling theorems and error estimates for random signals in the linear canonical transform domain, *Signal Process,* **111**(6): 31–38, (2015).

[137] D. F. James and G. Agarwal, The generalized Fresnel transform and its application to optics, *Optics Commun.,* **126**: 207–212, (1996).

[138] M. Ismail, Analytic properties of complex Hermite polynomials, *Trans. Amer. Math. Soc.,* **368**(2): 1189–1210, (2016).

[139] M. Ismail and A. Zayed, A q-analogue of the Whittaker-Shannon-Kotel'nikov Sampling theorem, *Proc. Amer. Math. Soc.,* **131**: 3711–3719, (2003).

[140] M. Ismail and D. Stanton, Addition theorems for the q-exponential function. (English summary) q-series from a contemporary perspective (South Hadley, MA, 1998), Contemp. Math., Amer. Math. Soc., Providence, RI, **254**: 235–245, (2000).

[141] M. Ismail and R. Zhang, Diagonalization of certain integral operators, *Advances in Math.*, **109**(1): 1–33, (1994).

[142] D. F. James and G. S. Agarwal, The generalized Fresnel transform and its application to optics, *Optics Commun.,* **126**: 207–212, (1996).

[143] A. J. Janssen, The Zak transform and sampling theorems for wavelet subspaces, *IEEE Trans. Signal Process.*, **41**(12): 3360–3364, (1993).

[144] A. J. Janssen, The Zak transform: A signal transform for sampled time-continuous signals, *Philips J. Research*, **43**: 23–69, (1988).

[145] A. J. Janssen, Bargmann transform, Zak transform, and coherent states, *J. Math. Phys.*, **23**: 720–731, (1982).

[146] A. J. E. M. Janssen, Gabor representations of generalized functions, *J. Math. Appl.,* **80**: 377–394, (1981).

[147] A. J. Janssen, Weighted Wigner distributions vanishing on lattices, *J. Math. Anal. Appl.*, **80**: 156–167, (1981).

[148] M. Jinming, M. Hongxial, S. Xinhua, G. Chang, K. Xuejing, and T. Ran, Research progressin theories and applications of the fractional Fourier transform, *J. Opto-Electronic Engineering*, **45**: 1129–1143 (1997).

[149] A. J. Jerri, The Shannon sampling theorem-its various extensions and applications, A tutorial review, *Proc. IEEE,* **11**: 1565–1596, (1977).

[150] G. Jumarie, Laplace's transform of fractional order via the Mittag- Leffler function and modified Riemann-Liouville derivative, *Appl. Math. Letters*, **22**: 1659–1664, (2009).

[151] G. Jumarie, Fourier's transform of fractional order via Mittag-Leffler function and modified Riemann-Liouville derivative, *J. Appl. Math.& Informatics*, **26**(5-6): 1101–1121, (2008).

[152] T. Kagawa and T. Suzuki, Characterization of the Gyrator Transform via the Fractional Fourier transform, *Integral. Transf. & Special Func.,* online (2022) https://www.tandfonline.com/doi/full/10.1080/10652469.2022.2138868.

[153] R. Kamalakkannan, R. Roopkumar, and A. Zayed, Quaternionic coupled fractional Fourier transform on Boehmians, to appear in Monograph, *Sampling, Approximation, and Signal Analysis,* Birkhauser, (2024).

[154] R. Kamalakkannan, R. Roopkumar, and A. Zayed, On the extension of the coupled fractional Fourier transform and its properties, *J. Integral Transforms & Special Functions,* **33**(1): 68–80, (2022).

[155] R. Kamalakkannan, R. Roopkumar, and A. Zayed, Short time coupled fractional Fourier transform and the uncertainty principle, *Frac Calculus & Applied Analys*, **24**(3): 667–688, (2021).

[156] A. Karoui, I. Mehrzi, and T. Moumni, Eigenfunctions of the Airy's integral transform: Properties, numerical computations, and asymptotic behavior, *J. Math. Anal. & Appls.,* **389**: 989–1005, (2012).

[157] F. H. Kerr, Fractional powers of Hankel transforms in the Zemanian spaces, *J. Math. Anal. Appl.*, **166**: 65–83, (1992).

[158] F. H. Kerr, A fractional power theory for Hankel transforms in $L2(R+)$, *J. of Math. Analysis and Applications*, **158**: 114–123, (1991).

[159] F. H. Kerr, A distributional approach to Namias' fractional Fourier transforms, *Proc. Roy. Soc. Edinburgh Sect. A*, **108**(1-2), 133–143, (1988)

[160] F. H. Kerr, Namias' fractional Fourier transforms on L^2 and applications to differential equations, *J. Math, Anal. Appl.*, **136**(2): 404–418, (1988)

[161] J. Klauder and B. S. Skagerstam, *Coherent States*, World Scientific, Singapore (1985).

[162] J. Klauder and E. C. Sudarshan, *Fundamentals of Quantum Optics*, W. A. Benjamin, New York (1968).

[163] H. Kober, Wurzeln aus der Hankel, Fourier, und anderen stetigen Transformationen, *Quarterly Journal Math.*, Oxford, **10**: 45–59, (1939).

[164] T. Koornwinder and R. Swarttouw, On q-analogues of the Fourier and Hankel transforms, *Trans. Amer. Math. Soc.*, **333**(1): 445–461 (1992).

[165] T. Koornwinder and G. Walter, The finite continuous Jacobi transform and its inverse, *J. Approx. Theory,* **60**: 83–100, (1990).

[166] K. I. Kou and R. H. Xu, Windowed linear canonical transform and its applications, *Signal Processing,* **92**(1): 179–188, (2012).

[167] V. Kotel'nikov, On the carrying capacity of the ether and wire in telecommunications, material for the first All-Union, Conference on Questions of Communications, Izd. Red. Upr. Svyazi RKKA, Moscow, Russia (1933).

[168] H. Kramer, A generalized sampling theorem, *J. Math. Phys.,* **38**: 68–72, (1959).

[169] P. Kraniauskas, G. Cariolaro, and T. Erseghe, Method for defining a class of fractional operations, *IEEE Transactions on Signal Processing,* **46**: 2804–2807, (1998).

[170] M. A. Kutay, H. M. Ozaktas, O. Arikan, and L. Onural, Optimal filtering in fractional Fourier domains, *IEEE Trans. Signal Process*, **45**(5): 1129–1143, (1997).

[171] G. Kutyniok and D. Labate, Editors, *Shearlets: Multiscale Analysis for Multivariate Data*, Birkhauser, Springer Science (2012).

[172] B. Lacaze, About sampling for band-limited linear canonical, *Signal Process,* **91**: 1076–1078, (2011).

[173] P. S. Laplace, *Théorie Analytique des Probabilités,* **I**, Part 2, Lerch, Paris (1820).

[174] K. G. Larkin, A beginner's guide to the fractional Fourier transform, Part 2: A brief history of time frequency distribution, *Australian Optical Society News* 13–17, (1995).

[175] N. Lebedev, *Special Functions and Their applications*. Dover Publ. New York (1972).

[176] A. J. Lee, Characterization of bandlimited functions and processes, *Inform. Control,* **31**: 258–271, (1976).

[177] N. Levinson, Gaps and density theprems, *Amer. Math. Soc. Colloq. Publ. Ser.*, Providence, RI, **26**, (1940).

[178] T. Levitina, On the eigenfunctions of the finite Hankel transform, *J. Sampling Theor, Signal & Image Proc.*, **11**: 55–79, (2012).

[179] B. Z. Li and T. Z. Xu, Sampling in the linear canonical transform domain, *Math. Probl. Eng.,* June 2012, https://www.hindawi.com/journals/mpe/2012/504580/.

[180] B. Z. Li, R. Tao, T-Z Xu, and Y. Wang, The Poisson sum formulae associated with the fractional Fourier transform, *Signal Process,* **89**: 851–856, (2009).

[181] B. Z. Li, R. Tao, and Y. Wang, New sampling formulae related to linear canonical transform, *Signal Process,* **87**(5): 983–990, (2007).

[182] M. Li and J. Yao, All-optical short-time Fourier transform based on a temporal pulse-shaping system incorporating an array of cascaded linearly chirped fiber bragg gratings, *IEEE Photonics Technology Letters.*, **23**(20): 1439–1441, (2011).

[183] X. Liu, J. Shi, W. Xiang, Q Zhang, and N. Zhang, Sampling expansion for irregularly sampled signals in fractional Fourier transform domain, *Digital Signal Prpoc.,* **34**: 74–81, (2014).

[184] Y. Liu, K. Kou, and I. Ho, New sampling formulae for non-bandlimited signals associated with linear canonical transform and non-linear Fourier atoms, *Signal Process,* **90**(5): 933–945, (2010).

[185] A. W. Lohmann, D. Mendlovic, and Z. Zalevsky, Fractional Hilbert transform, *Optics Letters*, **21**(4): 281–283, (1996).

[186] A. W. Lohmann and B. H. Soffer, Relationships between the Radon–Wigner and fractional Fourier transforms, *J. Opt. Soc. Amer. A*, **11**: 1798–1801, (1994).

[187] A. W. Lohmann, Image rotation, Wigner rotation and the fractional Fourier transform, *J. Opt. Soc. Amer. A*, **10**: 2181–2186, (1993).

[188] W. Lone, F. Shah, and A. Zayed, Two-dimensional fractional shearlet transforms in $L^2(R^2)$, *Frac. Calculus and Appl. Anal.,* November (2022), https://doi.org/10.1007/s13540-022-00109-0.

[189] L. Lu, Y. Lu, et al., Deriving the integral representation of a fractional transform from a fractional Fourier transform, *Optics Letts.*, **23**: 1158–1160, (1998).

[190] L. F. Ludwig, General thin-lens action on spatial intensity distribution behaves as non-integer powers of Fourier transform, *Special Light Modulators and Applications*, OSA Technical Digest Series (Opt. Soc. Amer.,) **8**: 173–176, (1988).

[191] S. Mallat, Multiresolution approximations and wavelet orthonormal bases in $L^2(R)$, *Trans. Amer. Math. Soc.,* **315**: 69–87, (1989).

[192] P. Massopust and A. I. Zayed, On the invalidity of Fourier series expansions of fractional order, *Fractional Calculus and Applied Analysis,* **18**(6): 1507–1517, (2015).

[193] A. McBride and F. Kerr, On Namias's fractional Fourier transforms, *IMA J. Appl. Math.,* **39**: 159–175, (1987).

[194] A. C. Mcbride and F. H. Kerr, On Namias's fractional Fourier transforms, *IMA J. Appl. Math.*, **39**(2): 159–175, (1987).

[195] J. E. Mebius, Derivation of the Euler-Rodrigues formula for three dimensional rotations from the general formula for four-dimensional rotations, 2007, https://arxiv.org/abs/math/0701759.

[196] H. Mellin, Über den Zusammenhang zwischen den Linearen Differential und Differenzengleichungen, *Acta Math.,* **25**: 139–164, (1902).

[197] D. Mendlovic, Z. Zalevsky, and H. M. Ozakdas, The applications of the fractional Fourier transform to optical pattern recognition, in *Optical Pattern Recognition,* Ch.3, New York: Academic, (1998).

[198] D. Mendlovic, Z. Zalevsky, D. Mas, J. Garcia, and C. Ferreir, Fractional wavelet transform, *Appl. Optics,* **36**(20): 4801–4806, (1997).

[199] D. Mendlovic, Z. Zalevsky, A. W. Lohmann, and R. G. Dorsch, Signal spatial-filtering using the localized fractional Fourier transform, *Optics Comm.*, **126**: 14–18, (1996).

[200] D. Mendlovic, H. M. Ozaktas, and A. Lohmann, Fractional correlation, *Appl. Optics,* **34**: 303–309, (1995).

[201] D. Mendlovic, Y. Bitran, R. G. Dorsch, and A. W. Lohmann, Optical fractional correlation: experimental results, *Opt. Soc. Amer. A.,* **12**: 1665–1670, (1995).

[202] D. Mendlovic, H. M. Ozaktas, and A. Lohmann, Graded-index fibers, Wigner distribution functions and the fractional Fourier transform, *Appl. Opt.,* **33**(26): 6188–6193, (1994).

[203] D. Mendlovic and H. M. Ozaktas, Fractional Fourier transformations and their optical implementation I, *J. Opt. Soc. Amer. A,* **10**: 1875–1881, (1993).

[204] P. Mikusinski and A. Zayed, On the extension of the Zak transform, *J. Methods and Appls. Anal.*, **2**: 160–172, (1995).

[205] P. Mikusinski and A. Zayed, An extension of the Radon Transform, *Generalized Functions and Their Applications,* R.S. Pathak, editor, Plenum Press, New York, 141–147, (1993).

[206] P. Mikusinski and A. Zayed, Radon transform of Boehmians, *Proc. Amer. Math. Soc.,* **118**: 561–570, (1993).

[207] P. Mikusinski, Boehmians and generalized functions, *Acta Math. Hungarica,* **51**: 159–179, (1988).

[208] P. Mikusinski, Convergence of Boehmians, *Japan J. of Math.,* **9**: 159–179, (1983).

[209] J. Mikusinski and P. Mikusinski, Quotients de suites et leurs applications dans l'analyse functionnelle, *Comptes Rendus,* **293**, Series I: 463–464, (1981).

[210] J. Morlet, G. Arens, E. Fourgeau, and D. Giard, Wave propagation and sampling theory, Part I: Complex signal and scattering in multilayered media, *J. Geophys.,* **47**: 203–221, (1982).

[211] J. Morlet, G. Arens, E. Fourgeau, and D. Giard, Wave propagation and sampling theory, Part II: Sampling theory and complex waves, *J. Geophys.,* **47**: 222–236, (1982).

[212] M. Moshinsky, Canonical transformations and quantum mechanics, *SIAM J. Appl. Math.,* **25**(2): 193–212, (1973).

[213] M. Moshinsky and C. Quesne, Linear canonical transformations and their unitary representations, *Math Phys.*, **12**: 1772–1780, (1971).

[214] T. Moumni and A. Karoui, Fourier and Hankel bandlimited signal recovery, *Integ. Transf. & Special Funct.,* **21**(5): 337–349, (2010).

[215] D. Mustard, The fractional Fourier transform and the Wigner distribution, *J. Austral. Math. Soc. B-Appl. Math.*, **38**: 209–219, (1996).

[216] D. Mustard, Uncertainty principles invariant under the fractional Fourier transform, *J. Austral. Math. Soc. Ser. B,* **33**: 180–191, (1991).

[217] V. Namias, The fractional order Fourier transforms and its application to quantum mechanics, *J. Inst. Math. Appl.,* **25**: 241–265, (1980).

[218] V. Namias, Fractionalization of Hankel transform, *J. Instit. Mathematics and Applications*, **26**: 187–197, (1980).

[219] F. S. Oktem and H. Ozaktas, Equivalence of linear canonical transform domains to fractional Fourier domains and the bicanonical width product: A generalization of the space-width product, *Opt. Soc. Amer. A,* **27**: 1885–1895, (2010).

[220] H. M. Ozaktas, Z. Zalevsky, and M. Kutay, *The Fractional Fourier Transform with Applications in Optics and Signal Processing*, John Wiley, New York (2001).

[221] H. M. Ozaktas, M. Kutay, and Z. Zalevsky, Applications of the fractional Fourier transform to matched filtering, detection, and pattern recognition. Proc. Fractional Fourier Transform with Appl. Opt., *Signal Process.* 455–460, (2000)

[222] H. M. Ozakdas, M. A. Kutay, and D. Mendlovic, Introduction to the fractional Fourier transform and its applications, in *Advances in Imaging Electronics and Physics*, Ch. 4, Academic Press, New York, (1999).

[223] H. M. Ozaktas, O. Arikan, M. Kutay, and G. Bozdagi, Digital computation of the fractional Fourier transform, *IEEE Trans. on Signal Proc.*, **44**: 2141–2149, (1996) .

[224] H. M. Ozaktas and D. Mendlovic, Fractional Fourier optics, *J. Opt. Soc. Amer. A*, **12**: 743–751, (1995).

[225] H. M. Ozaktas, B. Barshan, D. Mendlovic, and L. Onural, Convolution filtering, and multiplexing in fractional Fourier domains and their relation to chirp and wavelet transforms, *Opt. Soc. Amer. A,* **11**: 547–559, (1994).

[226] H. M. Ozaktas, B. Barshan, D. Mendlovic, and L. Onural, Convolution, filtering, and multiplexing in fractional Fourier domains and their relationship to chirp and wavelet transforms, *J. Opt. Soc. Amer. A*, **11**: 547–559, (1994).

[227] H. M. Ozaktas and D. Mendlovic, Fourier transforms of fractional order and their optical interpretation, *Opt. Commun.*, **101**: 163–169, (1993).

[228] R. Paley and N. Wiener, The Fourier transforms in the complex domain, *Amer. Math. Soc. Colloq. Publ. Ser.,* Providence, Rhode Island, **19**, (1934).

[229] A. Papoulis, *Signal Analysis*, McGraw-Hill, New York (1977).

[230] A. Papoulis, Ambiguity function in Fourier optics, *Optical Soc. Amer.,* **64**(6): 779–788, (1974).

[231] E. Parzen, A simple proof and some sxtensions of the sampling theorems, *Tech. Report*, Stanford University, Stanford, California, **7**, (1956).

[232] R. S. Pathak, A. Prasad, and M. Kumar, Fractional Fourier transform of tempered distributions and generalized pseudo-differential operator, *J. Pseudo-Diff. Oper. Appl.,* **3**(2): 239–254, (2012).

[233] S. C. Pei and J. J. Ding, Fractional Fourier transform, Wigner distribution, and filter design for stationary and non-stationary random processes, *IEEE Trans. Signal. Proc.,* **58**(8): 4079–4092, (2010).

[234] S. C. Pei and J. J. Ding, Eigenfunctions of Fourier and fractional Fourier transforms with complex offsets and parameters, *IEEE Trans. Circuits and Systems I,* **54**(7): 1599–1611, (2007).

[235] S. C. Pei and J. J. Ding, Generalized prolate spheroidal wave functions for optical finite fractional Fourier and linear canonical transforms, *J. Opt. Soc. Amer.,* **22**(3): 460–474, (2005).

[236] S. C. Pei and J. J. Ding, Eigenfunctions of the offest Fourier, fractional Fourier, and linear canonical transforms, *J. Opt. Soc. Amer. A,* **20**(3): 522–532, (2003).

[237] S. C. Pei and J. J. Ding, Eigenfunctions of linear canonical transform, *IEEE Trans. Signal. Process.* , **50**(1): 11–26, (2002)

[238] S. C. Pei, M. Yeh, and T. Luo, Fractional Fourier series expansion for finite signals and dual extension to discrete-time fractional Fourier transform, *IEEE Trans. Signal Proc.,* **47**: 2883–2888, (1999).

[239] S. C. Pei, M. H. Yeh, and C. C. Tseng, Discrete fractional Fourier transform based on orthogonal projections, *IEEE Trans. Signal Proc.,* **47**: 1335–1347, (1999).

[240] S. C. Pei and M. H. Yeh, Two dimensional discrete fractional Fourier transform, *Signal Proc.,* **67**: 99–108, (1998).

[241] S. C. Pei and M. H. Yeh, Improved discrete fractional Fourier transform, *Opt. Letts.,* **22**: 1047–1049, (1997).

[242] I. Pesenson and A. I. Zayed, Paley-Wiener subspace of vectors in a Hilbert space with applications to integral transforms, *J. Math. Anal. & Appls.,* **353**: 566–582, (2009).

[243] A. Poularikas, *The Handbook of Transforms and Applications,* CRC Press, Boca Raton, Florida (1996).

[244] A. Prasad and P. Kumar, Pseudo-differential operator associated with the fractional Fourier transform, *Math. Commun.*, **21**(1): 115–126, (2016).

[245] A. Prasad, S. Manna, A. Mahato, and V. K. Singh, The generalized continuous wavelet transform associated with the fractional Fourier transform, *J. Comput. Appl. Math.,* **259**: 660–671, (2014).

[246] X. Qiang, H. Q. Zhen, and Q. K. Yu, Multichannel sampling of signals band-limited in offset linear canonical transform domains, *Circ. Syst. Signal Process,* **32**(5): 2385–2406, (2013).

[247] J. Radon, Über die Bestimmung von Funktionen durch ihre Integralwerte längs gewisser Mannigfaltigkeiten", Berichte über die Verhandlungen der Königlich-Sächsischen Akademie der Wissenschaften zu Leipzig, Mathematisch-Physische Klasse *Reports on the Proceedings of the Royal Saxonian Academy of Sciences at Leipzig, Mathematical and Physical Section*, Leipzig: Teubner **69**: 262–277, (1917).

[248] Q. W. Ran, H. Zhao, L. Y. Tan, and J. Ma, Sampling of bandlimited signals in fractional Fourier transform domain, *Circut. Syst. Signal Process,* **29**: 459–467, (2010).

[249] T. Ran, D. Bing, and W. Yue, Research progress of the fractional Fourier transform in signal processing, *Science in China Series F,* **49**(1): 1–25, (2006).

[250] J. Rodrigo, T. Alieva, and M. Calvo, Programmable two-dimensional optical fractional Fourier processor, *Optics Express*, **17**(7): 4976–4983, (2009).

[251] J. A. Rodrigo, T. Alieva, and M. Calvo, Gyrator transform: properties and applications, *Optics Express,* **15**(5): 2190–2203, (2007).

[252] J. A. Rodrigo, T. Alieva, and M. L. Calvo, Optical system design for ortho-symplectic transformations in phase space, *J. Opt. Soc. Amer. A*, **23**: 2494–2500, (2006).

[253] A. Sahin, M. A. Kutay, and H. M. Ozaktas, Non-separable two-dimensional fractional Fourier transform, *Appl. Optics,* **37**: 5444–5453, (1998).

[254] A. Shain, H. Ozaktas, and D. Mendlovic, Optical implementations of two-dimensional fractional Fourier transforms and linear caconical transforms with arbitrary parameters, *Appl. Optics,* **37**: 2310–2141, (1998).

[255] S. Saitoh, One approach to some general integral transforms and its applications, *J. Integ. Transf. & Special Functs.*, **3**: 49–84, (1995).

[256] S. Saitoh, *The Theory of Reproducing Kernels and its Applications*, Pitman Research Notes in Math., **189**, Longman Scientific & Technical, United Kingdom, (1988).

[257] S. Saitoh, Hilbert spaces induced by Hilbert space valued functions, *Proc. Amer. Math. Soc.*, **89**: 74–78, (1983).

[258] L. Schwartz, *Theorie des Distributions,* Hermann, Paris, (1966).

[259] E. Sejdi'c, I. Djurovi'c, and L. Stankovi'c, Fractional Fourier transform as a signal processing tool: An overview of recent developments, *Signal Proc.,* **91**(6): 1351–1369, (2011).

[260] F. Shah, A. Tantary, and A. Zayed, Papoulis' sampling theorem: Revisited, *Appl. Comput. Harmonic Anal.,* **64**: 118–142 (2023).

[261] F. Shah, A. Tantary, and A. Zayed, A Convolution-based special affine wavelet transform, *J. of Integral Transforms and Special Functions,* **32**(10): 780–800, (2021).

[262] F. A. Shah and A. Y. Tantary, Non-isotropic angular Stockwell transform and the associated uncertainty principles, *Appl. Anal.,* (2019) doi.org/10.1080/00036811.2019.1622681.

[263] F. A. Shah and A. Y. Tantary, Polar wavelet transform and the associated uncertainty principles, *Int. J. Theor. Phys.,* **57**(6): 1774–1786, (2018).

[264] C. E. Shannon, A Mathematical Theory of Communication, *Bell Syst. Tech. J.*, **27**: 379–423, (1948).

[265] V. B. Shakhmurov and A. I. Zayed, Fractional Wigner distribution and ambiguity functions, *J. Frac. Calculus & Appl. Anal.*, **6**(4): 473–490, (2003).

[266] K. K. Sharma, Comments on generalized sampling expansion for bandlimited signals associated with the fractional Fourier transform. *IEEE Signal Process. Lett.*, **18**(12): 12, 761, (2011).

[267] C. J. R. Sheppard, Free-space diffraction and the fractional Fourier transform, *J. Mod. Optics,* **45**: 2097–2103, (1998).

[268] J. Shi, J. Zheng, X. Liu, and Q. Zhang, Novel short-time fractional Fourier transform: theory, implementations, and applications, *IEEE Trans. Signal Process.*, **68**: 3280–3295, (2020).

[269] J. Shi, X. Liu, F. G. Yan, and W. Song, Error analysis of reconstruction from linear canonical transform-based sampling, *IEEE, Trans. Signal Process.*, **66**(7): 1748–1760, (2018).

[270] J. Shi, X. Liu, X. Sha, Q. Zhang, and N. Zhang, A sampling theorem for fractional wavelet transfrom with error estimates, *IEEE Trans. Signal Process.*, **65**(18): 4797–4811, (2017).

[271] J. Shi, X. Liu, L. He, M. Han, Q. Li, and N. Zhang, Sampling and reconstruction in arbitrary measurement and approximation spaces associated with linear canonical transform, *IEEE, Trans. Signal Process*, **64**(24): 6379–6391, (2016).

[272] J. Shi, X. Liu, and N. Zhang, Multiresolution analysis and orthogonal wavelets associted with fractional wavelet transform, *Signal Image & Video Process.,* **9**: 211–220, (2015).

[273] J. Shi, X. Liu, and N. Zhang, Generalized convolution and product theorems associated with linear canonical transform, *Signal Image & Video Process.*, **8**: 967–974, (2014).

[274] J. Shi, X. Liu, X. Sha, and N. Zhang, Sampling and reconstruction of signals in function spaces associated with linear canonical transform, *IEEE, Trans. Signal Process.*, **60**(11): 6041–6047, (2012).

[275] J. Shi, N. T. Zhang, and X. P. Liu, A novel fractional wavelet transform and its applications, *Sci. China Inf. Sci.* **55**(6): 1270–1279, (2012).

[276] J. Shi, Y. Chi, and N. Zhang, Multichannel sampling and reconstruction of bandlimited signals in fractional Fourier domain, *IEEE Signal Process. Lett.*, **17**(11): 909–912, (2010).

[277] S. Shinde and M. G. Vikram, An uncertainty principle for real signals in the fractional Fourier transform domain, *IEEE Trans. Signal Process.*, **49**(11): 2545–2548, (2001).

[278] R. Simon and K. B. Wolf, Fractional Fourier transform in two dimensions, *J. Opt. Soc. Amer. A,* **17**(12): 2368–2381, (2000).

[279] R. Simon and K. B. Wolf, Structure of the set of paraxial optical systems, *J. Opt. Soc. Amer. A*, **17**(2): 342–355, (2000).

[280] N. Singh and A. Sinha, Optical image encryption using fractional Fourier transform and chaos, *Optics and Laser in Engineering,* **46**(2): 117–132, (2008)

[281] I. Sneddon, *Fourier Transforms,* McGraw Hill, New York, (1951).

[282] H. M. Srivastava, A. Y. Tantary, F. Shah, and A. Zayed, An interplay of ridgelet and linear canonical transforms, *J. of Mathematics: Special Issue on Mathematical Analysis and Analytic Number Theory*, June 2022, https://www.mdpi.com/journal/mathematics/special issues/Mathematical Analysis Analytic Number Theory 2022 1736788; doi: 10.3390/math10121986.

[283] H. M. Srivastava, Komal Khatterwani, and S. K. Upadhyay, A certain family of fractional wavelet transformations, *Math. Methods in Appl. Sci.*, **42**(9): 3103–3122, (2019).

[284] H. Srivastava and H. Manocha, *A Treatise on Generating Functions,* Ellis Horwood, New York (1984).

[285] L. Stanković, *Digital Signal Processing: with Selected Topics: Adaptive Systems, Time-Frequency Analysis, Sparse Signal Processing,* North Charleston, SC, USA (2015).

[286] A. Stern, Uncertainty principles in linear canonical transform domains and some of their implications in optics, *J. Opt. Soc. Amer. A.,* **25**(3): 647–652, (2008).

[287] A. Stern, Sampling of compact signals in offset linear canonical transform domains, *Signal Image Video Process.,* **1**(4): 359–367, (2007).

[288] A. Stern, Sampling of linear canonical transformed signals, *J. Signal Proc.,* **86**: 1421–1425, (2006).

[289] T. S. Stieltjes, Recherches sur les fractions continues, *Annales de la Faculté des Sciences de Toulouse*, **8**: 1–123, (1894).

[290] R. G. Stockwell, L. Mansinha, and R. P. Lowe, Localization of the complex spectrum: the S transform, *IEEE Trans. Signal Process.*, **44**: 998–1001, (1996).

[291] E. C. G. Sudarshan, N. Mukunda, and R. Simon, Realization of first order optical system using thin lenses, *Opt. Acta,* **32**: 855–872, (1985).

[292] S. Suslov, Addition theorems for some q-exponential and trigonometric functions, *Methods and Applications of Anal.,* **4**: 11–32, (1997).

[293] G. Szegö, *Orthogonal Polynomials,* Amer. Math. Soc., Providence, RI, (1975).

[294] R. Tao, Y. L. Li., and Y. Wang, Short-Time fractional Fourier transform and its applications, *IEEE Trans. Signal Process.*, **58**(5): 2568–2580, (2010).

[295] R. Tao, B. Z. Li, Y. Wang, and G. K. Aggrey, On sampling of bandlimited signals associated with the linear canonical transform, *IEEE Trans. Signal Proc.,* **56**(11): 5454–5464, (2008).

[296] R. Tao, B. Deng, W. Q. Zhang, and Y. Wang, Sampling and sampling rate conversion of bandlimited signals in the fractional Fourier transform domain, *IEEE Trans. Signal Proc.,* **56**(1): 158–171, (2008).

[297] E. Titchmarsh, *Eigenfunction Expansion I,* Oxford University Press (1962).

[298] V. K. Tuan and A. I. Zayed, Generalization of a theorem of Boas to a class of integral transforms, *Resultät der Mathematik,* **38**: 362–376, (2000).

[299] M. Unser, Sampling 50 years after Shannon, *Proc. IEEE,* **88**(4): 569–587, (2000).

[300] D. Urynbassarova, B. Z. Li, and R. Tao, Convolution and correlation theorems for Wigner-Ville distribution associated with the offset linear canonical transform, *Optik,* **157**: 455–466, (2018).

[301] Ville, Torie et applications de la notion de signal analytique, Cables et Transmissions, **2A**: 61-74, (1948). Translated from French by L. Selin, "Theory and applications of the notion of (2000). complex signal", RAND Corporation Technical Report T-92, Santa Monica, CA, 1958.

[302] G. G. Walter and A. I. Zayed, Multiresolution analysis with sampling subspaces, *Frac. Calc & Appl. Anal.,* **1**: 109–124, (1998).

[303] G. G. Walter, *Wavelets and Other Orthogonal Systems with Applications,* CRC Press, Boca Raton, FL (1994).

[304] G. G. Walter, A sampling theorem for wavelet subspaces, *IEEE Trans. Information Theory,* **38**(2): 881–884, (1992).

[305] G. G. Walter and A. I. Zayed, On the inversion of integral transforms associated with Sturm-Liouville problems, *J. Math. Analy. & Applications,* **164**(1): 285–306, (1992).

[306] G. G. Walter and A. Zayed, The continuous Jacobi transform and its inverse when $\alpha+\beta+1$ is a positive integer, *Trans. Amer. Math. Soc.,* **305**(2): 653–664, (1988).

[307] G. Watson, *A Treatise on the Theory of Bessel Functions,* Cambridge Univeristy Press (1944).

[308] G. Watson, Notes on generating functions of polynomials: (4) Jacobi polynomials, *J. London Math. Soc.,* **9**: 22–28, (1934).

[309] D. Wei and Y. M. Li, Generalized sampling expansions with multiple sampling rates for Low-pass and band-pass signals in the fractional Fourier transform domain, *IEEE Trans. Signal Proc.,* **64**(18): 4861–4874, (2016).

[310] D. Wei and Y. M. Li, Generalized wavelet transform based on the convolution operator in the linear canonical transform domain, *Optik,* **125**(16): 4491–4496, (2014).

[311] D. Wei, Q. Ran, and Y. Li, Generalized sampling expansion for bandlimited signals associated with the fractional Fourier transform, *IEEE Signal Proc. Letts.,* **17**(6): 595–598, (2010).

[312] D. Wei, Q. Ran, Y. Li, J. Ma, and L. Tan, A convolution and product theorem for the linear canonical transform, *IEEE Signal Proc. Letts.,* **16**: 853–856, (2009).

[313] J. L. Weiner and G. R. Wilkens, Quaternions and rotations in $\mathbb{E}^4$, *Amer. Math. Monthly,* **112**: 69–76, (2005).

[314] G. Weiss and E. Hernandez, *A First Course On Wavelets,* CRC Press, Boca Raton, FL (1996).

[315] E. T. Whittaker, On the functions which are represented by the expansion of interpolation theory, *Proc. Roy. Soc. Edinburgh, Ser. A,* **35**: 181–194,(1915).

[316] N. Wiener, Hermitian polynomials and Fourier analysis, *J. Math. Phys.,* **8**: 70–73, (1929).

[317] E. P. Wigner, On the quantum correction for thermodynamic equilibrium, *Physical Review,* **40**: 749–759, (1932).

[318] C. Wilcox, The synthesis problem for radar ambiguity functions, MRC Technical Report, **157**, Math. Research Center, U. S. Army, Univ. Wisconsin, Madison (1960).

[319] E. Wilczok, New uncertainty principles for the continuous Gabor transform and the continuous wavelet transform, *Doc. Math.,* **5**: 201–226, (2000).

[320] K. B. Wolf, *Geometric Optics on Phase Space,* Springer-Verlag, Berlin (2004).

[321] K. B. Wolf, *Integral Transforms in Science and Engineering*, Plenum Press, New York (1979).

[322] P. Woodward, *Probability and Information Theory with Applications to Radar*, McGraw-Hill, New York (1953).

[323] Q. Xiang and K. Qin, Convolution, correlation, and sampling theorems for the offset linear canonical transform, *Signal Image Video Process.,* **8**(3): 433–442, (2014).

[324] Q. Xiang, K. Qin, and Q. Z. Huang, Multichannel sampling of signals bandlimited in offset linear canonical transform domains, *Circuits Systems and Signal Proc.,* **32**(5): 2385–2406, (2013).

[325] X. Xia, On bandlimited signals with fractional Fourier transform, *IEEE Signal Proc. Lettrs.*, **3**: 72–74, (1996).

[326] S. Xu, L. Huang, Y. Chai, and Y. He, Nonuniform sampling theorems for bandlimited signals in the offset linear canonical transform, *Circ. Syst. Signal Process.*, **37**: 3227–3244, (2018).

[327] T. Z. Xu and B. Z. Li, *Linear Canonical Transform and Its Applications,* Science Press, Beijing, China (2013).

[328] R. Young, *An Introduction to Nonharmonic Fourier Series,* 2nd Edition, Elsevier (2001).

[329] W. H. Young, On infinite integrals involving a generalization of the Sine and Cosine functions, *Quart. J. Pure & Appl. Math.,* **43**: 161–177, (1912).

[330] F. T. S. Yu and G. Lu, Short-time Fourier transform and wavelet transform with Fourier-domain processing, *Applied Optics,* **33** (23), 5262–5270, (1994)

[331] J. Zak, Finite translations in solid state physics, *Phys. Rev. Lett.*, **19**: 1385–1397, (1967).

[332] Z. Zalevsky and D. Mendlovic, Fractional Wiener filter, *Appl. Opt.*, **35**: 3930–3936, (1996).

[333] Z. Zalevsky and D. Mendlovic, Fractional Radon transform, *Appl. Opt.,* **35**(23): 4628–4631, (1996).

[334] A. Zayed, Sampling theorem for two-dimensional fractional Fourier transform, *J. of Signal Processing,* **181**: 107902, (2021), https://www.sciencedirect.com/science/article/abs/pii/S0165168420304461.

[335] A. Zayed, A New Perspective on the Two-Dimensional Fractional Fourier Transform and its Relationship with the Wigner Distribution, *J. Fourier Anal. Appl.*, **25**(2): 460–487, (2019).

[336] A. Zayed, On multivariate sampling of a class of integral transforms, in *Classical and Modern Analysis-In Memory of Yingkang Hu,* Birkhauser Publ., Editors: M. Abell, E. Jacob, et al, 347–367, (2019).

[337] A. Zayed, Two-dimensional fractional Fourier transform and some of its properties. *Integral Transforms Spec. Funct.*, **29**(7), 553–570, (2018).

[338] A. Zayed, Sampling of signals bandlimited to a disc in the linear canonical transform domain, *IEEE Signal Processing Letters*, **25**: 1765–1769, (2018). https://ieeexplore.ieee.org/document/8488503.

[339] A. Zayed, On Sampling Theorems for Fractional Fourier Transforms and Series, *Proceedings of the SampTa 17, the 12th International Conference on Sampling Theory and Applications,* Tallinn, Estonia (2017), also appeared digitally on IEEE Xplore Digital Library, http://ieeexplore.ieee.org/document/8024396/

[340] A. Zayed, Maximally concentrated signals in the special affine Fourier transformation domain, *Proceedings of 2015 International Conference on Sampling Theory and Applications (SampTA)* July 2015, http://ieeexplore.ieee.org/stamp/stamp.jsp?tp= & arnumber=7148841.

[341] A. Zayed, Linear transformations in signal and optical systems, *Handbook of Operator Theory,* Springer-Verlag, D. Alpay, Editor, 833–874, (2015).

[342] A. Zayed, On the notion of bandlimitedness and its generalizations, *J. Argentinean Math. Soc.,* Revista De La Union Matematica Argentina, **49**: 99–109, (2008).

[343] A. Zayed, A class of fractional integral transforms: A generalization of the fractional Fourier transform, *IEEE Trans. Signal Process.,* **50**: 619–627, (2002).

[344] A. Zayed, On the relationship between the fractional Fourier transform and the Riemann-Liouville fractional integral, *Proc. International Assoc. Math. & Computers in Simulation (IMACS)*, May (2000).

[345] A. Zayed and A. Garcia, New sampling formulae for the fractional Fourier transform, *Signal Proc.,* **77**(1): 111–114, (1999).

[346] A. Zayed, Fractional Fourier transform of generalized functions, *Integral Transforms and Special Functions,* **7**: 299–312, (1998).

[347] A. Zayed, Hilbert transform associated with the fractional Fourier transforms, *IEEE Sig. Proc. Letters,* **5**: 206–208, (1998).

[348] A. Zayed, A convolution and product theorem for the fractional Fourier transform, *IEEE Sig. Proc. Letters,* **5**: 101–103, (1998).

[349] A. Zayed and A. Garcia, Sampling theorems associated with Dirac operator and the Hartley transform, *J. Math. Anal. Appls.*, **214**: 587–598, (1997).

[350] A. Zayed, On the relationship between the Fourier and fractional Fourier transforms, *IEEE Sig. Proc. Letters,* **3**: 310–311, (1996).

[351] A. Zayed, Sampling in a Hilbert space, *Proc. Amer. Math. Soc.,* **124**: 3767–3776, (1996).

[352] A. Zayed, *Function and Generalized Function Transformations,* CRC Press, Boca Raton, Florida (1996).

[353] A. I. Zayed and P. Mikusinski, On the extension of the Zak transform, *J. Methods Appl. Anal.,* **2**: 160–172, (1995).

[354] A. Zayed, Integral Transforms, **30**th Edition of *CRC Standard Tables and Formulae,* CRC Press, Boca Raton, FL (1995).

[355] A. Zayed, Sampling theorem for functions bandlimited to a disc, *J. Complex Variables: Theory and Applications,* **26**(3): 245–254, (1994).

[356] A. Zayed, A new role of Green's function in interpolation and sampling theory, *J. Math. Anal. Appl.*, **175**: 222–238, (1993).

[357] A. Zayed, *Advances in Shannon's Sampling Theory,* CRC Press, Boca Raton, Florida (1993).

[358] A. Zayed, Kramer's sampling theorem for multidimensional signals and its relationship with Lagrange-type interpolation, *Multidimen. Systems & Signal. Proc.,* **3**: 323–340, (1992).

[359] A. I. Zayed, On Kramer's sampling theorem associated with general Sturm-Liouville boundary-value problems and Lagrange interpolation, *SIAM J. Appl. Math.,* **51**(2): 575–604, (1991).

[360] A. Zayed, G. Hinsen, and P. Butzer, On Lagrange interpolation and Kramer-type sampling theorems assocaited with Sturm-Liouville problems, *SIAM J. Appl. Math.*, **50**: 893–909, (1990).

[361] A. Zemanian, *Generalized Integral Transformations,* Dover Publishing, New York (1987).

[362] Y. Zhang, T. Funaba, and N. Tanno, Self-fractional Hankel functions and their properties, *Optics Comm.*, **176**: 71–75, (2000).

[363] Y. Zhang, B. Y. Gu, and G. Z. Yang, A new kind of windowed fractional transforms, *Opt. Commun.*, **152**: 127–134, (1998).

[364] Z-C. Zhang, New Wigner distribution and ambiguity function based on the generalized translation in the linear canonical transform domain, *J. Signal Process*, **118**: 51–61, (2016).

[365] L. Zhao, J. J. Healy, and J. T. Sheridan, Two-dimensional nonseparable linear canonical transform: Sampling Theorem and unitary discretization, *Opt. Soc. Amer. A,* **31**(12): 2631–2641, (2014).

[366] H. Zhao, J. Zhang, L. Qiao, and Y. Q. Chen, A multichannel compressed sampling method for fractional bandlimited signals, *IEEE Trans. Signal Proc.,* **64**(18): 4861–4874, (2016).

[367] H. Zhao, R. Wang, D. Song, and D. Wu, Maximally concentrated sequences in both time and linear canonical transform domains, *Signal, Image and Video Processing*, **8**: 819–829, (2014).

[368] H. Zhao, Q. W. Ran, J. Ma, and L.Y. Tan, On bandlimited signals associated with linear canonical transform, *IEEE Signal Process. Lett.*, **16**(5): 343–345, (2009).

[369] H. Zhao, Q. W. Ran, L. T. Tan, and J. Ma, Reconstruction of bandlimited signals in linear canonical transform domain from finite nonuniformly spaced samples, *IEEE Signal Process. Lett.*, **16**: 1047–1050, (2009).

[370] X. Zhi, D. Wei, and W. Zhang, A generalized convolution theorem for the special affine Fourier transform and its application to filtering, *Optik,* **127**(5): 2613–2616, (2016).

[371] Z. H. Zhuo, Poisson summation formulae associated with the special affine Fourier transform and offset Hilbert transform, *Math. Prob. Engineer.,* Article ID 1354129 (2017).

[372] A. Zygmund, *Trigonometric Series,* 1,2, Cambridge University Press, Cambridge (1959).

Index